17 Springer Series in Chemical Physics
Edited by Robert Gomer

Springer Series in Chemical Physics
Editors: V. I. Goldanskii   R. Gomer   F. P. Schäfer   J. P. Toennies

Volume 1   **Atomic Spectra and Radiative Transitions**
By I. I. Sobelman

Volume 2   **Surface Crystallography by LEED**   Theory, Computation and Structural Results   By M. A. Van Hove, S. Y. Tong

Volume 3   **Advances in Laser Chemistry**   Editor: A. H. Zewail

Volume 4   **Picosecond Phenomena**
Editors: C. V. Shank, E. P. Ippen, S. L. Shapiro

Volume 5   **Laser Spectroscopy**   Basis Concepts and Instrumentation
By W. Demtröder

Volume 6   **Laser-Induced Processes in Molecules**   Physics and Chemistry
Editors: K. L. Kompa, S. D. Smith

Volume 7   **Excitation of Atoms and Broadening of Spectral Lines**
By I. I. Sobelman, L. A. Vainshtein, E. A. Yukov

Volume 8   **Spin Exchange**   Principles and Applications in
Chemistry and Biology
By Yu. N. Molin, K. M. Salikhov, K. I. Zamaraev

Volume 9   **Secondary Ion Mass Spectrometry SIMS II**
Editors: A. Benninghoven, C. A. Evans, Jr.,
R. A. Powell, R. Shimizu, H. A. Storms

Volume 10   **Lasers and Chemical Change**
By A. Ben-Shaul, Y. Haas, K. L. Kompa, R. D. Levine

Volume 11   **Liquid Crystals of One- and Two-Dimensional Order**
Editors: W. Helfrich, G. Heppke

Volume 12   **Gasdynamic Laser**   By S. A. Losev

Volume 13   **Atomic Many-Body Theory**   By I. Lindgren, J. Morrison

Volume 14   **Picosecond Phenomena II**
Editors: R. Hochstrasser, W. Kaiser, C. V. Shank

Volume 15   **Vibrational Spectroscopy of Adsorbates**   Editor: R. F. Willis

Volume 16   **Spectroscopy of Molecular Excitons**
By V. L. Broude, E. I. Rashba, E. F. Sheka

Volume 17   **Inelastic Particle-Surface Collisions**
Editors: E. Taglauer, W. Heiland

Volume 18   **Modelling of Chemical Reaction Systems**
Editors: K. H. Ebert et al.

Inelastic Particle–Surface Collisions

Proceedings of the Third International Workshop on
Inelastic Ion-Surface Collisions
Feldkirchen-Westerham, Fed. Rep. of Germany
September 17–19, 1980

Editors: E. Taglauer and W. Heiland

With 194 Figures

Springer-Verlag Berlin Heidelberg New York 1981

Dr. Edmund Taglauer

Max-Planck-Institut für Plasmaphysik
D-8046 Garching bei München, Fed. Rep. of Germany

Professor Dr. Werner Heiland

Fachbereich 4 der Universität Osnabrück
D-4500 Osnabrück, Fed. Rep. of Germany

Series Editors

Professor Vitalii I. Goldanskii

Institute of Chemical Physics
Academy of Sciences
Vorobyevskoye Chaussee 2-b
Moscow V-334, USSR

Professor Robert Gomer

The James Franck Institute
The University of Chicago
5640 Ellis Avenue
Chicago, IL 60637, USA

Professor Dr. Fritz Peter Schäfer

Max-Planck-Institut für
Biophysikalische Chemie
D-3400 Göttingen-Nikolausberg
Fed. Rep. of Germany

Professor Dr. J. Peter Toennies

Max-Planck-Institut für Strömungsforschung
Böttingerstraße 6-8
D-3400 Göttingen
Fed. Rep. of Germany

ISBN 3-540-10898-X Springer-Verlag Berlin Heidelberg New York
ISBN 0-387-10898-X Springer-Verlag New York Heidelberg Berlin

This work is subject to copyright. All rights are reserved, whether the whole or part of the material is concerned, specifically those of translation, reprinting, reuse of illustrations, broadcasting, reproduction by photocopying machine or similar means, and storage in data banks. Under § 54 of the German Copyright Law where copies are made for other than private use, a fee is payable to "Verwertungsgesellschaft Wort", Munich.

© by Springer-Verlag Berlin Heidelberg 1981
Printed in Germany

The use of registered names, trademarks, etc. in this publication does not imply, even in the absence of a specific statement, that such names are exempt from the relevant protective laws and regulations and therefore free for general use.

Offset printing: Beltz Offsetdruck, 6944 Hemsbach/Bergstr. Bookbinding: J. Schäffer OHG, Grünstadt.
2153/3130-543210

Preface

The interaction of particles and photons with solid surfaces is interdisciplinary in character, so that very recent developments in solid-state physics, surface physics and atomic physics stimulate progress in the field or profit from results of the "ion-solid" community. Technical interest in the field ranges from catalysis and semiconductor manufacturing to fusion research, for instance by surface analytical techniques, or interest in phenomena such as sputtering and radiation damage.

The Third International Workshop on Inelastic Ion-Surface Collisions, held at Feldkirchen-Westerham under the auspices of Max-Planck-Institut für Plasmaphysik, Garching, Fed. Rep. of Germany, brought together 63 scientists from 12 countries for three days of very involved discussions. As at the previous workshops at Bell Laboratories in 1976 and McMaster University in 1978, the experiment of gathering experts from seemingly different disciplines was very successful in promoting the basic physical ideas.

The proceedings contain the 14 major reviews and a smaller number of contributions presented at the workshop. All papers have been reviewed with little delay, and the reviewer's efforts are gratefully acknowledged. The first group of papers is concerned with theoretical and experimental aspects of secondary electron emission due to ion impact, including the potential emission caused by slow metastables. This is followed by reviews of experiments and recent theoretical developments of electron- and photon-induced desorption. Leading on to the processes governed by nuclear interactions such as reflection and sputtering are two particular reviews discussing congruent phenomena of atomic, solid-state and surface physics. The second half of the volume then comprises charged-state and excited-state formation in reflection and sputtering; the interpretations of many recent results led to lively discussions at the Workshop.

The programme resulted from discussions within the committee composed of G. Blaise, H.D. Hagstrum, W. Heiland, R. Kelly, R.J. MacDonald, E. Taglauer, G. Thomas, and N.H. Tolk. The Workshop was generously supported by the Max-Planck-Institut für Plasmaphysik, IBM Germany and the US Office of Naval Research through Dr. L.R. Cooper

Garching, Osnabrück *E. Taglauer*
March, 1981 *W. Heiland*

Contents

Electron Emission

Electron Emission from Solids During Ion Bombardment.
Theoretical Aspects. By P. Sigmund and S. Tougaard 2

Ion-Induced Auger Electron Emission from Solids
By R.A. Baragiola 38

Interaction Between Metastable Rare Gas Atoms and Surfaces
By C. Boiziau 48

Deexcitation of Metastable He-Atoms Interacting with Clean and
Adsorbate Covered Metal Surfaces
By H. Conrad, G. Ertl, J. Küppers, W. Sesselmann, and H. Haberland   73

Electron and Photon Impact

The Use of Angle-Resolved Electron and Photon Stimulated
Desorption for Surface Structural Studies
By T.E. Madey 80

Auger-Initiated Desorption from Surfaces: Review + Prospects
By P.J. Feibelman 104

Optical Radiation from Electron Sputtering of Alkali Halides
By N.H. Tolk, L.C. Feldman, J.S. Kraus, R.J. Morris, T.R. Pian,
M.M. Traum, and J.C. Tully 112

Electron Transfer

Electron Capture and Loss to Continuum States in Gases and Solids
By I.A. Sellin and R. Laubert 120

New Experiments with Electron Capture Spectroscopy at Ni, Gd and
Cr Surfaces. By C. Rau and S. Eicher 138

Electronic Excitation in Ion-Molecule Collisions. By S. Datz 142

Charge Fractions of Reflected Particles. By W. Eckstein 157

Plasma Effects in the Theory of Charge Fractions
By M. Kleber and J. Zwiegel 184

Polarized Light Emission

Polarization of Balmer Radiation from Grazing Incidence Collisions of Protons on Surfaces
By J.C. Tully, N.H. Tolk, J.S. Kraus, C. Rau, and R.J. Morris 196

LYα Polarization After Electron Capture in Ion-Surface Interaction
By H. Schröder and J. Burgdörfer 207

Light Emission and Circular Polarization by Ion Surface Scattering at Grazing Incidence. By W. Graser and C. Varelas 211

Orientation of Target and Projectile States After N^+-He-Collisions with Impact Parameter Selection. By H. Winter 216

Excited Particle Emission

Outer Shell Excitation During Sputtering and Low Energy Ion Scattering. By R.J. MacDonald, C.M. Loxton, and P.J. Martin 224

Comment on the Energy Distribution of Excited Recoil Atoms
By P. Sigmund 251

Quantitative Aspects of Outer-Shell Excitation in Ion-Surface Collisions. By I.S.T. Tsong 258

Theory of Charge States in Sputtering. By Z. Šroubek 277

Boundary Conditions for Models of Ion and Excited-State Formation in the Sputtering Process. By R. Kelly 292

On the Velocity Measurements of Sputtered Excited Atoms
By M. Szymoński, A. Paradzisz, and L. Gabła 322

Index of Contributors 329

Electron Emission

Electron Emission from Solids During Ion Bombardment. Theoretical Aspects

Peter Sigmund and Sven Tougaard
Fysisk Institut, Odense Universitet, DK-5230 Odense M, Denmark

This contribution is devoted to the theory of ion-induced electron emission from solids, with the emphasis on bulk as opposed to surface processes. With the aim of describing yields and energy spectra of emitted electrons, we discuss pertinent elementary processes, i.e., primary and secondary excitation, electron transport, and escape. With regard to the primary process, proper distinction is made between high- and low-velocity ions, and reference is made to the physics of ion-atom collisions as well as to dielectric stopping theory. The discussion of secondary excitation makes reference to the theory of ionization cascades as well as atomic recoil cascades. Electron transport is considered first in an infinite medium and subsequently, within the diffusion approximation, in a semi-infinite medium. Special attention is paid to pertinent electron mean free paths, inelastic processes being estimated on the basis of the free-electron-gas model of a solid. We find that the neglect of elastic scattering of electrons in most existing treatments of electron emission has severe consequences on the interpretation of structure in measured electron spectra. Comments are made on the escape depth of electrons.

1. Introduction

The interaction of ionizing radiation with solids is known to give rise to electron emission. In particular, the interaction of an energetic ion beam with a solid may generate free electrons by a number of mechanisms:

I) Ionizing collisions in the target liberate electrons of which some may be energetic enough to reach the surface and overcome the surface barrier. Such ionizing collisions may be undergone by the primary ion, by energetic recoil atoms, by secondary electrons, and possibly by bombardment-induced photons.

II) Surface processes may give rise to electron ejection into the vacuum. Such processes may be undergone by bombarding ions approaching the surface as well as reflected or transmitted ions (or sputtered atoms) while leaving the surface.

III) Heavy particles ejected from the surface (reflected or transmitted ions, sputtered or desorbed atoms or molecules) may be in metastable states and deexcite via electron emission.

The very fact that energetic ions cause electron emission has been utilized for particle detection, although detection efficiencies are often small, in case of low electron yield. On the other hand, electron emission is a major source of error in measurements of particle current that is hard to be corrected for. Note that the total electron yield (number of emitted electrons per incident ion) does not differentiate between the three types of processes mentioned above, i.e. is not easily predicted theoretically; moreover, it is a common experience that electron yields depend sensitively on surface conditions and, therefore, are not always reproducible.

Energy spectra of emitted electrons, integrated over all directions or angular-resolved, contain substantial information about the processes giving rise to electron emission. In fact, experimental efforts have been directed increasingly toward the development of ion-induced electron emission as a spectroscopic tool along with other electron spectroscopies. Indeed, the three types of processes mentioned above indicate that ion-electron spectroscopy should yield information on both bulk and surface properties of solids, as well as atomic and molecular transitions.

Theoretical efforts directed specifically toward ion-electron emission have been rather sparse in the past. In particular, to our knowledge no summary has yet been written on that particular subject. However, the field has considerable overlap with a vast variety of topics, some of which have been extensively investigated:

a) Secondary electron emission by electron bombardment;

b) Photoelectron spectroscopy;

c) Ionization phenomena in gases and solids;

d) Electron diffraction;

f) Penetration of ions and electrons;

g) Sputtering.

Moreover, the underlying collision and transport theory is common in general terms to an even wider range of physical phenomena. It is not the least important goal of the present work to collect some of the pertinent information available from these fields.

In view of the general outline of this volume, we have concentrated on bulk processes (group I of the processes specified above), while surface processes (group II) have been covered in other contributions (SELLIN et al., 1981; C. BOIZIAU, 1981). The formation of metastable states belongs to the prime topic of this workshop; therefore, processes belonging to group III will, explicitly or implicitly, be included in most contributions of this volume.

Even though we restrict attention to bulk processes, we deal with a wide variety of bombardment conditions including light and heavy ions of both low and high velocities, corresponding to an energy range from a few hundred eV to several hundred MeV. Correspondingly, we deal with a broad range of ejected-electron energies from less than 1 eV up to several hundred keV. A comprehensive description of the processes on an atomic scale is hardly achievable within the near future. Therefore, even though this article is aimed at theoretical aspects, experimental observations will have to be used when appropriate to support theoretical steps.

Comprehensive reviews of the experimental literature on ion-electron emission date back more than ten years ago (ARIFOV, 1961; KAMINSKY, 1965; MEDVED et al., 1965; ABROYAN et al., 1967; KREBS, 1968; CARTER et al., 1968), but shorter, more recent summaries are available (PETROV, 1974; KREBS, 1976; BARAGIOLA et al., 1979). These summaries focus on electron yields. For the spectroscopic aspects, the reader is referred to recent contributions (CLERC et al., 1973; FOLKMANN et al., 1975; VIEL et al., 1976; GROENEVELD, 1976; BENAZETH et al., 1978; MISCHLER et al., 1979; HENNEQUIN et al., 1981).

2. Qualitative Considerations

According to the classification of processes mentioned in the introduction, we concentrate on the emission of electrons from the bulk of the solid. From a qualitative point of view, an emission event splits into four steps.

- A. Primary ionization by the bombarding ion, i.e., liberation of target and/or projectile electrons during penetration;

- B. Secondary ionization by energetic recoil atoms, energetic electrons and, possibly, photons;

- C. Transport of liberated electrons toward the target surface;

- D. Ejection of electrons through the potential barrier at the surface.

We go here through a qualitative evaluation of the electron yield and spectrum in order to find the pertinent parameters. Let $\nu(\varepsilon,x)dx$ be the mean number of electrons liberated per bombarding ion in a layer (x,dx) with an initial kinetic energy exceeding some arbitrary value ε; both primary and secondary processes (A&B) enter into $\nu(\varepsilon,x)$. Here, x is a depth coordinate; the target surface (which is assumed to be planar) is taken at $x=0$ (for emission in the backward direction).

Assume that $\nu(\varepsilon,x)$ varies sufficiently slowly with x so that within the depth of origin of emitted electrons, we can set $\nu(\varepsilon,x) \simeq \nu(\varepsilon,0)$. Then, for an impinging ion current ψ[no. ions/time], we have a stationary density of liberated electrons $G(\varepsilon)d\varepsilon$ [no. electrons/depth] given by

$$G(\varepsilon)d\varepsilon = \psi \nu(\varepsilon,0) d\tau , \qquad (1)$$

where $d\tau$ is the mean time spent by an electron in slowing down from $\varepsilon+d\varepsilon$ to ε,

$$d\tau \simeq \frac{d\varepsilon}{|d\varepsilon/dt|} = \frac{d\varepsilon}{|d\varepsilon/dx|v} , \qquad (2)$$

$d\varepsilon/dx$ is the stopping power of an electron at energy ε, and v its velocity. (Note that an electron can only slow down through the interval $(\varepsilon,d\varepsilon)$ if its initial energy exceeds ε, hence the occurrence of $\nu(\varepsilon,0)$ in (1).)

Eqs. (1) and (2) are based on the assumption of continuous slowing down of electrons in an infinite medium; we shall go beyond this simplified approach in section 5.

Under certain conditions, the angular distribution of the electrons is isotropic, in which case (1) can be expanded into

$$G(\varepsilon,\vec{\Omega})d\varepsilon d^2\Omega \simeq \psi\nu(\varepsilon,0) \cdot \frac{d\varepsilon}{|d\varepsilon/dx|} \cdot \frac{1}{v} \cdot \frac{d^2\Omega}{4\pi} \quad, \tag{3}$$

where $d^2\Omega$ is an element of solid angle, and (2) has been inserted. Any of the following features may contribute to assure isotropy of the velocity distribution:

a) the primary ionization process itself may lead to a near isotropic ejected-electron distribution;

b) the primary ion may undergo a large number of scattering processes on target nuclei during slowing down;

c) electrons may predominantly originate from secondary or higher order events (B), and

d) electrons may undergo a large number of (almost) elastic scattering events (electron-phonon interactions).

If acting, the last process may easily be dominating and sufficient to ensure isotropy (SCHOU, 1980a, 1980b; GANACHAUD et al., 1979). The process has usually been ignored in the theory of secondary electron emission (WOLFF, 1954; DEKKER, 1958; HACHENBERG et al., 1959; and many others); the reason for this is illuminating. Elastic scattering, i.e., potential scattering of electrons on the ion cores of the solid, was considered equivalent with electron-phonon scattering and, therefore, asserted to be temperature dependent. In view of the absence of a substantial temperature dependence of measured electron yields, electron-phonon scattering was discarded as a significant process. On the other hand, (3) shows that even though the cross section for elastic scattering does not enter explicitly into the yield formula for an infinite medium, the process may be crucial to provide the necessary degree of isotropy of the electron flux in order to make (3) a valid estimate. Indeed, this isotropy, though well-established experimentally, did not find a satisfactory explanation in early treatments (HACHENBERG et al., 1959).

The mean number of electrons passing the plane at x=0 in a time interval dt with energy $(\varepsilon,d\varepsilon)$ in a direction $(\vec{\Omega},d^2\Omega)$ is given by $G(\varepsilon,\vec{\Omega})d\varepsilon d^2\Omega v|\cos\theta|dt$, where θ is the angle between the unit vector $\vec{\Omega}$ and the surface normal. Dividing by ψdt, we obtain the number of electrons passing through x=0 per incident ion,

$$F(\varepsilon,\vec{\Omega})d\varepsilon d^2\Omega \simeq \nu(\varepsilon,0) \cdot \frac{d\varepsilon}{|d\varepsilon/dx|} \cdot \frac{|\cos\theta|}{4\pi} \cdot d^2\Omega \quad. \tag{4}$$

In passing a metal surface, an electron has to overcome a potential barrier which, as a first approximation, may be assumed to be planar with a height U where

$$U \simeq E_F + \phi \quad, \tag{5}$$

E_F being the Fermi energy and ϕ the work function. This gives rise to refraction according to

$$\varepsilon'\cos^2\theta' = \varepsilon\cos^2\theta - U; \quad \varepsilon' = \varepsilon - U \quad, \tag{6}$$

where ε' and θ' denote the energy and direction of motion of the electron in vacuum. When (6) is valid, (4) reads

$$F'(\varepsilon',\vec{\Omega}')d\varepsilon'd^2\Omega' \simeq \left(\frac{\nu(\varepsilon,0)}{\varepsilon|d\varepsilon/dx|}\right)_{\varepsilon=\varepsilon'+U} \cdot \varepsilon'd\varepsilon' \cdot \frac{|\cos\theta'|}{4\pi} \cdot d^2\Omega' \quad . \tag{4'}$$

The total electron yield γ (from processes under group I) is given by

$$\gamma \simeq \frac{1}{4\pi}\int d\varepsilon \frac{\nu(\varepsilon,0)}{|d\varepsilon/dx|}\int d^2\Omega|\cos\theta| \tag{7}$$

which, if (6) holds, reduces to

$$\gamma \simeq \frac{1}{4}\int_U^\infty d\varepsilon \frac{\nu(\varepsilon,0)}{|d\varepsilon/dx|}(1-U/\varepsilon) \quad . \tag{7'}$$

The argument leading to (7) is a slight generalization of a calculation of the sputtering yield by ion bombardment (SIGMUND, 1977; cf. also SIGMUND, 1969). With appropriate specifications for $\nu(\varepsilon,0)$ and $d\varepsilon/dx$, (7) and (7') reduce to the result of a recent transport calculation (SCHOU, 1980 b).

It is instructive to compare this result with that of a frequently applied argument in elementary theories of secondary electron emission (BRUINING, 1954; CHUNG et al., 1974, 1977). In the present notation, the production rate of liberated electrons per energy, depth and bombarding ion is $|\partial\nu(\varepsilon,x)/\partial\varepsilon|$. Then, one may write

$$\gamma' \simeq \int_0^\infty dx\int d\varepsilon|\partial\nu(\varepsilon,x)/\partial\varepsilon|\int \frac{d^2\Omega}{4\pi} e^{-x/\lambda(\varepsilon)\cos\theta}, \tag{8}$$

where $\lambda(\varepsilon)$ is a mean free path. Eq. (8) implies that an electron has a probability $\exp(-x/\lambda\cos\theta)$ to move along a straight path from its point of origin, and to be ejected with its original energy. If it undergoes a collision, it will not be ejected, regardless of energy or direction of motion. Approximating again $\nu(\varepsilon,x)\simeq\nu(\varepsilon,0)$, one finds

$$\gamma' \simeq \frac{1}{4}\int_U^\infty d\varepsilon|\partial\nu(\varepsilon,0)/\partial\varepsilon|\cdot\lambda(\varepsilon)\cdot(1-U/\varepsilon) \tag{8'}$$

if (6) is valid.

Although formally very similar, the two expressions (7') and (8') differ substantially in physical content. While (7') contains only quantities that are in principle well-defined, (8') contains a mean free path, the significance of which is not clear in general. It is usually implied that $\lambda(\varepsilon)$ is the total inelastic mean free path[1].

Depending on what mechanism of inelastic scattering is involved, the discrepancy between (7') and (8') may be partially reconciled if $\lambda(\varepsilon)$ is identified as a <u>transport</u> mean free path. Even then, however, there is a major difference between (7') and (8') that is most readily revealed in

---

[1] For electronic collisions, the terms "elastic" and "inelastic" usually refer to the laboratory frame of reference, unlike for heavy-particle collisions where they conveniently refer to the centre-of-mass frame.

the spectra. Indeed, if the primary production rate $|\partial\nu(\varepsilon,o)/\partial\varepsilon|$ shows a peak structure due to e.g. Auger electrons, this structure will be recovered in the ejected-electron spectrum corresponding to (8') (shifted by U according to (6)), while it will show up as a <u>step structure</u> in the spectrum corresponding to (7'), due to occurrence of the integral production rate $\nu(\varepsilon,o)$. Thus, (8') describes a physical situation where a liberated electron loses essentially all its energy in the first collision, while (7') applies to the situation of (more or less) continuous slowing down. While we believe that (8') is inadequate to determine total yields, we will show in section 5 that this relation has some significance in the description of discrete features of ejected-electron spectra.

The quantity $\nu(\varepsilon,o)$ entering into (4), (4') and (7') is assumed to contain all primary and secondary ionizations (A & B) caused by the bombarding ion; electron transport (C) enters into the term $d\varepsilon/dx$, and the ejection process (D) is symbolized by the barrier height U, eq. (5). Each of these items will be discussed in some detail in the following sections. For very rough orientation, let us consider a fast light ion propagating along a straight line. It generates primary electrons with initial energy (T,dT) at a density of $Nd\sigma_i(E,T)$ [electrons/unit depth] where N is the number of target atoms per volume, E the ion energy, and $d\sigma_i(E,T)$ the differential cross section per target atom. Assuming the number of secondaries ($>\varepsilon$) per primary (T) to be given by

$$n(T,\varepsilon) \simeq \text{const} \cdot T/\varepsilon \quad \text{for } T \gg \varepsilon, \tag{9}$$

we find

$$\nu(\varepsilon,o) \simeq \text{const} \cdot \frac{N}{\varepsilon} \int T d\sigma_i(E,T) \tag{10}$$

or, with (7'),

$$\gamma \simeq \frac{1}{4} \cdot \text{const} \cdot |dE/dx|_e \int_U^\infty \frac{d\varepsilon}{\varepsilon |d\varepsilon/dx|} (1-U/\varepsilon), \tag{11}$$

where $(dE/dx)_e$ is the electronic stopping power of the ion,

$$(dE/dx)_e = -N \int T d\sigma_i(E,T). \tag{12}$$

The constant in (9) can be determined experimentally or from the theory of ionization cascades (ICRU, 1979; FOWLER, 1923). For binary collisions, one finds (SIGMUND, 1969) a value

$$\text{const} = \Gamma_m = m/(\psi(1) - \psi(1-m)); \quad 0 < m < 1 \tag{13}$$

which is $\sim 1/2$, and also emerges from a recent transport theory (SCHOU, 1980 b). Here, $\psi(x) = d \log \Gamma(x)/dx$, and $\Gamma(x)$ the gamma function. Except for a depth-dependent factor β in front of the stopping power, (11) is strictly equivalent with SCHOU's result. Eq. (11) predicts the electron yield to increase with <u>increasing</u> electronic stopping power of the ion, and with <u>decreasing</u> stopping power of the electron. However, while the latter feature is inherent in the rather general statement, (7'), and reflects the depth of origin of ejected electrons, the former feature relies heavily on the specification that $T \gg \varepsilon$ in (9), i.e., that the majority of all ejected electrons are due to <u>secondary ionization.</u> That condition is by no means generally fulfilled. Thus, although (11) gives a rough estimate of the dependence of γ on ion type and energy, it is at best qualitative. As a matter of fact, in the opposite case of primary ionization only -- still assuming isotropy -- we find

$$\nu(\varepsilon,0) \simeq N \int_{\varepsilon'=\varepsilon}^{\infty} d\sigma_i(E,\varepsilon') \tag{10'}$$

instead of (10). Then, (7') reads

$$\gamma \simeq \frac{1}{4} N \int_U^{\infty} d\varepsilon \frac{\int_\varepsilon^{\infty} d\sigma_i(E,\varepsilon')}{|d\varepsilon/dx|} \cdot (1-U/\varepsilon) , \tag{11'}$$

a relation that is in general quite different from (11).

Had we started from (8') instead of (7'), appropriate specification of $d\dot{\sigma}_i(E,T)$ in (10) and a more accurate estimate of $n(T,\varepsilon)$ than (9) would have led to STERNGLASS's (1957) well-known expression for the electron yield.

3. Primary Ionization Processes

Liberation of electrons can occur in collisions initiated by the bombarding ion itself or by the action of an energetic recoil atom, electron or photon. In this section, we consider the first category of processes. The investigation of this very wide range of phenomena constitutes a large subfield in atomic collision and radiation physics.

Ideally, one might wish to know ionization rates doubly differential in electron energy and angle to the beam <u>for penetration in a solid</u>. A less ambitious and more promising approach would be to apply (measured or calculated) ionization cross sections for ion-atom collisions in the gas phase, in so far as inner-shell excitations are concerned. For outershell processes, theoretical estimates based on quasi-free electrons seem to make up an acceptable approach. For the mere purpose of yield calculations, even Thomas-Fermi type estimates might be sufficient. In either case, (4) indicates that single-differential spectra in energy may be satisfactory for many processes.

A convenient reference standard is the cross section for Coulomb scattering with an atom containing Z_2 weakly bound electrons,

$$d\sigma_c(v,\varepsilon) = \frac{2\pi e_1^2 e^2}{mv^2} Z_2 \frac{d\varepsilon}{\varepsilon^2} , \quad 0 \leq \varepsilon \leq 2mv^2 , \tag{14}$$

where e_1 and v denote the charge and initial velocity of the ion, e and m the elementary charge and electron mass, and ε the kinetic energy of the recoiling electron. We note that in this limit the electron is ejected at an angle

$$\phi = \arccos (\varepsilon/2mv^2)^{1/2} \tag{14a}$$

to the ion beam. Note the peak in (14) toward low electron energies and the associated anisotropy.

Deviations from (14) occur
  i)   when the projectile is not a point charge,
  ii) at electron energies where the binding and mutual interaction of target electrons is important, and
  iii) in the presence of solid-state effects.

Unlike in case (14), essential differences occur between classical and quantal predictions, once those effects become important. Most available experimental data as well as many calculations refer to light ions, in particular protons, in the range of ion velocities where the Bethe theory (BETHE, 1930) should apply. For recent summaries cf.TOBUREN et al.(1979) on experiments and KIM (1975) and INOKUTI (1971) for theory.[2] Figs. 1a and 1b show results from a recent systematic study of H^+-He collisions(MANSON et al., 1975). It appears that the single-differential energy spectrum is reasonably well understood while there is somewhat greater uncertainty with regard to angular distributions.

For gas targets bombarded with high-velocity protons, deviations from the spectrum (14) occur both at the lower and the upper end and are conveniently described by the "Platzman plot"(PLATZMAN, 1961, KIM 1975) which shows the ratio of $d\sigma(v,\varepsilon)/d\sigma_c(v,\varepsilon)$ vs. ε. Unlike for free electrons, (14a), electron emission is predicted and measured also at backward angles ($\phi > 90^\circ$); at ejected energies comparable to atomic binding energies, the angular distributions are close to isotropic. However, at angles near 0° and 180°, major discrepancies occur between predicted and measured distributions. They are, though, hardly important with regard to integrated quantities like stopping power. The latter is well described by the Bethe theory (INOKUTI, 1971).

These results can to some extent be scaled to heavier projectiles at comparable or higher ion velocities, i.e., higher ion energies, but the limits of validity of this strategy have been little investigated. The main uncertainty is the replacement of an ion core by an effective point charge. Some success has been achieved in the scaling of stopping powers (BROWN et al., 1972).

A most severe limitation on the validity of the cross section (14) is the neglect of electronic motion; this restriction determines the range $0 \leq \varepsilon \leq 2mv^2$ of ejected electron energies. A conceptually simple procedure to remove this limitation is the so-called binary-encounter approximation (BEA) in which target electrons are considered free but given an initial distribution in velocity $\vec{v}_e$ (GRYZINSKI, 1965; GERJUOY, 1966). Conservation laws show that the maximum transferred energy becomes, then,

$$\varepsilon_{max} = 2mv(v+v_e) , \qquad (15)$$

if all directions of $\vec{v}_e$ are represented. For ion velocities $\lesssim v_e$, this decreases much more slowly with decreasing v than the upper limit in (14). The main domain of applicability of the BEA appears to be in innershell excitation by moderate- velocity light ions (GARCIA, 1970), in particular with regard to indirect processes (e.g. emission of Auger electrons).

---

[2] The fact that much theoretical work refers to incident electrons rather than protons is of minor significance in this context.

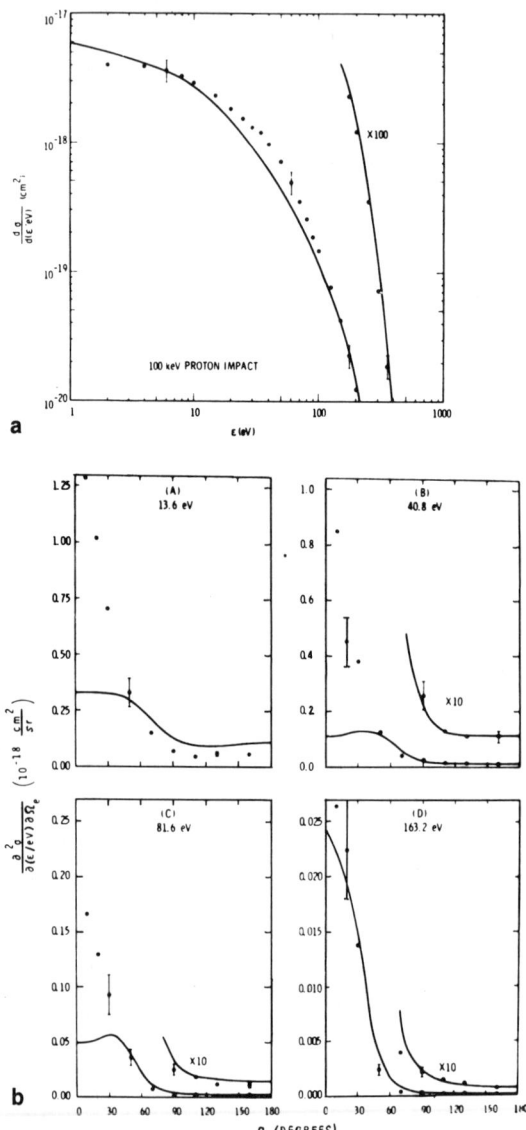

Fig. 1(a) Single differential cross section for electrons ejected from He by 100-keV proton impact as a function of ejected electron energy ε, (b) Angular distributions for electrons ejected at 13.6, 40,8, 81.6, and 163.2 eV by 100 keV proton impact on He. (———) theoretical results based upon the Born approximation. (·····) experimental results. From MANSON et al. (1975). By courtesy of the American Institute of Physics

A theoretical scheme suitable to treat electronic excitation by light ions at all velocities, in particular of outer shells and to some extent inner shells, is the dielectric theory (LINDHARD, 1954; RITCHIE, 1959). In comparison with (14), this theory takes into account the internal motion of target electrons as that of a degenerate Fermi gas, the screening of a point charge by target electrons as well as, to some extent, the mutual interaction between target electrons. The latter effect causes collective excitations (plasmons) that may decay by emission of electrons (FERRELL, 1958; PINES, 1960). The cross section for direct excitation of electrons has been evaluated in this model by BRICE et al. (1980). At high velocities, the cross section resembles (14) with three pronounced differences (Fig. 2a),

a) near the upper bound on ejected-electron energy, (14), $d\sigma$ falls continuously to zero between $2mv(v-v_F)$ and $2mv(v+v_F)$, v_F being the Fermi velocity,

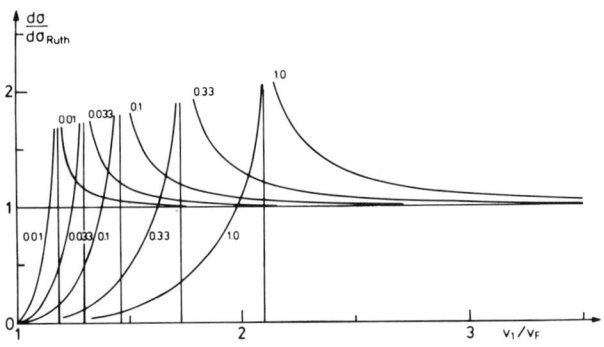

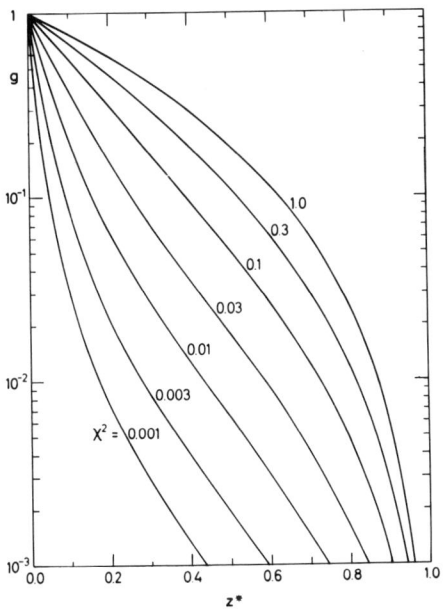

Fig. 2a Calculated energy spectra of primary electrons excited by fast moving point charge in Fermi gas. Abscissa: $v_1 = \sqrt{2(\varepsilon-\varepsilon_F)/m}$; v_F = Fermi velocity. Ordinate: $d\sigma$ = differential cross section; $d\sigma_{Ruth}$ = free Coulomb cross section, eq. (14). Numbers on curves indicate the value of the density parameter $\chi^2 = e^2/\pi\hbar v_F$. Singularities originate in the transition region between plasmon and single-particle excitations in the $(\vec{k},\omega)$ plot (LINDHARD et al., 1964). The curves are independent of projectile velocity for $v \gg v_F$, except for the upper cut-off around $\varepsilon = 2m(v\pm v_F)^2$

Fig. 2b Function $g(z^*)$ determining low-velocity-ion-excited electron spectra according to (16).

From BRICE et al. (1980). By courtesy of Det Kongelige Danske Videnskabernes Selskab

b) the singularity near $\varepsilon = 0$, eq. (14), becomes much less pronounced because the Pauli principle allows only a small fraction of electrons to undergo low excitations, and

c) a singularity develops instead at an electron velocity between $\sim v_F$ and $\sim 2v_F$, dependent on electron density, which is connected with the propagation of plasmons (BRICE et al., 1980; a somewhat related effect was discussed by SCHÄFER et al., 1978, 1980).

At low velocities, for $v \ll v_F$, the cross section takes on the form

$$d\sigma(\varepsilon_1) \simeq \text{const} \cdot g(\frac{\varepsilon_1}{2mv_Fv}) \, d(\frac{\varepsilon_1}{2mv_Fv}) \quad , \tag{16}$$

where $\varepsilon_1 = \varepsilon - \varepsilon_F$ and g is a function that depends on electron density and decreases from 1 to zero for ε_1 going from zero to $2mv_Fv$, eq. (15) (Fig. 2b).

When applied to atomic systems in accordance with the Thomas-Fermi principle, i.e., the local-density approach (GOMBAS,1956), the associated total cross section becomes comparable to the geometric cross section, hence the contribution of this type of excitation to electron emission is determined mainly by the magnitude of the maximum energy $2mvv_F$, which will be sizable in particular for the deeper-lying shells. There are indications from ion-atom collisions (ØSTGAARD OLSEN et al, 1976) that the (dominating) continuum part of measured electron spectra scales indeed like predicted by (16). Predictions on angular distributions are also available (BRICE et al., 1980).

Electrons can also be emitted in low-velocity collisions as a consequence of inner-shell excitation followed by an Auger transition. The excitation mechanism is considered to be electron promotion (FANO et al., 1965); the probability of electron emission increases with decreasing atomic mass, i.e., decreasing fluorescence yield. Auger electrons constitute a discrete portion of the ejected-electron spectrum at all projectile velocities, but there is little experimental evidence (OGURTSOV et al., 1970; STOLTERFOHT et al., 1974; DAHL et al., 1976) to support the notion (PARILIS et al., 1960) that the majority of ejected electrons -- even of high-energy electrons -- should have that origin (fig. 3).

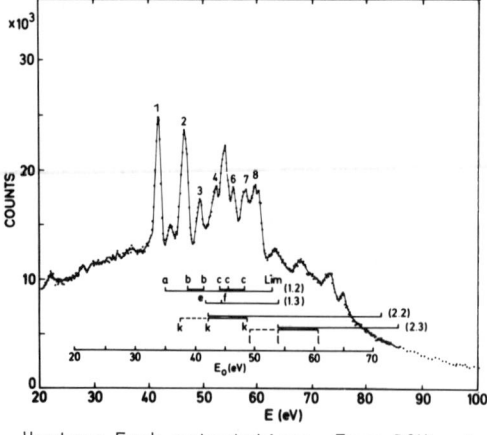

Fig. 3 Spectrum of electrons ejected from argon during 20.9 keV Mg$^+$ bombardment. The energies E and E_0 are the electron energies in the laboratory and source frame (Mg) respectively. Note that the Auger electrons are superimposed on an intense continuum. The initial electronic state has been characterized by the set (λ,γ) where λ is the number of $L_{2,3}$ vacancies and γ the number of outer electrons. In the figure the energies of the corresponding Auger groups of transitions $[(\lambda,\gamma) \to (\lambda-1,\gamma-2)]$ are shown. Transitions a,b, etc. are the electron energies obtained by Hartree-Fock calculations. From DAHL et al. (1976). By courtesy of The Institute of Physics

Rough estimates can be based on the Kessel model for inner-shell excitation cross sections (CACAK et al., 1970; KESSEL et al., 1973) which implies a step-function type of behaviour for each excitation cross section when the pertinent electron shells overlap . Estimated and measured fluorescence yields exist for a great number of systems (MCGUIRE, 1969, 1970; WALTERS et al., 1971, CHEN et al., 1973). For a comprehensive survey cf. MEHLHORN (1978).

A qualitative estimate of outer-shell electron promotion is due to FIRSOV (1959), based on geometric arguments. This model has been quite successful in predicting differential (MORGAN et al., 1962) and integrated energy losses (stopping power) in ion-atom collisions (for a summary cf. SIGMUND, 1975). PARILIS et al. (1960) extended this model by asserting that a number of valence electrons

$$\sim T(p)/I \qquad (17)$$

are excited in an ion-atom collision, where $T(p)$ is the mean electronic energy loss at impact parameter p and I the average ionization potential for outer shells, and that the deexcitation takes place predominantly through Auger processes. Eq. (17) implies that the majority of ion-atom collisions at impact parameters where $T(p) \gg I$ lead to multiple ionization. This is hardly supported by experimental evidence in the energy range which the theory was designed for, i.e., low-keV ions.

This section closes with a brief mentioning of various effects that influence the ionization rate of particles penetrating <u>solids</u>.

First, penetrating ions are scattered by nuclear collisions with target atoms. As a result, the ionization rate per unit depth is dependent on multiple scattering phenomena; in particular for light ions at low and moderate energies, energy deposition takes place in a much smaller depth range than what one would estimate by assuming projectile motion along a straight line. A rough estimate of the enhancement of ionization rates can be found by comparing tabulated depth distributions of deposited energy in ionization (BRICE, 1975; WINTERBON, 1975) with the underlying electronic stopping powers (HOLMEN et al., 1979; SCHOU, 1980b).

Ionization rates depend in general on the (charge) state of the bombarding ion. To the extent that processes near the target surface contribute to electron emission, this dependence should show up directly in measured electron yields and spectra. For inner-shell excitation by electron promotion, such effects may be pronounced (FASTRUP et al., 1971). The state of moving ions is known to be influenced by the nature of the solid, mainly because of the high frequency of collisions involving inner-shell excitation as compared to gases (BOHR et al., 1954). Although some aspects have been discussed extensively (BRIGGS et al., 1977; BETZ, 1976) little appears to be known about the implications on ejected-electron spectra.

In case of single-crystal bombardment, proper consideration needs to be made of the anisotropy of the ionization rate. Close-encounter processes are known to be suppressed for bombardment along channeling directions (LINDHARD, 1965). This can influence the electron yields both directly (through a change in the spectrum for primary ionization) and indirectly (through changes in ionization density because of changes in ion penetration). The anisotropy of ion-electron emission has long been known (MAGNUSON et al., 1963; MASHKOVA et al., 1963; FONTBONNE et al., 1970) and has been utilized successfully as a monitor for crystal lattice defects

(HOLMEN et al., 1972). Quantitative understanding is complicated due to the competition of bulk and surface processes.

4. Secondary Ionization Processes

4.1 Electrons

If the kinetic energy of a primary electron exceeds the lowest ionization potential of the target, secondary electrons will be generated. In case of high primary-electron energy, such an ionization cascade may involve a great number of electrons, given approximately by the relation

$$n(T) = \frac{T}{W}, \tag{18}$$

where $n(T)$ is the total number of electron-ion pairs (in a gas) or electron-hole pairs (in a solid) generated by a primary electron of energy T. The quantity $W = W(T)$ is known to approach a constant value for $T \gg W$. Experimental values of W have been tabulated in a recent compilation (ICRU, 1979).

We note that (18) does not contain any spectral information about the kinetic energy of the secondary electrons. Therefore, we need to briefly consider a formalism that allows to extract such information. Let $n(T,\varepsilon)$ be the mean number of secondary electrons with initial kinetic energy exceeding ε, generated in an ionization cascade initiated by a primary electron with initial energy T. Then, within the general framework of cascade theory (FOWLER, 1923; LINDHARD et al., 1963; SIGMUND, 1972; INOKUTI et al., 1976), we find the following integral equation for $n(T,\varepsilon)$,

$$\sum_i d\sigma_i(T,\Delta T)\left(n(T,\varepsilon) - n(T-\Delta T,\varepsilon) - n(\Delta T - I_i,\varepsilon) - \theta(\Delta T - I_i - \varepsilon)\right) = 0, \tag{19}$$

where $\theta(\xi)$ is a step function, $\theta(\xi) = \int_{-\infty}^{\xi} d\eta\, \delta(\eta)$, ΔT is the energy lost by the primary electron in an individual electron-atom collision, i indicates the various modes of excitation of the target atom, $d\sigma_i(T,\Delta T)$ the excitation cross section for energy transfer ΔT, and I_i the respective ionization potential. Obviously, one must require that

$$n(T,\varepsilon) = 0 \text{ for } T < \varepsilon; \tag{19a}$$

therefore, those modes of excitation that do not lead to ionization yield zero contribution to the last two terms in the brackets of (19).

Let first ε be much greater than the highest ionization potential so that I_i can be ignored. Then, disregarding boundaries in (19), we note that the asymptotic solution must be of the form

$$n(T,\varepsilon) \sim \Gamma \cdot \frac{T}{\varepsilon} - 1 \quad \text{for } T \gg \varepsilon \gg I_i \tag{19b}$$

with the constant Γ determined by the parameters in the cross section $d\sigma_i(T,\Delta T)$ only. Because of energy conservation, Γ cannot exceed 1, thus

$$0 < \Gamma < 1. \tag{19c}$$

For special cross sections, one obtains results like (13) for Γ (SIGMUND, 1969). For $\varepsilon=0$, $n(T,\varepsilon)$ has to go over into $n(T)$, (18). Keeping T large, we can accomplish this with the expression

$$n(T,\varepsilon) \sim \frac{T}{W + \varepsilon/\Gamma} - 1 \quad \text{for } T \gg \varepsilon, I_i \tag{20}$$

which is a reasonable interpolation. A similar expression has been derived for a specific model by ANDERSEN et al. (1974).

Detailed theoretical evaluations of Γ and W require a considerable amount of cross sectional data for the pertinent scattering processes (DOUTHAT, 1979). In effect, straightforward evaluation of $n(T,\varepsilon)$ from (19) can be avoided since the electron *flux* can be found directly from a similar equation (cf. below).

We need, however, an estimate of the behaviour of $n(T,\varepsilon)$ for small primary energy T. As a first approximation, (20) yields

$$n(T,\varepsilon) = 0 \quad \text{for } T \leq W + \varepsilon/\Gamma , \tag{21}$$

while an accurate solution of the integral equation (19) would predict a smooth decrease toward zero as T approaches $I_0+\varepsilon$, with I_0 the lowest ionization potential. More detailed calculations are available (INOKUTI, 1975).

It is of interest, then, to evaluate the relative contributions of primary and secondary ionization processes to the electron spectrum $\nu(\varepsilon,x)$. For a travelled path-length element dx, that spectrum is

$$\nu(\varepsilon,0)dx \simeq Ndx \left\{ \int_{T \geq \varepsilon} d\sigma(\nu,T) + \int d\sigma(\nu,T)n(T,\varepsilon) \right\} \tag{22}$$

in the present notation.

For the case where (14) is valid, i.e., for high-velocity ions and $\varepsilon \gg W$, (22) reads

$$\nu(\varepsilon,0) \simeq N \frac{2\pi e_1^2 e^2}{mv^2} \frac{Z_2}{\varepsilon} \left(1 + \frac{\varepsilon}{W+\varepsilon/\Gamma} \left(\log \frac{2mv^2}{W+\varepsilon/\Gamma} - 1 \right) \right) . \tag{23}$$

The first term between the brackets represents the primary electrons while the second, due to cascades, is usually smaller, even for high v, because Γ is $\ll 1$ for m close to 1 (free Coulomb-scattering, cf. (13) and (14)).

To take the other extreme, consider generation of primary electrons by Auger processes only, at a primary energy E_A, i.e., $d\sigma(\nu,T) = \sigma_A(\nu)\delta(T-E_A)dT$, with σ_A the cross section for Auger excitation. Then, (22) reads

$$\nu(\varepsilon,0) \simeq N\sigma_A(\nu) \left(1 + \left[\frac{E_A}{W+\varepsilon/\Gamma} - 1 \right] \cdot \theta(\Gamma \cdot (E_A-W)-\varepsilon) \right) \tag{23'}$$

for $\varepsilon < E_A$, where $\theta(\xi)$ is the step function. Obviously, for $\varepsilon, W \ll E_A$, the cascade contributions play the dominating role in the liberation of electrons.

As an illustration of a continuous transition between the two extremes, we may evaluate the spectrum (16) approximated by an exponential (ØSTGAARD OLSEN et al., 1976; WATANABE et al., 1979; WOERLEE et al., 1980),

$$d\sigma(v,T) \simeq \sigma(v) \, e^{-\beta T/v} \, d(\beta T/v), \quad 0 < T < \infty \tag{16'}$$

with an atomic parameter β; then, (22) reads

$$\nu(\epsilon,0) \simeq N\sigma(v) \, e^{-\beta\epsilon/v}(1 + \frac{v}{\beta}(W+\epsilon/\Gamma))^{-1} \, e^{\frac{\beta}{v}(\epsilon-W-\epsilon/\Gamma)} \;). \tag{23''}$$

We note that the total ionization cross section following from (16') is $\sigma(v)$ while the ionization stopping cross section is

$$\int_0^\infty T d\sigma(v,T) = \frac{v}{\beta}\sigma(v) \; ; \tag{24}$$

Thus, at low velocities, (23") is proportional to the ionization cross section while a proportionality with the ionization stopping cross section is approached at higher velocities, in agreement with the arguments presented in sec.2.

4.2 Recoil Atoms

At projectile velocities far above typical velocities of target electrons ($v \gg Z_2^{2/3} e^2/\hbar$), the specific energy loss is dominated heavily by electronic stopping (BOHR, 1948); recoil atoms receive less than 1/1000 of the total energy loss. Thus, secondary excitation by recoiling atoms is a negligible effect.

At lower projectile velocities, this is different. Indeed, for $v \ll Z_2^{2/3} \cdot e^2/\hbar$, for not too light ions, nuclear stopping is competitive, and at the low-energy end, defined by

$$\epsilon = \frac{M_2 E}{M_1+M_2} / (Z_1 Z_2 e^2/a) \lesssim 1 \; , \tag{25}$$

nuclear stopping shows a more or less pronounced dominance (Fig. 4)

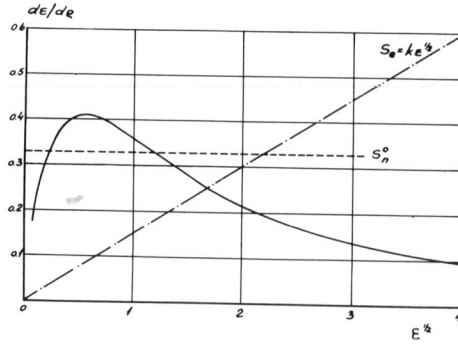

Fig. 4 Theoretical stopping cross sections $d\epsilon/d\rho$. Here ϵ and ρ are dimensionless measures of ion energy and path length, respectively. The horizontal dashed line and the full-drawn curves are the nuclear stopping cross section for Thomas-Fermi and inverse-square interaction, respectively. The dot-dashed line is a representative electronic stopping cross section. From LINDHARD et al. (1963). By courtesy of Det Kongelige Danske Videnskabernes Selskab

(LINDHARD et al., 1963). Here a is the Thomas-Fermi screening radius,

$$a = 0.8853\, a_o (Z_1^{2/3} + Z_2^{2/3})^{-1/2} \tag{25a}$$

with $a_o = 0.529$ Å.

The question arises, then, to what extent recoil atoms contribute a substantial amount of electronic excitation. A qualitative criterion can be based on the following parameter,

$$\zeta = \frac{\int d\sigma_{12}(E,T) \eta_2(T)}{S_{1e}(v)}, \tag{26}$$

where S_{1e} is the electronic stopping cross section of the ion, $d\sigma_{12}(E,T)$ the differential elastic scattering cross section for ion-target collisions, and

$$\eta_2(T) \simeq \int_0^T dE' \frac{S_{2e}(E')}{S_{22}(E')+S_{2e}(E')} \tag{26a}$$

the energy ending up in electronic excitation from a recoil atom slowing down from energy T to zero, where S_{2e} and S_{22} are the electronic and nuclear stopping cross section of a recoil atom, respectively. Eq. (26) can be evaluated on the basis of conventional cross sections for elastic scattering (LINDHARD et al., 1968, WINTERBON et al., 1970),

$$d\sigma_{ij} = C_{ij} E^{-m} T^{-1-m} dT; \quad 0 \leq T \leq \gamma_{ij} E, \tag{27a}$$

and the low-velocity stopping cross section (LINDHARD et al., 1961)

$$S_e = \xi_e \cdot 8\pi e^2 a_o \cdot \frac{Z_1 Z_2}{Z} \cdot \frac{v}{v_o}, \tag{27b}$$

with all parameters properly defined in the respective references. Then, one finds

$$\zeta = \Delta_m \cdot \frac{S_{12}(E)}{S_{1e}(v)} \cdot \frac{S_{2e}(v_{max})}{S_{22}(T_{max})}, \tag{28}$$

with $\Delta_m = (1-m)(\frac{1}{2} + m)^{-1}(\frac{1}{2} + 2m)^{-1}$. Specifically, for $m = \frac{1}{2}$ one obtains the simple result

$$\zeta = \frac{\sqrt{2}}{3}(\frac{Z_2}{Z_1})^{1/6} \frac{Z}{Z_2}(\frac{M_1}{M_1+M_2})^2 \quad ; \quad Z = (Z_1^{2/3} + Z_2^{2/3})^{1/2} \tag{26'}$$

which indicates that nuclear stopping is negligible for $Z_1 \ll Z_2$ (i.e., $M_1 \ll M_2$), causes about 1/4 of the electronic excitation for $Z_1 \sim Z_2$, and dominates for $Z_1 \gg Z_2$.

The importance of this effect was pointed out on the basis of measured electron yields as a function of Z_1 (DOROZHKIN et al., 1974). Theoretical estimates are available (HOLMEN et al., 1979; VINOKUROV et al., 1976). The recoil effect shows up directly in the energy dependence of the total secondary electron yield (Fig. 5).

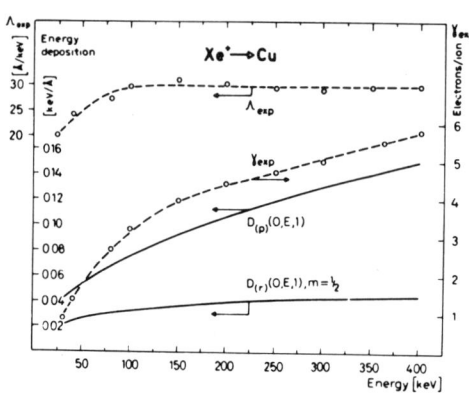

Fig. 5 Contribution of target recoils to electron emission yield. $D_{(p)}$ and $D_{(r)}$ indicate the energy deposited per unit depth in electronic motion (cf. (63)) by primary ions (Xe^+) and recoil atoms (Cu), respectively. E is the initial ion energy; $1 = \cos\theta$ indicates perpendicular incidence. γ_{exp} is the measured electron yield, and Λ_{exp} the material parameter extracted from (63), $\Lambda_{exp} = \gamma_{exp}/(D_{(p)} + D_{(r)})$. From HOLMÉN et al. (1979). By courtesy of the American Institute of Physics

4.3 Photons

While light is an important source in electron emission studies (ultraviolet or x-ray photoelectron spectroscopy), it appears less important in general to consider electron emission due to excitation by secondary photons. Indeed, primary photon yields are comparable in magnitude with primary electron yields, but free paths for photons are much longer, hence secondary excitation in the near-surface region pertinent to electron emission is a comparatively rare event. Exceptions to this rule ought to be looked for in insulators with large electron mean free paths.

For relativistic projectiles, energy loss due to bremsstrahlung may be dominating over ionization losses. Thus, electron emission caused by bremsstrahlung photons may eventually be competitive.

5. Electron Slowing-Down

According to (4), the slowing-down properties of target electrons enter into the electron spectrum primarily through the stopping-power term $|d\varepsilon/dx|$ in the denominator. Thus, the minimum knowledge needed is an estimate of $|d\varepsilon/dx|$ in the range of electron energies where spectral information is desired. In particular, for a calculation of the total electron yield, the region around the low-energy peak is crucial. Hence, $|d\varepsilon/dx|$ will be needed even at energies as low as ~ U, cf. (5). This range of energies is far below that where energy loss measurements can be made by transmission through thin foils. Indeed, experimental determinations of stopping parameters for low-energy electrons rely on electron emission spectra from pure and covered surfaces. It is important, then, to provide a precise analysis of the measured spectra.

5.1 Processes

Electrons moving in a solid are scattered either elastically on the atoms or ion cores of the target, or inelastically by exciting target electrons. At high electron energies (>> 1 keV), a description in terms of atomic cross sections, combined with theoretical estimates concerning plasmon and valence electron excitation, should provide an adequate description (RITCHIE et al., 1975; TUNG et al., 1977; ASHLEY et al., 1979), comparable with that used for electron slowing-down in gases (DOUTHAT, 1975, 1979). Increasing uncertainties arise the more the electron energy drops below 1 keV. With regard to inelastic scattering, most existing estimates have been based on the free-electron-gas (jellium) model which disregards band structure effects. Scattering mean-free-paths for elastic scattering on target atoms decrease with decreasing electron energy, such as to cause diffraction phenomena of the type known from low-energy electron diffraction to become noticeable.

Scattering processes in solids are conveniently characterized by a transition probability per unit time, $W(\vec{k},\vec{k}')d^3k'$, with initial and final wave vectors $\vec{k}$ and $\vec{k}'$, respectively; the relation to a description in terms of cross sections is provided by

$$Nv \, d\sigma(\vec{v},\vec{v}') \Rightarrow W(\vec{k},\vec{k}')d^3k' \; ; \tag{29}$$

In this manner, band gaps and density-of-states effects are readily included in the description. Eq. (29) determines various inelastic scattering cross sections or differential inverse mean free paths, $\tau(\varepsilon,T)dT$. The latter is the probability per unit path length for an electron with energy ε to lose energy (T,dT). Figs. 6 and 7 show $\tau(\varepsilon,T)$ for excitation of single electron-hole pairs (τ_{ee}) and for plasmon excitation (τ_p), respectively. The evaluation is due to RITCHIE et al. (1975) and is based on LINDHARD's (1954) dielectric function for the electron gas. Most notable for single-particle excitation (Fig. 6) is the finite maximum at some intermediate energy and the lack of a singularity at $T = 0$ as would be expected for pure Coulomb scattering, cf. (14). Thus, the total mean free path is nonzero for these processes, and the spectrum is of a quasi-discrete nature. This feature is even more pronounced for plasmon losses (Fig. 7).

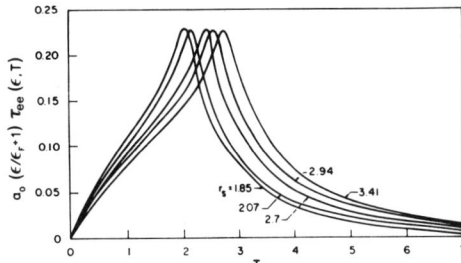

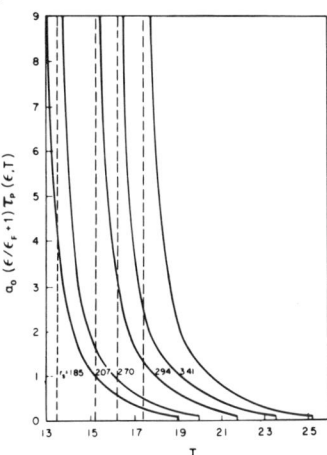

Fig. 6 Inverse differential mean free path of an electron in a free electron gas at various electron densities ($r_s=(4\pi \cdot a_0^3 \cdot n/3)^{-1/3}$), where n is the electron density, a_0 the Bohr radius, and ε_F the Fermi energy. $\tau_{ee}(\varepsilon,T)dT$ is the probability per unit path length that an electron with energy ε shall lose energy (T,dT) to electron-hole pair excitation. From RITCHIE et al. (1975). By courtesy of Academic Press, Inc.

Fig. 7 Same as Fig. 6 for plasmon creation. From RITCHIE et al. (1975). By courtesy of Academic Press, Inc.

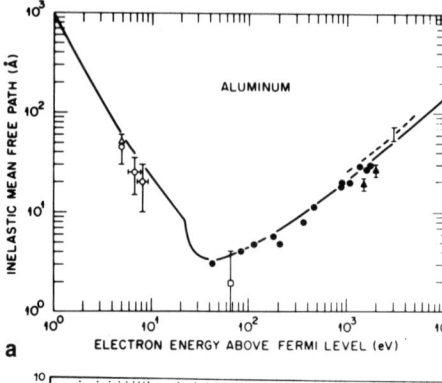

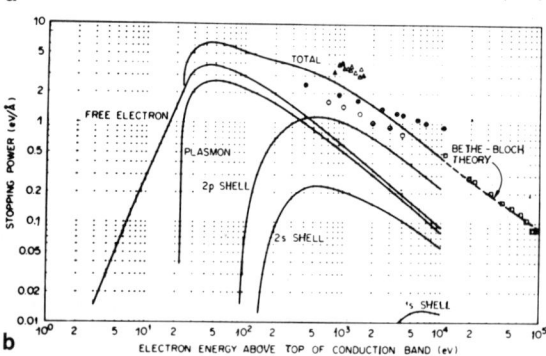

Fig. 8 Calculated inelastic mean free path (a) and stopping power (b) of aluminium for electrons as a function of electron energy. The points are experimental data taken from various sources. From ASHLEY et al. (1979). By courtesy of North-Holland Publishing Company

This type of calculation has been elaborated with increasing detail over a number of years. Figures 8a and 8b show results for the total inelastic mean free path and the stopping power for electrons in aluminium (ASHLEY et al., 1979). Note the minimum in the former, ~3Å for $\varepsilon \approx 40$ eV.

Similar considerations have been made by GANACHAUD et al. (1979). Figure 9 shows their results for the mean free path for inelastic as well as elastic scattering, the latter being evaluated on the basis of phase shifts for a muffin-tin potential. One notes the dominating role of elastic scattering.

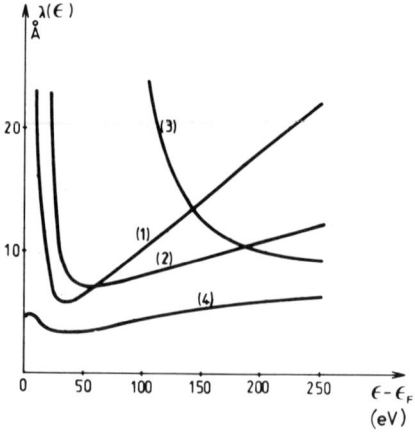

Fig. 9 Calculated mean free paths of electrons in aluminium: (1) electron-electron; (2) electron-plasmon; (3) ionization (x 1/10); (4) elastic. Note that elastic collisions dominate over inelastic processes at all energies studied. From GANACHAUD et al. (1979). By courtesy of North-Holland Publishing Company

5.2 Homogeneous Source in Infinite Medium

Let $G(\vec{v}_0,\vec{v},\vec{r},t)d^3r d^3v$ be the probability that a primary electron, moving with velocity $\vec{v}_0$ in $\vec{r} = 0$ at $t = 0$, be found in $(\vec{r},d^3r)$ at time t with a velocity $(\vec{v},d^3v)$. This function obeys the Boltzmann transport equation

$$(-\frac{1}{v_0}\frac{\partial}{\partial t} - \frac{\vec{v}_0}{v_0}\cdot\vec{\nabla})G_0 = N\int d\sigma(G_0-G') \quad , \tag{30}$$

with $G_0 = G(\vec{v}_0,\vec{v},\vec{r},t)$, and $d\sigma = d\sigma(\vec{v}_0,\vec{v}')$ the cross section for scattering from $\vec{v}_0$ to $(\vec{v}',d^3v')$.

Moreover,

$$G(\vec{v}_0, \vec{v}, \vec{r}, 0) = \delta(\vec{v}_0-\vec{v})\delta(\vec{r}). \tag{30a}$$

For a constant source of ψ [electrons/unit time], the (stationary) velocity distribution, integrated over all space and directions of motion, thus obeys the equation

$$\frac{\psi}{V}\delta(E-\varepsilon) = N\int d\sigma\cdot(G(E,\varepsilon) - G(E',\varepsilon)) \quad , \tag{31}$$

where

$$G(E,\varepsilon) = \psi\int d^3r\int d^3v\,\delta(\varepsilon - \frac{m}{2}v^2)\int_0^\infty dt\,G(\vec{v}_0,\vec{v},\vec{r},t) \quad , \tag{31a}$$

and $E = \frac{m}{2}v_0^2$.

The present form of Boltzmann's equation, eq. (30), called backward equation, is more convenient for analytical evaluation than the conventional forward form (SPENCER et al., 1954), but the two forms are strictly equivalent (cf. LINDHARD et al., 1971).

Now, let $E' = E-T$, with T the energy loss in an individual collision, and expand

$$G(E',\varepsilon) = G(E,\varepsilon) - T\frac{\partial G}{\partial E} + \frac{1}{2}T^2\frac{\partial^2 G}{\partial E^2} \dots ; \tag{32}$$

then, to first order in T, (31) reads

$$\frac{\psi}{V}\cdot\delta(E-\varepsilon) \simeq NS(E)\frac{\partial G}{\partial E} \quad , \tag{32'}$$

or

$$G(E,\varepsilon) \simeq \frac{\psi}{VNS(\varepsilon)}\cdot\theta(E-\varepsilon) \quad , \tag{33}$$

where $NS(E) = N\int T\,d\sigma = |dE/dx|$, i.e., the stopping power of the electron and θ is the step function. Eq. (33) is consistent with (3).

For a number of reasons, the continuous-slowing-down-approximation (csda) leading to (32') does not describe adequately the behavior of $G(E,\varepsilon)$ near the edge $\varepsilon \simeq E$. Here, small energy losses are decisive. Guided by figs. 6 and 7, let us solve (31) for a peaked energy loss spectrum,

$$d\sigma(E,T) = \sigma(E)\delta(T-\hbar\omega)dT \quad , \tag{34}$$

with som fixed energy loss quantum $\hbar\omega$; here we find

$$G(E,\varepsilon) = \frac{\psi}{vN\sigma(\varepsilon)} \{\delta(\varepsilon-E)+\delta(\varepsilon-E+\hbar\omega)+ \ldots\} \quad , \tag{35}$$

for $\hbar\omega \ll E$, i.e., a peaked structure. Both features, the continuous one below the leading edge, (33), and an oscillatory structure corresponding to (35), show up in nummerical solutions of the Boltzmann equation (Fig. 10) (Tung et al., 1977).

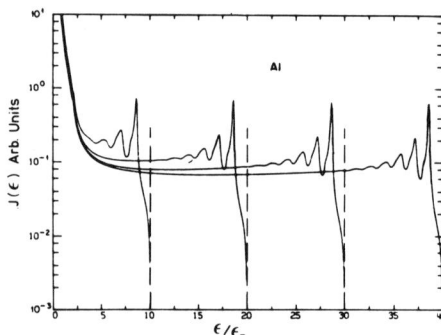

Fig. 10 Electron slowing-down flux $J(\varepsilon)$ in aluminium due to monoenergetic sources of electrons with energies $\varepsilon/\varepsilon_F$ = 10, 20, 30 and 40 (in units of the Fermi energy ε_F). From TUNG et al. (1977). By courtesy of the American Institute of Physics

Note that $S(\varepsilon) = \hbar\omega \cdot \sigma(\varepsilon)$ for the spectrum (34). When this is inserted into (33), one finds that (33) correctly describes the <u>average</u> of $G(E,\varepsilon)$, (35), taken over each interval $\hbar\omega$.

It is easily seen that inclusion of higher order terms in the expansion (32) does not affect the spectrum in the limit of $\varepsilon \ll E$. Indeed, take the equation of next higher order,

$$\frac{\psi}{v}\delta(E-\varepsilon) = NS(E)\frac{\partial G}{\partial E} - \frac{1}{2}NW(E)\frac{\partial^2 G}{\partial E^2} \quad , \tag{36}$$

with $W = \int T^2 d\sigma$. Eq. (36) can be solved iteratively, by insertion of (33) into the second derivative. The first iteration yields

$$\frac{\partial G}{\partial E} \simeq \frac{\psi}{vNS(\varepsilon)}\left(\delta(E-\varepsilon) + \frac{W(\varepsilon)}{2S(\varepsilon)}\delta'(E-\varepsilon)\right)$$

i.e.,

$$G(E,\varepsilon) \simeq \frac{\psi}{vNS(\varepsilon)}\left(\theta(E-\varepsilon) + \frac{W(\varepsilon)}{2S(\varepsilon)}\delta(E-\varepsilon)\right) \quad , \tag{37}$$

which differs from (33) only near the upper bound. Eq. (37) shows that G is insensitive to higher moments of the cross section for $\varepsilon \ll E$.

Even though we assumed a monochromatic source according to (30a), secondary electrons will be generated according to sect. 4.1. In order to take those into account, we amplify (31) by a term analogous to the third term in (18), i.e.,

$$\frac{\psi}{v}\delta(E-\varepsilon) = N \sum_i \int d\sigma_i(E,T) \{G_{(1)}(E,\varepsilon) - G_{(1)}(E-T,\varepsilon) - G_{(1)}(T-I_i,\varepsilon)\} \tag{38}$$

where $G_{(1)}(E,\varepsilon)$ is defined as $G(E,\varepsilon)$, but takes into account the ionization cascade. We split $G_{(1)}(E,\varepsilon)$ into G, (33), and a cascade correction ΔG,

$$G_{(1)}(E,\varepsilon) = G(E,\varepsilon) + \Delta G(E,\varepsilon) \quad . \tag{39}$$

Inserting (39) into (38) and taking into account (31), we find the following integral equation for ΔG,

$$\sum_i \int d\sigma_i(E,T)\{\Delta G(E,\varepsilon) - \Delta G(E-T,\varepsilon) - \Delta G(T-I_i,\varepsilon) - G(T-I_i,\varepsilon)\} = 0 \quad . \tag{40}$$

Now, according to (33), we have

$$G(T-I_i,\varepsilon) \simeq \frac{\psi}{vNS(\varepsilon)} \cdot \theta(T-I_i-\varepsilon) \quad , \tag{41}$$

hence (40) becomes identical with (19) apart from a factor, and we find

$$\Delta G(E,\varepsilon) = \frac{\psi}{vNS(\varepsilon)} n(E,\varepsilon) \quad , \tag{42}$$

or

$$G_{(1)}(E,\varepsilon) = \frac{\psi}{vNS(\varepsilon)} (1 + n(E,\varepsilon)) \quad \text{for } \varepsilon \lesssim E. \tag{43}$$

ΔG approaches zero for $\varepsilon \simeq E$, but for $\varepsilon \ll E$, (20) yields

$$G_{(1)}(E,\varepsilon) \simeq \frac{\psi}{vNS(\varepsilon)} \cdot \frac{E}{W + \varepsilon/\Gamma} \quad \text{for } E \gg \varepsilon, I_i \quad . \tag{44}$$

In practice, we may apply (44) for $\varepsilon \lesssim \Gamma \cdot (E-W)$, and

$G_{(1)} = G$, eq. (33), for $\varepsilon \gtrsim \Gamma \cdot (E-W)$, in complete analogy with (21).

5.3 Homogeneous Source in Semi-Infinite Medium

In the preceding paragraph, we were dealing with the energy spectrum of moving particles for a source of ψ [primary electrons/unit time], regardless of the distribution in space and direction of motion. In an infinite medium, with an isotropic source of electrons of initial energy E at a source strength Ψ [primary electrons / unit time and volume], the mean number of electrons at energy $(\varepsilon,d\varepsilon)$ in the volume element $[\vec{r},d^3r]$ is given by

$$G(E,\varepsilon,\vec{\Omega},\vec{r}) \, d\varepsilon d^2\Omega d^3r = \frac{\Psi}{\psi} G(E,\varepsilon) \, d\varepsilon \frac{d^2\Omega}{4\pi} d^3r \quad , \tag{45}$$

with $G(E,\varepsilon)$ being the quantity defined by (31). In the presence of secondary ionization, G has to be replaced by $G_{(1)}$, cf. (38). From (45), we obtain the flux (electrons per unit time and area) through the plane $x = 0$,

$$J d\varepsilon d^2\Omega = v \cos\theta \ G(E,\varepsilon,\vec{\Omega},0,y,z) \ d\varepsilon d^2\Omega , \tag{45'}$$

where θ is the angle between the electron velocity and the outward surface normal. As was observed already in sect. 2, the cross section for angular deflection of the electrons does not enter into this expression.

Let us, next, consider a semi-infinite medium, but disregard all angular deflection of moving electrons. Thus electrons originating at depth x slow down continuously along straight lines. In this case,

$$J = \Psi \ \frac{1}{4\pi} \int_0^\infty dx \ \delta(\varepsilon - \varepsilon(E, \frac{x}{|\cos\theta|})) , \tag{46}$$

where $\varepsilon = \varepsilon(E,s)$ is the energy of an electron at path length s for initial energy E. The integral is easily evaluated and yields

$$J(E,\varepsilon,\vec{\Omega}) = \Psi \cdot \frac{1}{4\pi} \ \frac{|\cos\theta|}{|d\varepsilon/dx|} \cdot \theta(E-\varepsilon) . \tag{47}$$

The same result is achieved when (45) and (33) are inserted into (45').

Of course, in the absence of angular deflection, the same electron spectrum is obtained for a semi-infinite as for an infinite medium.

Let us, next, consider the case of strongly dominating elastic scattering. If an electron undergoes many elastic scattering events before losing a substantial portion of its energy, a diffusion approximation is justified to describe electron transport (BETHE et al., 1938).

We consider first the spatial distribution of electrons from a monochromatic isotropic source in $x=0$ as a function of time, disregarding all energy loss. Boltzmann's equation can be shown to yield the following (rigorous) expression for the width of the depth profile in an infinite medium,

$$<x^2> = \frac{2v}{3N\sigma_1} \ [t + \frac{1}{vN\sigma_1} \ (e^{-vN\sigma_1 t} -1)], \tag{48}$$

where σ_1 is a transport cross section,

$$\sigma_1 = \int d\sigma(\phi) \ (1-\cos\phi). \tag{48a}$$

For small t, (48) yields

$$<x^2> \simeq \frac{1}{3}(vt)^2 , \tag{49}$$

corresponding to rectilinear motion, while for large t, we find

$$<x^2> \simeq 2 Dt \tag{50}$$

with

$$D = \frac{1}{3} \ \frac{v}{N\sigma_1} , \tag{50a}$$

i.e., a diffusion-like behavior.

The one-dimensional diffusion equation has the following solution for the density of electrons [electrons/unit depth] at time t in a semi-infinite medium ($0<x<\infty$),

$$\frac{Qdx'}{\sqrt{4\pi Dt}} \left(\exp\left(-\frac{(x-x')^2}{4Dt}\right) - \exp\left(-\frac{(x+x')^2}{4Dt}\right)\right), \tag{51}$$

if Qdx' electrons started from x' at $t = 0$ with an isotropic velocity distribution and speed v. The flux through $x = 0$ at time t is then found to be

$$-\frac{Q\, x'dx'}{\sqrt{4\pi Dt^3}} e^{-x'^2/4Dt}, \tag{52}$$

by means of Fick's first law. For a homogeneous distribution of sources, integration over x' yields a total current density

$$\sqrt{\frac{D}{\pi t}}\, Q \tag{53}$$

of electrons crossing the surface at time t. Within the range of validity of the diffusion approximation, this flux is distributed according to the cosine law. Hence, with $Q=\Psi dt$, we find a stationary current density

$$dJ = \Psi \frac{d^2\Omega}{\pi} |\cos\theta| \cdot \frac{d(Dt)}{\sqrt{\pi Dt}} \tag{54}$$

of electrons which cross the surface in the direction $(\vec{\Omega}, d^2\Omega)$ after having been in motion for a time (t,dt). Eq. (54) holds for $t \gtrsim t_1$, where t_1 is the crossover between the straight-line solution (49) and the diffusion solution (50), i.e.,

$$t_1 = \frac{6D}{v^2} = \frac{2}{vN\sigma_1}; \tag{55}$$

in the other case, for $t \lesssim t_1$, the assumption of straight-line motion yields

$$dJ = \Psi \frac{d^2\Omega}{4\pi} |\cos\theta| vdt, \tag{54a}$$

instead of (54).

We can now convert the time spectrum into an energy spectrum by means of $G(E,\varepsilon,t)$ which is the function introduced in (30) integrated over spatial and angular variables. In doing so, we include the energy dependence of σ_1, i.e., of D, by way of the age approximation (BETHE et al., 1938). This has already been accounted for in the notation of (54). Thus, we find an electron spectrum

$$Jd\varepsilon d^2\Omega \simeq \Psi \frac{d^2\Omega}{\pi} |\cos\theta|\, d\varepsilon \left\{ \frac{1}{4} \int_0^{t_1} vdt + \int_{t_1}^{\infty} \frac{d(Dt)}{\sqrt{\pi Dt}} \right\} \cdot G(E,\varepsilon,t). \tag{56}$$

Within the continuous slowing-down approximation, i.e. for

$$G(E,\varepsilon,t) = \delta(\varepsilon-\varepsilon(E,t)),$$

(56) yields

$$J(E,\varepsilon,\vec{\Omega}) = \Psi \frac{|\cos\theta|}{4\pi} \cdot \frac{1}{NS(\varepsilon)} \quad \text{for } \varepsilon_1 \lesssim \varepsilon \lesssim E, \tag{57a}$$

and

$$J(E,\varepsilon,\vec{\Omega}) = \Psi \cdot \frac{|\cos\theta|}{\pi} \cdot \left(\frac{1}{\sqrt{\pi Dt}} \cdot \frac{1}{\partial\varepsilon/\partial(Dt)}\right)_\varepsilon =$$

$$= \Psi \frac{|\cos\theta|}{\pi} \cdot \frac{f(\varepsilon)}{(\pi \int_\varepsilon^E d\varepsilon' f(\varepsilon'))^{\frac{1}{2}}} \quad \text{for } 0<\varepsilon \leq \varepsilon_1 , \qquad (57b)$$

where

$$f(\varepsilon) = (3 N^2 \sigma_1(\varepsilon) S(\varepsilon))^{-1} \qquad (57')$$

and

$$\varepsilon_1 = \varepsilon(E, t_1).$$

As it must be, we retain the solution neglecting scattering, (47), in (57a), but at energies below ε_1, the calculated spectrum falls below by a factor of

$$\frac{4}{3 N \sigma_1(\varepsilon) [\pi \int_\varepsilon^E d\varepsilon' f(\varepsilon')]^{\frac{1}{2}}}, \qquad (58)$$

which is of the order of $\sim(S(\varepsilon)/\varepsilon\sigma_1(\varepsilon))^{\frac{1}{2}}$ for very low electron energies ε. Obviously, the square-root dependence prevents a drastic reduction except in extreme cases. In the case of a narrow edge, $E-\varepsilon_1 \ll E$, the behavior of (57b) near $\varepsilon \approx \varepsilon_1$ is given by

$$J \simeq \Psi \frac{|\cos\theta|}{\pi} \left(\frac{f(E)}{\pi(E-\varepsilon)}\right)^{\frac{1}{2}} , \qquad (59)$$

i.e., an inverse square-root type behavior for $\varepsilon \leq \varepsilon_1$, with

$$E - \varepsilon_1 \simeq \frac{16 S(E)}{3\pi \sigma_1(E)} , \qquad (59a)$$

and a constant for $\varepsilon_1 \leq \varepsilon \leq E$.

For the region near the edge, we could also apply the discrete energy-loss spectrum (34) instead of the continuous slowing-down approximation. Boltzmann's equation leads then to

$$G(E,\varepsilon,t) \simeq e^{-N\sigma vt} \sum_\nu \delta(\varepsilon-E+\nu\hbar\omega) \frac{1}{\nu!} (N\sigma vt)^\nu$$

for $|\varepsilon-E| \ll E$, and, hence, to

$$J \simeq \Psi \frac{|\cos\theta|}{4\pi} \cdot \frac{1}{N\sigma} \sum_\nu \frac{1}{\nu!} \delta(\varepsilon-E+\nu\hbar\omega) \cdot g_\nu(s_1), \qquad (60)$$

where

$$g_\nu(s_1) \simeq \int_0^{s_1} ds\, s^\nu e^{-s} + s_1^{\frac{1}{2}} \int_{s_1}^\infty ds\, s^{\nu-\frac{1}{2}} e^{-s}, \qquad (60a)$$

and $s_1 \simeq 2\sigma/\sigma_1$. The integral (60a) could be expressed in terms of incomplete gamma functions (ABRAMOWITZ et al.). For $s_1 \ll 1$, we find $g_\nu(s_1) \simeq \Gamma(\nu+\frac{1}{2}) s_1^{\frac{1}{2}}$ while for $s_1 \gg 1$, $g_\nu(s_1) \simeq \Gamma(\nu+1)$, i.e., again a ratio of the order of $(\sigma/\sigma_1)^{\frac{1}{2}}$ between the solutions with and without scattering.

5.4 Lattice Structure Effects

The application of transport theory in electron slowing down, as manifested by the occurrence of various mean free paths, implies a random distribution of scattering centers in the target. The validity of this picture is tested most directly by measurement of angular distributions of electrons emitted from single crystals. Some structure is well-documented experimentally in the energy-resolved angular distribution (HLICS et al., 1970; MISCHLER et al., 1973; BENAZETH et al., 1976); such structure (cf. fig. 11) is presumably related to diffraction effects familiar in low-energy electron diffraction, angular-resolved photoemission and alike. Theoretical descriptions are available (ABERDAM et al., 1976, 1977; NÉGRE et al. 1978; MISCHLER et al., 1979), but predict much more pronounced structure than what is measured (Fig. 12).

Taken as a correction to transport theory, diffraction effects appear to constitute an effect of minor significance (<10 pct.) in comparison with other uncertainties inherent in the theory. Note in particular that diffraction effects tend to cancel in polycrystals.

Taken as a subject matter of its own, the investigation of diffraction effects under ion bombardment may eventually reveal important information on the spatial and temporal structure of atomic collision cascades, as well as the details of the excitation mechanism. Angular distributions of Auger electrons are distinctly different from those of so-called true secondary electrons; the analysis of this difference must involve the depth as well as the lattice location of the Auger-emitting atom.

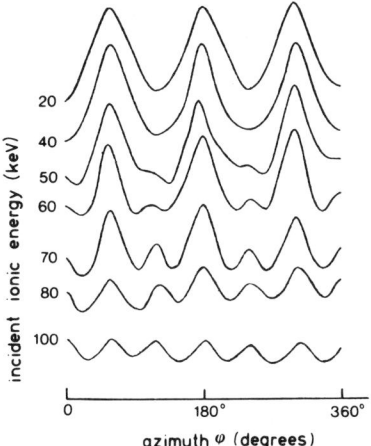 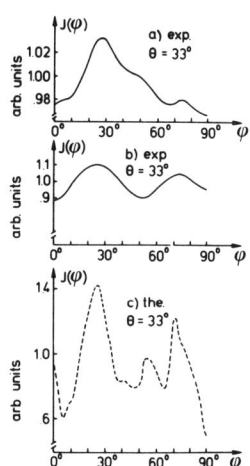

Fig. 11 Angular distribution of 15 eV secondary electrons emitted from a Cu(111) crystal under Ar$^+$-ion bombardment. The polar angle of emission is fixed ($\theta=33°$) while the azimuthal angle is varied. From BENAZETH et al. (1976). By courtesy of North-Holland Publishing Company

Fig. 12 Angular distribution $J(\phi)$ of 63 eV Auger electrons from Al (110); azimuthal angle ϕ; fixed value of the polar angle ($\theta=33°$); a: experimental curve for 50 keV Ar$^+$ bombardment; b: experimental curve for electron bombardment; c: calculated curve. From MISCHLER et al. (1979). By courtesy of North-Holland Publishing Company

For the true secondary electrons with energies just above the vacuum level the density of electronic states in the emitting crystal may be of importance. When this is the case one will expect the energy distribution of emitted electrons to directly reflect the crystal density of states. This effect has been demonstrated experimentally (WILLIS et al., 1978).

6. Electron Escape

Within the scope of this article, we are dealing with bulk processes; hence, electrons are emitted after having been excited into the conduction band. This implies that the pertinent surface barrier is given by (5), with ϕ the work function as determined by conventional means (CARDONA et al., 1978, HÖLZL et al., 1979), and E_F the Fermi energy of the target material. A distinction between the surface potential for a core electron and a valence electron should be necessary only to the extent that the presence of the hole modifies the work function; this ought to be an effect of higher order.

The surface barrier $U=E_F+\phi$ enters the energy spectrum and angular distribution of ejected electrons via (6). It follows, by comparison of (4) and (4'), that for $U \neq 0$, the spectrum in general approaches zero for ε' going to zero, i.e., there is a maximum in the energy spectrum at an energy of the order of $\sim U$ (AMELIO, 1970; CHUNG et al., 1974). Provided that U is well known for a given target surface under the particular experimental conditions, one might invert the step from (4) to (4'), i.e., determine the flux function $\nu(\varepsilon,o)/|d\varepsilon/dx|$ from a measured energy spectrum of ejected electrons. It follows from (4) and (4') that in case of an isotropic velocity distribution in the target, the angular distribution of ejected electrons follows the Knudsen cosine law, regardless of the magnitude of U. On the other hand, in the presence of diffraction phenomena, measured angular distributions must depend on U.

7. Discussion

While most of the elements pertinent to a theory of electron emission have been discussed in some detail in the foregoing sections, we refrain from actually collecting them to make predictions on yields and energy spectra in view of the limited space as well as the deadline posted by the editors of this volume. Instead, we provide a brief summary of the theoretical literature in context with the present outline, as well as a qualitative discussion of physically observable effects.

7.1. Yields

In an assessment of the validity of existing yield calculations, it appears important to estimate the difference between (8'), upon which most existing analytical models rely (WOLFF, 1954; STERNGLASS, 1957; PARILIS et al.,1960; GHOSH et al., 1962, 1963; KANAYA et al., 1972, 1974; BEUHLER et al., 1977) and (7') which is implicit in the model of SCHOU (1980).

Consider first the case of dominating Auger excitation at an energy $\varepsilon=\varepsilon_A$. Then, for $\varepsilon_A \gg U$, and an energy-loss spectrum according to (34) for liberated electrons, we find a ratio of

$$\frac{\gamma}{\gamma'} \simeq \frac{\sigma(\varepsilon_A)}{\hbar\omega} \int_U^\varepsilon \frac{d\varepsilon}{\sigma(\varepsilon)} \left(1 - \frac{U}{\varepsilon}\right) \tag{61}$$

from (7') and (8'). The value of the integral (61) is often dominated by contributions from $\varepsilon \sim U$; in that case one finds $\gamma/\gamma' \sim (U/\hbar\omega) \cdot (\sigma(\varepsilon_A)/\sigma(U))$. It is obvious that the ratio (61) depends on how strongly $\sigma(E)$ varies with ε. If the variation is stronger than linear, γ' underestimates the yield.

When the excitation function is a power law, $|d\nu(\varepsilon,o)/d\varepsilon| \propto \varepsilon^{-\alpha}$, we find

$$\frac{\gamma}{\gamma'} \simeq [\int_U^\infty \frac{d\varepsilon}{\varepsilon^\alpha \sigma(\varepsilon)} \cdot \frac{\varepsilon}{|\alpha-1|\hbar\omega} (1-\frac{U}{\varepsilon})] \cdot [\int_U^\infty \frac{d\varepsilon}{\varepsilon^\alpha \sigma(\varepsilon)} (1-\frac{U}{\varepsilon})]^{-1}, \qquad (62)$$

which is of the order of $\sim U/\hbar\omega|\alpha-1|$) if the dominating contributions stem from $\varepsilon \sim U$.

This ratio differs less drastically from 1 than does (61); it is, however, reasonable to conclude that a factor of two disagreement between γ and γ' should be the rule rather than the exception.

Although (7') ignores the effect of the target surface while (8') does not, we assert that (7') is a better starting point for a yield calculation than (8').

Eq. (11) suggests the electron yield to be written in the form (HOLMEN et al., 1979; SCHOU, 1980),

$$\gamma = \Lambda \cdot D(x=0,E), \qquad (63)$$

where Λ is a material constant determined by U and $d\varepsilon/dx$, and D the energy deposited per unit depth by the bombarding ion, i.e., the stopping power corrected for multiple scattering of the ion as well as energy transport by liberated electrons and recoil atoms. The discussion in Sect. 4.1 indicates that (63) should be valid to the extent that high-energy electrons ($\Gamma \gg U$) dominate the spectrum of primary excitation. This assumption is rarely fulfilled. Deviations from (63) can be estimated in specific cases on the basis of (23), (23'), and (23").

For evaluation of (63), the reader is referred to SCHOU (1980b).

Most existing analytical models reduce the energy dependence of the electron yield to that of the electronic stopping power of the ion, with various corrections applied.

Finally, mention should be made of numerical approaches to the problem, both by straight Monte Carlo simulation of the slowing-down process (KOSHIKAWA et al., 1974; GANACHAUD et al., 1979), and by brute-force solution of the (forward) Boltzmann equation (BINDI et al., 1980). A theoretical formalism based on the invariant-embedding approach has been set up (MITITEL, 1976) but apparently not yet been evaluated.

Some of the above-quoted papers refer to electron rather than ion bombardment. While there are both quantitative and qualitative differences between the two situations, this limitation usually reflects the research interest of the respective authors, rather than an inherent limitation of the theoretical scheme.

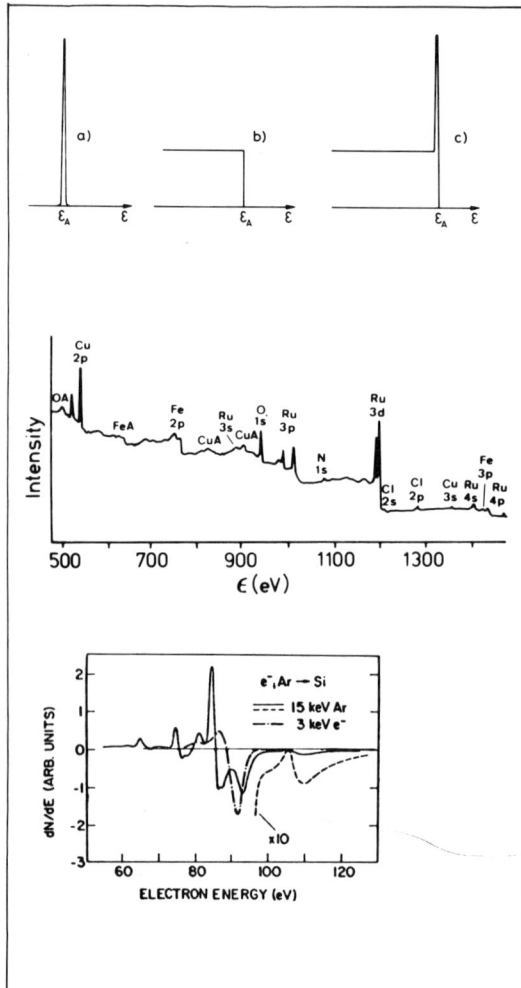

Fig. 13 Schematic illustration of the energy distribution of electrons detected outside the crystal. a) and b) refer to the two models discussed in the text. A monoenergetic, homogeneously distributed source of electrons at energy ε_A is assumed in both cases. Because of the presence of a target surface and inelastic scattering, measured spectra should exhibit both features as shown in c), according to the discussion in sect. 5.4

Fig. 14 Example of a photoelectron spectrum of a mixed metal catalyst. The emission intensity is shown as a function of photo-electron energy. FeA, CuA and OA indicate Auger electrons. From SHEPARD et al. (1977). By courtesy of the American Society for Testing and Materials

Fig. 15 Differentiated Auger electron spectra of silicon bombarded with electrons and argon, respectively. Note that the ion-excited Auger spectrum differs markedly from the electron-excited spectrum, which indicates that the density and/or population of electronic states in a collision cascade differs considerably from that of a perfect crystalline structure. From WITTMAACK(1979). By courtesy of North-Holland Publishing Company

7.2 Spectra

The striking difference between the spectra predicted by the conventional approach (WOLFF, 1954; AMELIO, 1970; CHUNG et al., 1974), i.e. the expression equivalent to (8),

$$F'(\varepsilon',\vec{\Omega}')d\varepsilon'd^2\Omega' \simeq \left|\frac{\lambda(\varepsilon)}{\varepsilon}\frac{d\nu(\varepsilon,o)}{d\varepsilon}\right|_{\varepsilon=\varepsilon'+U} \cdot \varepsilon'd\varepsilon' \cdot \frac{|\cos\theta'|}{4\pi} \cdot d^2\Omega' \quad (64)$$

and the spectrum (4') has already been discussed in sect. 2. This difference shows up most pronouncedly in case of a quasi-monoenergetic source, i.e., in photoemission and Auger electron emission where (64) predicts a sharp <u>peak</u>

at $\varepsilon = \varepsilon_A$ -- if ε_A is the source energy -- and (4') a step function for $\varepsilon \lesssim \varepsilon_A$ (Fig.13). The arguments put forward in sect. 5.4 suggest that, because of the presence of a target surface and elastic scattering events, measured spectra ought to exhibit *both* features, i.e., a peak at ε_A superimposed a step function, and measured photoelectron spectra seem to corroborate this conclusion (Fig. 14). Similar structures have been observed in ion-induced electron spectra (Fig. 15), but an (almost) undeformed Auger peak can in that case as well be due to Auger electrons emitted from *sputtered* atoms. (HENNEQUIN et al., 1967, 1974; VRAKKING et al. 1979; WITTMAACK, 1979; METZ et al., 1980).

The relative significance of the peak is related to the frequency of elastic scattering events. Indeed, in the absence of elastic scattering of source electrons, an Auger peak can be observed if the energy-loss spectrum shows a clear peak structure. Conversely, in case of continuous stopping, the step-like shape is deformed due to elastic scattering into a more or less pronounced maximum. Weak singularities of this type have been observed in high-energy ion bombardment (GROENEVELD et al., 1974; FOLKMANN et al., 1975).

While we believe that the cases discussed in Sect. 5.3 are sufficiently different to bracket most realistic situations, we are aware of the fact that none of the proposed spectra are suited to provide an overall description of a realistic electron spectrum, especially in the important case where elastic and inelastic scattering cross sections have comparable magnitude. The authors are working on this problem.

Several isolated features emerge, however, quite readily from (4'). The presence of a maximum in the spectrum around $\varepsilon \sim U$ has been mentioned previously. More striking perhaps is the behaviour at electron energies near the maximum of the electronic stopping power $d\varepsilon/dx$. Above this energy, $d\varepsilon/dx$ bends over and approaches an approximate $\sim \varepsilon^{-1}$ behaviour. Thus, for $\nu(\varepsilon) \sim \varepsilon^{-1}$, the spectrum becomes almost constant, in agreement with experimental observations (FOLKMANN et al., 1975; PFERDEKÄMPER et al., 1977). Note that the energy dependence of $d\varepsilon/dx$ follows approximately that of the *electron-induced secondary-electron yield* of the material.

We note that the most uncertain quantity in both yield and spectra is the low-energy stopping power $d\varepsilon/dx$ near $\varepsilon \sim U$. To the extent that it is this quantity which determines the yield, a feasible approach is to extract the energy dependence of $d\varepsilon/dx$ from measured electron spectra, to determine a normalizing constant at a reasonably high ε, and on this basis to evaluate the material constant Λ in (63). This strategy has to some extent been followed by SCHOU (1980).

7.3 Escape Depth

A quantity of prime importance in surface analysis is the escape depth of secondary electrons, especially of Auger electrons. It is measured conventionally by recording the attenuation of an Auger signal as a function of the thickness of an overlayer (MAYER et al., 1966; BAER et al., 1970; TARNG et al., 1973; BATTYE et al., 1974; POWELL et al., 1974, 1977; SEAH et al., 1979). While there is a practical problem of background subtraction connected with this procedure (TOUGAARD et al., 1981), there is also the basic problem of interpreting the extracted escape depth in theoretical terms. Following the conventional description, eq. (8) -- which has also been applied in photoemission (BERGLUND et al., 1964) one usually interprets this quantity as being proportional with the inelastic mean free path. The discussion in Sect. 5.3 suggests the escape depth to be related both to the transport mean free path for elastic collisions and the inelastic mean free path, the details being somewhat dependent on the model. From (50), one finds an escape depth Δ_s to be

$$\Delta_s \simeq \sqrt{2Dt_s} \quad , \tag{65}$$

with t_s being the maximum slowing-down time for an electron *recorded as an Auger electron*. For discrete energy loss, we have

$$t_s \simeq \frac{1}{N\sigma v} \quad , \tag{66}$$

where σ is the cross section for inelastic scattering. Hence, from (50a), (65), and (66),

$$\Delta_s \simeq \sqrt{\frac{2}{3} \Lambda_1 \Lambda_i} \quad , \tag{67}$$

where $\Lambda_1 = 1/N\sigma_1$ and $\Lambda_i = 1/N\sigma$ are the transport and inelastic mean free paths, respectively.

On the other hand, for continuous slowing-down we find

$$t_s \simeq \frac{1}{v|d\varepsilon/dx|} \cdot \frac{16}{3\pi} \Lambda_1 \cdot NS(\varepsilon) = \frac{16}{3\pi} \frac{\Lambda_1}{v} \tag{68}$$

by means of (58), i.e.,

$$\Delta_s \simeq \frac{4}{3}\sqrt{\frac{2}{\pi}} \Lambda_1 \simeq \Lambda_1 \quad . \tag{69}$$

The seeming contradiction between (67) and (69) is easily reconciled by noting the different definition of the width of an Auger peak in the two cases.

7.4 Molecular, Charge State and Isotope Effects

Considerable interest has been devoted to secondary electron yields in case of molecular-ion bombardment (PETROV, 1963; PROPST et al., 1963; LARGE, 1963; STAUDENMAIER et al., 1976; NIKOLAEV et al., 1978; THUM et al., 1978; BARAGIOLA et al., 1979; ALONSO et al., 1980). It is customary to compare the electron yield per constituent atom, γ/n for an n-atomic molecular ion, with the yield for an atomic ion at the same *velocity*. As a first approximation, one expects these yields to be similar because molecular ions tend to dissociate and slow down independently (TEL'KOVSKII, 1956).[3]

There is ample evidence to support the conclusion that the electron yield $\gamma/2$ for H_2^+ is smaller than that for H^+ at the same velocity. For low-velocity ions, such a difference has been asserted to be caused by a difference in potential electron emission (Propst et al., 1963). At velocities

---

[3] Notable exceptions to the latter statement occur for high-speed bombardment under channeling conditions, cf. GEMMELL et al., 1975).

$v \simeq e^2/\hbar$, this difference would be too small and would decrease with increasing velocity. Even at 100 keV, a substantial difference has been found experimentally (Large, 1963). A plausible explanation (SVENSSON, 1980) is based on the incident charge per atom which is halved in case of H_2^+ bombardment. At high velocity, proton backscattering is insignificant, hence the pertinent excitations take place very close to the target surface where the molecule has not yet dissociated.

Enhanced electron yields have been looked for (THUM et al., 1979) in case of heavy-ion bombardment, but were not found. The motivation was the early picture of KAPITZA (1923) of electron emission from a bombardment-induced hot zone (spike). While the excistence of such hot zones is well documented (SIGMUND, 1974; ANDERSEN et al., 1974, 1975), it is obvious from a consideration of the time scale of slowing down that a local increase of ion temperature above ambient temperature cannot be shared by (metallic) electrons. This does not preclude the possibility of spike effects on electronic excitation phenomena: Indeed, any electronic effect with a measurable cross section at energies corresponding to spike temperatures ($\lesssim$ 10 eV) should be enhanced under spike conditions.

Charge-state effects on electron emission by *atomic* ions have been investigated mainly at low velocities where potential emission dominates (for reviews cf. KREBS, 1968; PARILIS, 1980). With increasing velocity, charge equilibration takes place at increasingly large depth, hence effects of the instantaneous charge state on kinetic secondary emission should be visible (PARILIS, 1974).

Electron emission yields have been extensively measured for different isotopes of bombarding ions. Most early investigations (for a review cf. KREBS, 1968) report comparisons at constant ion *energy* rather than ion *velocity*. No measurable difference has been found between D^+ and H^+ yields at equal velocity ($v \simeq e^2/\hbar$ (BARAGIOLA et al., 1979; ALONSO et al., 1980). Increased emission is expected for H^+ at lower velocity because of a higher reflection coefficient (VUKANIC et al., 1976). Isotope effects should also be observable in high-precision measurements with heavy ions when emission by recoil atoms becomes important.

Acknowledgement

Discussions with David Adams, Nils Andersen, David Brice, Gillis Holmén, Mitio Inokuti, Joseph Macek, Per Morgen, Jens Onsgaard, Rufus Ritchie, Arnold Russek, Jørgen Schou, Bengt Svensson and Harold Winters about various aspects of electron emission have proved valuable during the writing up of this article.

References

D. Aberdam, R. Baudoing, E. Blanc, and C. Gaubert, Surf. Sci 57, 306 (1976).
D. Aberdam, R. Baudoing, E. Blanc, and C. Gaubert, Surf. Sci. 65, 77 (1977).
M. Abramowitz and I.A. Stegun, Handbook of Mathematical Functions, Dover.
I.A. Abroyan, M.A. Eremeev, N.N. Petrov, Usp. Fiz. Nauk. 92, 105 (1967);
    engl. transl. Sov. phys. Uspekhi 10, 332 (1967).
E.V. Alonso, R.A. Baragiola, J. Ferron, M.M. Jakas, A. Oliva-Florio, Phys. Rev. B 22, 80 (1980).
G.F. Amelio, J. Vac. Sci. Technol. 7, 593 (1970).

H. H. Andersen and H.L. Bay, J. Appl. Phys. 45, 953 (1974); ibid. 46, 2416 (1975).
N. Andersen and P. Sigmund, Mat. Fys. Medd. Dan. Vid. Selsk. 39, no. 3 (1974).
U.A. Arifov, Interaction of Atomic Particles with a Solid Surface, Tashkent (1961); engl. transl. AEC Washington (1963).
J.C. Ashley, C.J. Tung, and R.H. Ritchie, Surf. Sci. 81, 409 (1979).
Y. Baer, P.F. Hedén, J. Hedmann, M. Klason, and C. Nordling, Sol. St. Commun. 8, 1479 (1970).
R.A. Baragiola, E.V. Alonso, J. Ferron, and A. Oliva-Florio, Surf. Sci. 90, 240 (1979).
R.A. Baragiola, E.V. Alonso, and A. Oliva-Florio, Phys. Rev. B 19, 121 (1979).
F.L. Battye, J.G. Jenkin, J. Liesegang, and R.C.G. Leckey, Phys. Rev. B 9, 2887 (1974).
N. Benazeth, J. Agusti, C. Benazeth, J. Mischler, and L. Viel, Nucl. Instr. Meth. 132, 477 (1976).
C. Benazeth, N. Benazeth and L. Viel, Surf. Sci. 78, 625 (1978).
C.N. Berglund and W.E. Spicer, Phys. Rev. 136, A 1030 (1964).
H.A. Bethe, Ann. Physik 5, 325 (1930).
H.A. Bethe, M.E. Rose, and L.P. Smith, Proc. Am. Phil. Soc. 78, 573 (1938).
H.D. Betz, Nucl. Instrum. Meth 132, 19 (1976).
R.J. Beuhler and L. Friedman, J. Appl. Phys. 48, 3928 (1977).
R. Bindi, H. Lanteri, and P. Rostaing, J. Phys. D 13, 267 (1980); ibid. 13, 461 (1980).
N. Bohr, Mat. Fys. Medd. Dan. Vid. Selsk. 18, no. 8 (1948).
N. Bohr and J. Lindhard, Mat. Fys. Medd. Dan. Vid. Selsk. 28 no. 7 (1954).
C. Boiziau, this volume.
D.K. Brice, Ion Implantation Range and Energy Deposition Distributions, Plenum Press, Vol. 1 (1975).
D.K. Brice and P. Sigmund, Mat. Fys. Medd. Dan. Vid Selsk. 40, no. 8 (1980).
J.S. Briggs and K. Taulbjerg, in Structure and Collisions of Ions and Atoms, I. Sellin, ed., Springer (1978), p. 105.
M.D. Brown and C.D. Moak, Phys. Rev. B 6, 90 (1972).
H. Bruining, Physics and Applications of Secondary Electron Emission, McGraw-Hill (1954).
R.K. Cacak, Q.C. Kessel, and M.E. Rudd, Phys. Rev. A 2, 1327 (1970).
M. Cardona and L. Ley, in Photoemission in Solids I, M. Cardona et al., eds., Springer 1978, p. 1.
G. Carter and J.S. Colligon, Ion Bombardment of Solids, Heineman (1968).
M.H. Chen and B. Craseman, Phys. Rev. A 8,7 (1973).
M.S. Chung and T.E. Everhart, J. Appl. Phys. 45, 707 (1974).
M.S. Chung and T.E. Everhart, J. Appl. Phys. 15, 4699 (1977).
H.G. Clerc, H.J. Gehrhardt, L. Richter, and K.H. Schmidt, Nucl. Instrum. Meth. 113, 325 (1973).
P. Dahl, M. Rødbro, G. Hermann, B. Fastrup, and M.E. Rudd, J. Phys. B 9, 1581 (1976).
A.J. Dekker, in Solid State Physics, F. Seitz et al., eds., Acad. Press. Vol 6, 251 (1958).
A.A. Dorozhkin and N.N. Petrov, Fiz. Tverd. Tela 16, 947 (1974); engl. transl. Sov. Phys. Solid State 16, 611 (1974).
D.A. Douthat, Radiat. Res. 61, 1 (1975).
D.A. Douthat, J. Phys. B 12, 663 (1979).
D.A. Douthat, in Radiation Research, S. Okada et al., eds., Tokyo (1979) p.89.
U. Fano and W. Lichten, Phys. Rev. Lett. 14, 627 (1965).
B. Fastrup, G. Hermann, and Q.C. Kessel, Phys. Rev. Lett. 27, 771 (1971).
R.A. Ferrell, Phys. Rev. 111, 1214 (1958).

O.B. Firsov, Zh. Eksp. Teor. Fiz. 36, 1517 (1959), engl. transl. Sov. Phys. JETP 9 1076 (1959).
F. Folkmann, K.O. Groeneveld, R. Mann, G. Nolte, S. Schumann, and R. Spohr, Z. Physik A 275, 229 (1975).
J.P. Fontbonne, N. Colombié, and B. Fagot, Compt. Rend. Acad. Sci. B 270, 1573 (1970).
R.H. Fowler, Proc. Cambr. Phil. Soc. 21, 531 (1923).
J.P. Ganachaud and M. Cailler, Surface Sci. 83, 498 (1979); ibid. 83, 519 (1979).
J.D. Garcia, Phys. Rev. A 1, 280 (1970).
D.S. Gemmell, J. Remillieux, J.C. Poizat, M.J. Gaillard, R.E. Holland, and Z. Vager, Phys. Rev. Lett. 34, 1420 (1975).
E. Gerjuoy, Phys. Rev. 148, 54 (1966).
S.N. Ghosh and S.P. Khare, Phys. Rev. 125, 1254 (1962).
S.N. Ghosh and S.P. Khare, Phys. Rev. 129, 1638 (1963).
P. Gombas, in Atoms II, Encyclopedia of Physics, Vol. XXXVI, ed. by S. Flügge (Springer Berlin, Göttingen, Heidelberg 1956) p. 109
K.O. Groeneveld, in Beam-Foil Spectroscopy, I.A. Sellin et al., eds., Plenum Press, Vol. 2, 593 (1976).
K.O. Groeneveld, R. Mann, W. Meckbach, and R. Spohr, Vacuum 25, 9 (1974).
M. Gryzinski, Phys. Rev. 138A, 305, 322, 336 (1965).
O. Hachenberg and W. Brauer, Adv. Electronics and Electron Phys. 11, 413 (1959).
J.F. Hennequin, P. Joyes, and R. Castaing, Compt. Rend. Acad. Sci. 265, 312 (1967).
J.F. Hennequin and P. Viaris de Lesegno, Surf. Sci. 42, 50 (1974).
J.F. Hennequin and P. Viaris de Lesegno, in Physics of Ionized Gases 1980, B. Cobic, ed., Boris Kidrich Institute, Belgrade, in press.
R. Hlics and H.J. Binder, phys. stat. sol. 38, K 27 (1970).
G. Holmén and P. Högberg, Radiat. Eff. 12, 77 (1972).
G. Holmén, B. Svensson, J. Schou, and P. Sigmund, Phys. Rev. B 20, 2247 (1979).
J. Hölzl and F. K. Schulte, Springer Tracts in Modern Physics, G. Höhler, ed., Vol. 85, 1 (1979).
M. Inokuti, Rev. Mod. Phys. 43, 297 (1971).
M. Inokuti, Radiat. Res. 64, 6 (1975).
M. Inokuti, D.A. Douthat, and A.R.P. Rau, in Proc. 5. Sympos. on Microdosimetry, J. Booz et al., eds. Commiss. Europ. Communities (1976), p. 977.
International Commission on Radiation Units and Measurements, ICRU Report 31, Washington, D.C. (1979).
M. Kaminsky, Atomic and Ionic Impact Phenomena on Metal Surfaces (Springer Berlin, Heidelberg, New York 1965)
K. Kanaya and H. Kawakatsu, J. Phys. D 5, 1727 (1972).
K. Kanaya and S. Ono, Jap. J. Appl. Phys. 13, 944 (1974).
P. Kapitza, Phil. Mag. 45, 989 (1923).
Q. C. Kessel and B. Fastrup, in Case Stud. Atomic Phys. 3, 137 (1973).
Y.K.Kim, in Radiat.Research,Chemical, Physical and Biomedical Perspectives, O. Nygaard et al., eds., Acad. Press, p. 741 (1975).
T. Koshikawa and R. Shimizu, J. Phys. D 7, 1303 (1974).
K.H. Krebs, Fortschr. Physik 16, 419 (1968).
K.H. Krebs, in Physics of Ionized Gases 1976, B. Navinsek, ed., Ljubljana, p. 379 (1976).
L.N. Large, Proc. Phys. Soc. 81, 1101 (1963).
J. Lindhard, Mat. Fys. Medd. Dan. Vid. Selsk. 28, no. 8 (1954).
J. Lindhard, Mat. Fys. Medd. Dan. Vid. Selsk. 34, no. 14 (1965).
J. Lindhard and V. Nielsen, Mat. Fys. Medd. Dan. Vid. Selsk. 38, no. 9 (1971).
J. Lindhard, V. Nielsen, and M. Scharff, Mat. Fys. Medd. Dan. Vid. Selsk.

36, no. 10 (1968).
J. Lindhard, V. Nielsen, M. Scharff, and P.V. Thomsen, Mat. Fys. Medd. Dan. Vid. Selsk. 33, no. 10 (1963).
J. Lindhard and M. Scharff, Phys. Rev. 124, 128 (1961).
J. Lindhard, M. Scharff, and H.E. Schiøtt, Mat. Fys. Medd. Dan. Vid. Selsk. 33, no. 14 (1963).
G.D. Magnuson and C.E. Carlston, Phys. Rev. 129, 2409 (1963).
S.T. Manson, L.H. Toburen, D.H. Madison, and N. Stolterfoht, Phys. Rev. A 12, 60 (1975).
E.S. Mashkova, V.A. Molchanov, and D.D. Odintsov, Fiz. Tverd. Tela 5, 3426 (1963); engl. transl. Sov. Phys. Sol. State 5, 2516 (1964).
H. Mayer and J. Hölzl, phys. stat. sol. 18, 779 (1966).
E.J. McGuire, Phys. Rev. 185, 1 (1969).
E.J. McGuire, Phys. Rev. A 2, 273 (1970).
D.B. Medved and I.S. Strausser, Adv. Electronics and Electron Phys. 21, 101 (1965).
W. Mehlhorn, Electron Spectroscopy of Auger and Auto-ionizing States: Experiment and Theory. Lectures at Univ. Aarhus (1978).
W.A. Metz, K.O. Legg, and E.W. Thomas, J. Appl. Phys. 51, 2888 (1980).
J. Mischler and N. Colombié, Surf. Sci. 40, 311 (1973).
J. Mischler, M. Nègre, N. Benazeth, D. Spanjaard, C. Gaubert, and D. Aberdam, Surf. Sci. 82, 453 (1979).
E.G. Mititel, Izv.Akad. Nauk SSSR, Ser. Fiz. 40, 2628 and 2646 (1976); engl. transl. Bull Acad. Sci. USSR, Phys. Ser. 40, 157 and 173 (1977).
G.H. Morgan and E. Everhart, Phys. Rev. 128, 667 (1962).
M. Nègre, J. Mischler, N. Bénazeth, C. Noguera, and D. Spanjaard, Surf. Sci. 78, 174 (1978).
E.N. Nikolaev, G.D. Tantsyrev, and V.A. Saraev, Zh. Tekh. Fiz. 48, 406 (1978); engl. transl. Sov. Phys. Tech. Phys. 23, 241 (1978).
G.N. Ogurtsov, I.P. Flaks, and S.V. Avakyan, Zh. Tekn. Fiz. 40, 2124 (1970); engl. transl. Sov. Phys. Techn. Phys. 15, 1656 (1971).
J. Østgaard Olsen and N. Andersen, unpublished (1976).
E.S. Parilis, Izv. Akad. Nauk SSSR 37, 2565 (1973); engl. transl. Bull. Acad. Sci. USSR 37, 83 (1974).
E.S. Parilis, in Proc. Symposium on Sputtering, P. Varga et al., eds., Inst. Allg. Phys. Vienna (1980) p. 664.
E.S. Parilis and L.M. Kishinevskii, Fiz. Tverd. Tela 3, 1219 (1960); engl. transl. Sov. Phys. Solid State 3, 885 (1960).
N.N. Petrov, Izvest. Akad. Nauk SSSR, Ser. Fiz. 26 (1963); engl. transl. Bull. Acad. Sci. USSR, Phys. Ser. 26, 1350 (1963).
N.N. Petrov, in Physics of Ionized Gases 1974, V. Vujnovic, ed., Zagreb, p. 533 (1974).
K.E. Pferdekämper and H.G. Clerc, Z. Phys. A 280, 155 (1977).
D. Pines, Physica 26, S 103 (1960).
R.L. Platzman, Int. J. Radiat. and Isotopes 10, 116 (1961).
C.J. Powell, Surf. Sci. 44, 29 (1974).
C.J. Powell, R.J. Stein, P.B. Needham, and T.J. Driscoll, Phys. Rev. B 16, 1370 (1977).
F.M. Propst and E. Lüscher, Phys. Rev. 132, 1037 (1963).
R.H. Ritchie, Phys. Rev. 114,644 (1959).
R.H. Ritchie, C.J. Tung, V.E. Anderson, and J.C. Ashley, Radiat. Res. 64, 181 (1975).
W. Schäfer, H. Stöcker, B. Müller, and W. Greiner, Z. Physik A 288, 349 (1978).
W. Schäfer, H. Stöcker, B. Müller, and W. Greiner, Z. Physik B 36, 319 (1980).
J. Schou, Nucl. Instrum. Meth. 170, 317 (1980 a).

J. Schou, Phys. Rev. B 22, 2141 (1980 b).
M.P. Seah & U.A. Dench, Surf. Interf. Anal. 1, 2 (1979).
I. Sellin & R. Laubert, this volume.
P. Sigmund, Appl. Phys. Lett. 14, 114 (1969).
P. Sigmund, Phys. Rev. 184, 383 (1969); ibid. 187, 768 (1969).
P. Sigmund, Rev. Roum. Phys. 17, 823, 969, 1079 (1972).
P. Sigmund, Appl. Phys. Lett. 25, 169 (1974); ibid. 27, 52 (1975).
P. Sigmund, in Radiation Damage Processes in Materials, C.H.S. Dupuy, ed., Noordhoff, Leiden, p. 3 (1975).
P. Sigmund, in Inelastic Ion-Surface Collisions, N.H. Tolk et al., eds., Academic Press, p. 121 (1977).
L.V. Spencer & U. Fano, Phys. Rev. 93, 1172 (1954).
G. Staudenmaier, W.O. Hofer & H. Liebl, Int. J. Mass Spect. Ion Phys. 11, 103 (1976).
E.J. Sternglass, Phys. Rev. 108, 1 (1957).
N. Stolterfoht, D. Schneider, D. Burch, H. Wieman & J.S. Risley, Phys. Rev. Lett. 33, 59 (1974).
N. Stolterfoht, D. Schneider & H. Gabler, Phys. Lett. 47A, 271 (1974).
B. Svensson, Thesis, Chalmers University of Technology, Göteborg (1980).
M.L. Tarng & G.K. Wehner, J. Appl. Phys. 44, 1534 (1973).
V.G. Tel'kovskii, Izv. Akad. Nauk SSSR, Ser. Fiz., 20 (1956); engl. transl. Bull. Acad. Sci. USSR, Phys. Ser. 20, 1070 (1956).
F. Thum & W. Hofer, Surf. Sci. 90, 331 (1979).
L.H. Toburen & W.E. Wilson, in Radiation Research, S. Okada et al., eds., Jap. Assoc. Rad. Research Tokyo, p. 80 (1979).
S. Tougaard & P. Sigmund, to be published (1981).
C.J. Tung & R.H. Ritchie, Phys. Rev. B 16, 4302 (1977).
L. Viel, C Benazeth & N. Benazeth, Surf. Sci. 54, 635 (1976).
Ya. A. Vinokurov, L.M. Kishinevskii & E.S. Parilis, Izv. Akad. Nauk SSSR, Ser. Fiz 40, 1745 (1976); engl. transl. Bull. Acad. Sci. USSR, Phys. Ser. 40, 166 (1977).
J.J. Vrakking & A. Kroes, Surf. Sci. 84, 153 (1979).
J. Vukanic & P. Sigmund, Appl. Phys. 11, 265 (1976).
D.L. Walters & C.P. Bhalla, Atomic Data 3, 301 (1971).
T. Watanabe, P.H. Woerlee & Yu.S.Gordeev, in VIth International Seminar on Ion-Atom Collisions, Abstracts of Contributed Papers, Japan Atomic Energy Research Institute (1979) p. 65.
R.F. Willis & N.E. Christensen, Phys. Rev. B 18, 5140 (1978).
K.B. Winterbon, Ion Implantation Range and Energy Deposition Distributions, Plenum Press, Vol. 2 (1975).
K.B. Winterbon, P. Sigmund & J.B. Sanders, Mat. Fys. Medd. Dan. Vid. Selsk. 37, no. 14 (1970).
K. Wittmaack, Surf. Sci. 85, 69 (1979).
P.H. Woerlee, Yu. S. Gordeev, H. De Waard & F.W. Saris, J. Phys. B (in press).
P.A. Wolff, Phys. Rev. 95, 56 (1954).

ADDENDUM: The authors' attention has been drawn to an important paper by W.PLOCH (Z.Physik 130,174(1951) which offers a detailed interpretation of the isotope effect in emission yields as well as an outline of the role of electron promotion in the excitation process.

Ion-Induced Auger Electron Emission from Solids [1]

Raúl A. Baragiola
Centro Atómico Bariloche[2] and Instituto Balseiro[3]
8400 - S.C. de Bariloche, Argentinia

1 Introduction

In the tail of the energy distribution of electrons emitted from solids bombarded by keV ions one can notice, in certain cases, Auger electrons from the filling of inner-shell (core) holes of target atoms.

Interest in studying this ion-excited Auger electron emission arises for several reasons. One is the information carried by the yield of these electrons on core excitation collisions at impact velocities much smaller than the average velocity of the core electrons, excitations which are more important in solids than in gases due to, other things being equal, a larger accessible density of target atoms.

Another point of interest, and one which attracted the early attention of several research groups, is the question of the effect of inner-shell excitations and subsequent Auger emission on other observables occurring during ion bombardment. For instance, in 1960 Parilis and Kishinevskii [1] proposed a theory of ion-induced electron emission in which the main process determining total electron yields was core excitation and Auger decay. This theory survived for nearly two decades until it was examined and discussed in detail recently [2-5]. On the contrary, the emission of Auger electrons and of multiply charged ions from some solids under ion impact are highly correlated, an observation which prompted early research in ion-induced Auger electron emission [6]. A major role is likely played also by the Auger effect (this time in the projectile) in determining charge states of backscattered heavy ions in particular ion-target combinations [7,8].

A third region of interest for research in this field, and where we will mostly dwell in this paper, is the possibility of deriving not only from the yields but from spectroscopic details, information about the main steps of the process, i.e., about how the core hole is formed and what is its environment when it decays.

In this communication we will report on our measurements of yields and spectra of Auger electrons from clean samples of Mg, Al, and Si bombarded with 0.7-15 keV noble gas ions. In the light of these results we will show that it is possible to solve the two basic controversies which have persisted since the very early days of this field:
  - Are core excitations produced mainly from asymmetric collisions between

---

[1] Partially supported by the International Atomic Energy Agency
[2] Comisión Nacional de Energía Atómica
[3] Comisión Nacional de Energía Atomica and Universidad Nacional de Cuyo

the incident ion and target atoms or from symmetric collisions between target atoms in the collision cascade?
- Are atomic-like Auger lines in the electron spectra caused by atoms decaying inside or outside the solid?

2 Experimental Details

The apparatus used in this work [9] is shown schematically in Fig.1. The targets to be ion bombarded (of purity better than 99.999%) were mounted inside a Mu-metal ultrahigh vacuum chamber routinely operated at a base pressure of less that 3×10^{-11} Torr. The samples were cleaned by sputtering with low energy noble gas ions from the ion gun AG2 and the cleanliness of the surface layers monitored by means of Auger electron spectroscopy induced by electrons from the LEG61 electron gun. The surface contamination in all our studies amounted to less than the equivalent of 1% of a monolayer.

The accelerated ions used in these studies were produced by means of the electron-bombardment-type ion gun AG5. They were mass analyzed using a Wien filter and deflected electrostatically onto the target to separate them from non mass-analyzed neutrals formed by electron capture from gas atoms issuing from the ion source region.

The energy of the ejected electrons was determined with the aid of a hemispherical electrostatic analyzer working at typically 0.2-0.6% energy resolution. Electrons were detected by means of a channel electron multiplier and the data acquisition was performed either manually, or under computer control using signal averaging techniques.

2.1 Calibration of the Electron Energy Scale

It is important in our work to accurately determine electron energies since we will make comparisons with absolute Auger energies measured by other workers in gas-phase collisions. The calibration was performed from the observation of the elastic reflection from the target of electrons accelerated through an accurately known potential V_f. This is depicted schematically in Fig.2 (all potentials are measured between the Fermi levels of the different electrodes). We can then determine the work function of the spectrometer, ϕ_{sp} by measuring the most probable (peak) energy of the elastically scattered electrons:

$$\phi_{sp} = eV_f - eV_{sp} + \phi_f + E_{kT}$$

Here e is the electron charge, and $eV_{sp} = eV_r + CV_p$ the pass energy of the spectrometer where V_r is the voltage by which electrons are retarded prior to analysis, V_p the voltage between the hemispheres and C the geometrical constant of the analyzer. The work function of the filament ϕ_f and the most probable excess thermal energy E_{kT} were taken to be 4.55 eV and 0.23 eV respectively, which should correspond closely (to within 0.1 eV) with our experimental situation with a tungsten emitter at T$\sim$2700°K. The constant of the analyzer, C, could also be accurately determined by tuning on the peak of elastically reflected electrons and measuring the required inter-hemisphere voltage as a function of the retarding potential V_r.

As an independent check, we have compared our measured Fermi edges in the derivative electron excited $L_{2,3}$VV Auger spectra of Mg and Al with the binding energies of the $L_{2,3}$ shells as determined by X-ray emission spectros

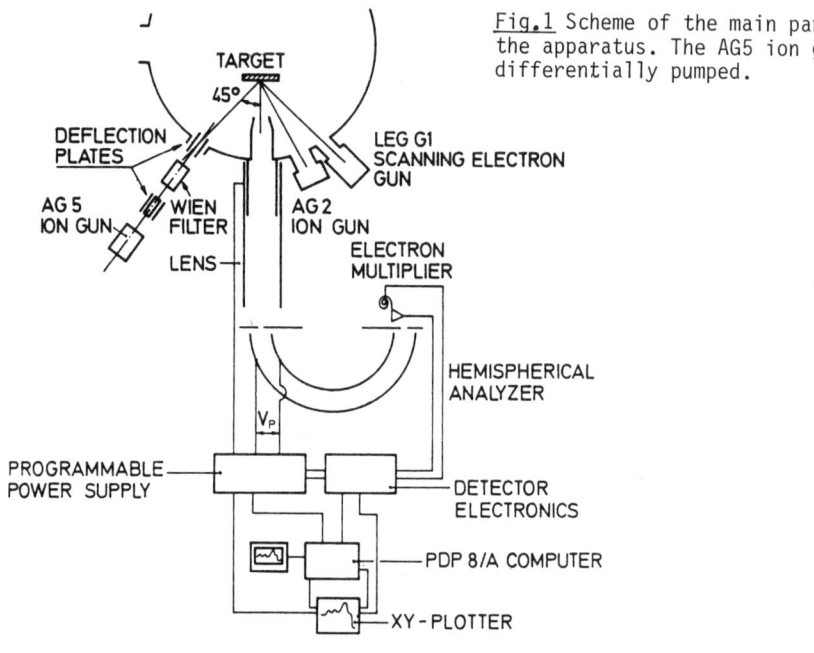

Fig.1 Scheme of the main part of the apparatus. The AG5 ion gun is differentially pumped.

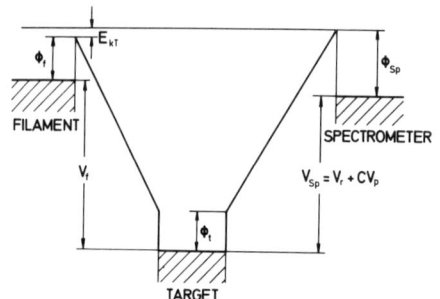

Fig.2 Electron energy diagram appropriate to the calibration of the energy scale.

copy [10]. We found the agreement to be excellent, to within 0.1 eV.

3 Gross Spectral Features

Typical ion-excited Auger spectra are shown in Fig.3 in comparison with electron excited bulk $L_{2,3}VV$ transitions. Besides the broad feature which can be associated with transitions involving the valence electrons from the solid, like in the electron-impact case, one can observe sharp, atomic-like peaks in the ion induced spectra. The origin of the larger of these atomic peaks has been attributed in the past either to excited sputtered ions [6,11-13], or to excited atoms moving inside the solid carrying outer electrons dynamically "decoupled" from the normal band [14] or from a drastically modified band of valence electrons in the region of the collision cascade [12,15].

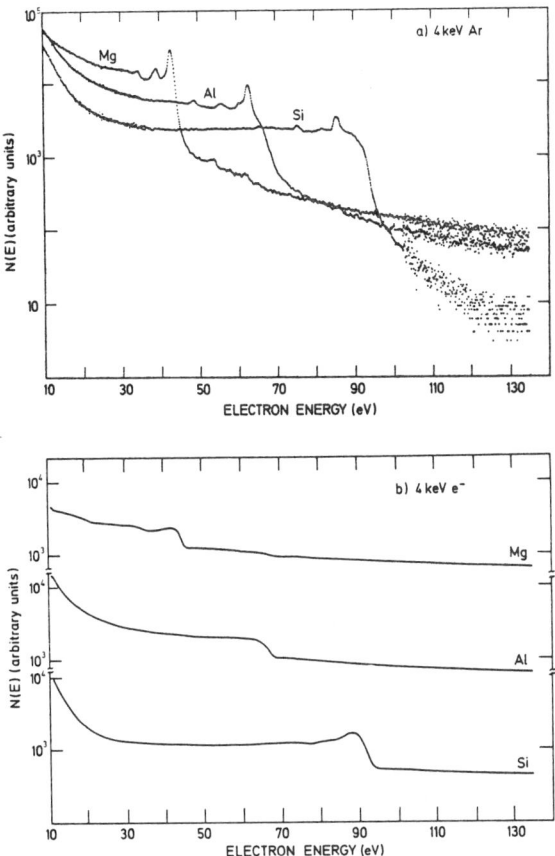

Fig.3 Auger spectra of Mg, Al and Si induced by 4 keV ion: a) and electron: b) bombardment.

The smaller atomic peaks have been attributed to different sources, like plasmon losses [11,16,17], modified plasmon losses [15], excitation of 3s electrons [18,19], and atomic transitions from different initial and final states [12-14,19].

It was also observed previously that the peaks are broader than those obtained under electron bombardment of gaseous targets and this has been attributed to the Doppler effect from atoms moving inside the solid [12-14]. With the experimental geometries used in these past studies, however, it was not possible to derive any more conclusions other than the widths of the atomic peaks increase with impact energy, since the analyzers looked at electrons emitted over a wide angular range with respect to the surface normal. In our experiments, with a simple geometry, it is possible to examine this broadening in detail. As shown in Fig.4, which is representative of all the studied projectile-target combinations, the broadening occurs only towards high energies, indicating that the electron source is moving towards the analyzer,

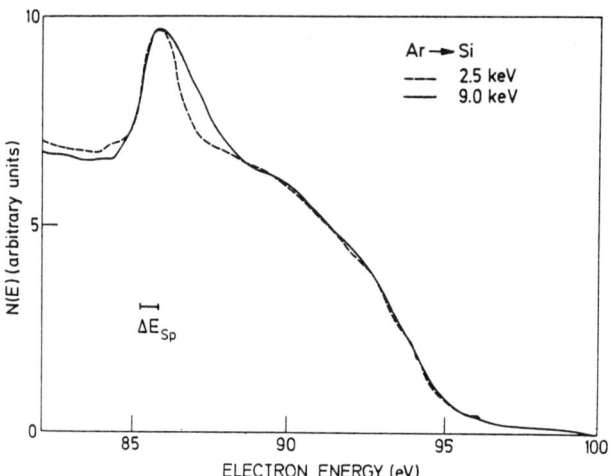

Fig.4 Showing the effect of the kinetic energy of the Argon projectile on the Auger spectrum of Silicon.

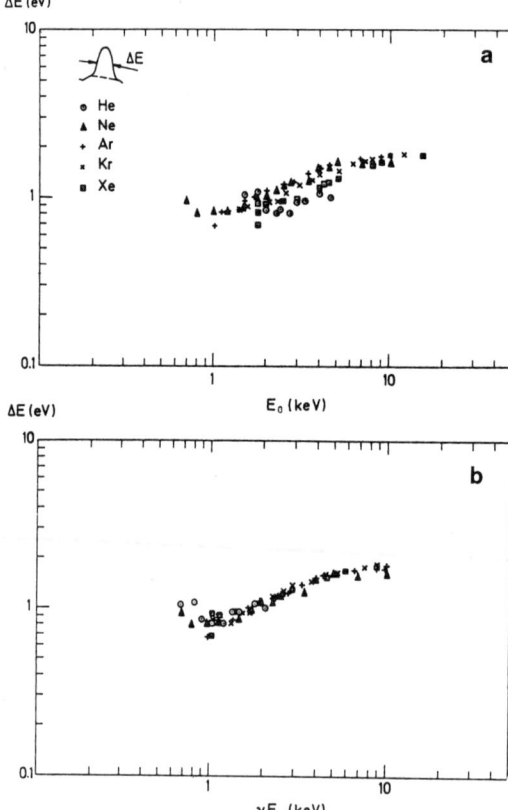

Fig.5 Full width at half maximum of the main atomic line of Al vs. the energy of the projectile E_0: a) and vs. γE_0, the maximum energy transfer in a collision between the projectile and a target atom: b).

that is, away from the solid [20]. Thus the sharp atomic lines are due to sputtered atoms since if one could accept that narrow lines exist for atoms which move inside the solid with outer shell wavefunctions overlapping those of the valence band, the lines would be broadened also towards low energies, because in that case one would certainly find in the collision cascade produced by the projectile, atoms moving with velocity components towards the interior of the solid. We should mention that a broadening towards the high energy side could also result from different decay distances of the excited atoms from the surface and therefore different shifts in energy levels due to the proximity to the solid. The impact energy dependence of the broadening as shown in Fig.5 for the case of Al, suggests that this effect is relatively small, since higher impact energies imply higher mean energies of sputtered atoms with a resulting larger mean distance of decay from the surface and less energy shift of the atomic levels.

For the case of Al shown in Fig.5 and for our geometry, the Doppler shift is $E_D=0.07$ eV $\sqrt{E_p(eV)}$, with $E_p=Mv_p^2/2$ where M and v_p are the mass and velocity component perpendicular to the surface, of the sputtered excited Al atom. Then, mean shifts of 0.35-0.9 eV ($\sim 1/2$ of the FWHMs of Fig.5 after taking care of the 0.4 eV energy resolution used in this case) correspond to "transverse" energies E_p between 24.5 eV and 162 eV which are of the order but somewhat higher than those expected for sputtered atoms (the short lifetime of the 2p-vacancy, $\sim 2\times 10^{-14}$ sec, lowers the probability of decay of low energy excited atoms outside the solid and causes a larger mean energy than that which would result for sputtered atoms of longer lifetime). For the bulk-type Auger transitions, no impact-energy dependent broadening was observed indicating that many electrons take part in the sharing of the momentum involved in the Auger process.

4 Yields

For the case of Aluminum, we have studied the impact-energy dependence of the yield of the larger atomic line [21] after substracting the bulk-like $L_{2,3}VV$ feature as shown in Fig.6. This substraction is approximate, since it is difficult to estimate the "background", but uncertainties are thought to be below 30%. With the purpose of giving to the reader an idea of total yields, we have multiplied our results by 2500, before plotting them in Fig.6. This number is based on the assumption of a cosine distribution of ejected electrons and on an estimate of the transmission of the electron spectrometer, and should be accurate to within a factor of 2.

One can observe from Fig.6a that the yields increase extremely fast with ion energy. This stresses the necessity of working with monoenergetic, mass-analyzed ion beams [22]. Our results become nearly independent of projectile type, except for He and low energy Ne, if plotted as a function of γE_0, the maximum energy transfer between a projectile of mass m and energy E_0, and a target atom of mass M, where $\gamma=4mM(m+M)^{-2}$. This energy scaling was first found by Wittmaack [12] for his "Auger production efficiency" in Si targets and suggests that collisions between target atoms dominate the production of target 2p-vacancies. We have found, however, that it is not necessary to scale the yields by dividing them by the sputtering yields, as done by Wittmaack, a procedure which may also be misleading since, and particularly at low impact energies, a collision cascade in which one of the atoms suffered a very hard, inner-shell excitation encounter is not a typical collision cascade leading to sputtering. The scaling with γE was also found to be accurate for the case of the widths of the atomic Auger lines, as exemplified by Fig.5b.

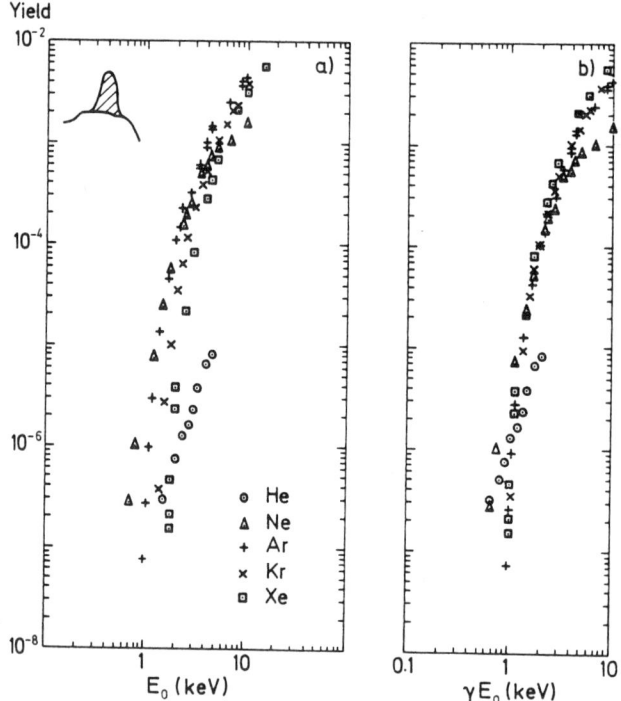

Fig.6 Yields of the main atomic line of Al vs. the energy of the projectile E_0: a) and vs. γE_0, the maximum energy transfer in a collision between the projectile and a target atom: b). The yield scale is an estimate for the total yield assuming a cosine distribution of ejected electrons and should be accurate to within a factor of 2.

The data for Ar, Kr, and Xe projectiles point to an extrapolated "threshold" at $\gamma E_t = 0.9$ keV. After receiving this energy, an Al atom can collide with another reaching a distance of closest approach of 0.58Å [26], close to 0.61Å, twice the average radius of the 2p-shell of Al [27], which is the distance at which one can expect a level overlap sufficient to promote the $4f\sigma$ orbital to an energy where it can be ionized by crossing empty levels. The closeness of the agreement may be partly accidental, but it strongly suggests the physical origin of the threshold.

It is apparent from Fig.6b that for He^+ and Ne^+ projectiles different excitation mechanisms are effective, involving projectile-Al collisions, at least below $\gamma E_0 = 0.9$ keV. It can easily be shown that Al 2p-shell excitation is extremely unlikely in projectile-Al collisions at these energies and for the one-electron transitions of the Fano-Lichten-Barat's molecular orbital model [28,29]. Alternatively it is possible, as suggested to the author by N.Stolterfoht, that two-electron transitions may be important, wherein in the incoming part of the collision $He^+(Ne^+)$ is ionized to He^{++} (Ne^{++}) and upon separation it captures an Al 2p-electron in a transition with a relatively small energy defect. Further work is needed to ascertain this new inner-shell excitation mechanism at such low velocities.

What we have just discussed were the yields of the most intense atomic Auger lines. Roughly the same energy dependence was observed for the smaller atomic lines, though this could not be ascertained near the threshold due to signal to noise considerations. We have observed that the ratios of the atomic to bulk Auger intensities have an energy dependence similar in shape to that shown for the line widths in Fig.5. This ratio is a measure of the chance that a 2p-excited Al atom decays outside rather than inside the solid; the main reason it increases with ion energy is the small value of the lifetime of the core hole: the larger the impact energy, the larger the mean energy of 2p-excited Al atoms and the larger the chance that they will survive as excited during transit to the surface, and that they decay outside rather than inside the solid.

5 Identification of the Atomic Auger Lines

As mentioned above, there have been many different assignments made by different workers, of the sharp lines in the spectra. In this report we shall restrict our discussion to lines due to single 2p-excitation of Mg, for which there are measurements in the gas phase under electron bombardment [30].

Main differences between electron bombardment of Mg vapour and our work are the absence in the first case of significant Doppler broadening and the visibility of many more lines, partially due to a different excitation mechanism but also due to a much larger peak/background ratio. We will only consider initial states of the excited sputtered atom which cannot undergo the very fast surface ionization process, i.e. outer-shell electrons in levels lying below the top of the filled valence band of the solid.

Line C in the spectra of Fig.7 can be assigned to the transition Mg^+ $(2p^5 3s^2) \rightarrow Mg^{++}(2p^6)$, which is the most intense line in the electron excited spectra of atomic Mg. The line position in this work is shifted 0.5 eV due partly to the surface interaction and partly to the Doppler effect.

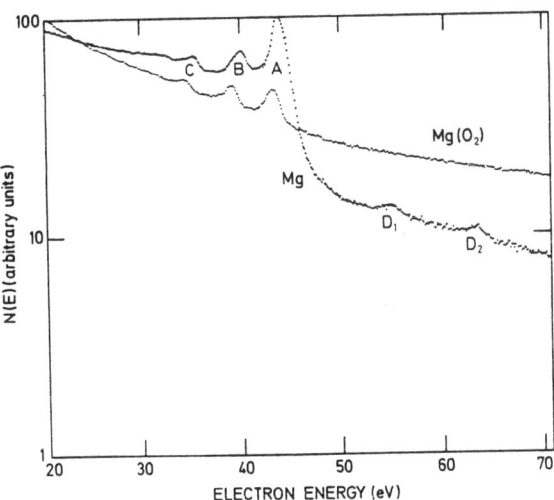

Fig.7 The Auger spectrum of magnesium under 4.0 keV argon bombardment and the effect of exposure to 10^{-3} Torr sec of oxygen. The energy scale is relative to the vacuum level of clean magnesium. Lines A, B and C are due to singly 2p-excited sputtered Mg, lines D_1 and D_2 are for sputtered Mg with two 2p-shell holes.

Our most intense line, A, is assigned to neutral Mg $(2p^53s^23p)$ decaying into $Mg^+(2p^63s)$. We calculate the mean energy of this transition using the Z+1 equivalence method [31] to be 44.0 eV. The other decay channel, leading to $Mg^+(2p^53p)$, is expected to yield 39.5 eV electrons with an estimated intensity of about 10% of that of line A [32]. Line B is found centered at 40.3 eV, and with intensity about 20% of that of line A. Another possible assigment for line B is the transition $Mg^+(2p^53s3p) \rightarrow Mg^{++}(2p^6)$ for which part of the intensity will fall in the region of peak A, the estimated energies being 39.5 eV and 42.3 eV (other estimates yield 38.6 eV and 41.4 eV [31]).In an attempt to distinguish between these two possibilities, we have measured spectra after exposing the surface to pure oxygen to a dose of 10^{-3} Torr sec. For oxidized Mg there is less chance that the sputtered Mg, originally in an ionic state in the solid, captures an electron from a negative oxygen ion to become neutral. This is confirmed in the spectrum shown in Fig.7 for oxyde-covered Mg: peak A due to neutral 2p-excited Mg decreases in intensity upon oxidation. From this reasoning it follows that peak B belongs to the initial state of Mg^+ just mentioned, since it was not much affected by oxidation.

It is also of interest to note in Fig.7 that, after oxidation, the bulk Auger transitions have disappeared; the same has occurred with the high energy atomic lines due to double 2p-vacancies.

6 Epilogue

In this work we have been able to explain controversial points in studies of ion-excited Auger emission from light element solids. We have found that for the heavier noble gas ions violent collisions between target atoms dominate the Auger electron production, in our energy range. For low energy Ne, and for He projectiles, projectile-target collisions, probably involving two-electron transitions, are more important. The sharp lines in the Auger spectra have been identified as originating from 2p-excited neutral and ionized sputtered atoms, with energies shifted by the Doppler effect. We leave for a future publication the discussion of other interesting subjects such as the differences between electron- and ion-excited bulk $L_{2,3}VV$ spectra, many-body effects, and the origin of Auger spectra from the projectile, as well as a more detailed treatment of the effect of oxidation.

7 Acknowledgements

The author thanks here the invaluable help of many colleagues who have contributed in different aspects of this work, in particular, to Eduardo Alonso, Humberto Raiti, Mario Jakas, Nikolaus Stolterfoht and Neil Callwood. A special acknowledgement goes to Edmund Taglauer and Werner Heiland for their kind invitation to a very stimulating workshop.

References

1. E.S.Parilis, L.M.Kishinevskii: Sov.Phy.Sol.State 3, 885 (1960)
2. J.Ferrón: Trabajo Especial, Univ.Nacional de Cuyo (1977)
3. R.A.Baragiola, E.V.Alonso, J.Ferrón, A.Oliva-Florio: Phys.Rev. B 19, 121 (1979)
4. R.A.Baragiola, E.V.Alonso, J.Ferrón, A.Oliva-Florio: Surface Sci.90,245 (1979)
5. E.V.Alonso, R.A.Baragiola, J.Ferrón, M.M.Jakas, A.Oliva-Florio: Phys.Rev. B22, 80 (1980)
6. J.F.Hennequin, P.Joyes, R.Castaign: C.R.Acad.Sci.Paris 265B, 312 (1967)
7. L.M.Kishinevskii, E.S.Parilis, V.K.Verleger: Rad.Effects 29, 215 (1976)

8. R.A.Baragiola: Abstracts of the VII International Conference on Atomic Collision in Solids (Moscow State Univ.Publishing House, 1977) p.151.
9. Made by VG Scientific, U.K.
10. J.A.Tagle, E.T.Arakawa, and T.A.Callcott: Phys.Rev.B 21, 4552 (1980)
11. J.T.Grant, M.P.Hooker, R.W.Springer, T.W.Haas: J.Vac.Sci,Technol.12,481 (1975)
12. K.Wittmaack, Surface Sci.85, 69 (1979)
13. W.A.Metz,K.O.Legg, E.W.Thomas: J.Appl.Phys.51, 2888 (1980)
14. C.Benazeth, N.Benazeth, L.Viel: Surface Sci.78, 625 (1978)
15. J.J.Vrakking, A.Kroes: Surface Sci.84, 153 (1979)
16. L.Viel, N.Colombie, B.Fagot, C.Fert: C.R.Acad.Sci.Paris 271B, 239 (1970)
17. F.Louchet, L.Viel, C.Benazeth, B.Fagot, N.Colombie: Rad.Effects 14, 123 (1972)
18. C.Benazeth, L.Viel, N.Colombie: C.R.Acad.Sci.Paris 276B, 863 (1973)
19. J.-F.Hennequin, P.Viaris de Lesegno: Surface Sci.42, 50 (1974)
20. It can be easily seen that the fact that our analyzer is not aimed along the surface normal but 15° away from it is unimportant (cos 15°=0.97, and very few energetic atoms are sputtered at angles larger than 75° from the surface normal)
21. R.A.Baragiola, E.V.Alonso: to be published.
22. The reader may readily estimate the effect in the shape of the yield vs. energy curve of a contamination of a few percent of doubly charged ions with energy $2E_o$ on a beam of single charged ions of energy E_o, a situation found in the work of references 15, 23, 24 and 25.
23. J.Ferrante, S.V.Pepper: Surface Sci.58, 613 (1976)
24. J.Kempf, G.Kaus: Appl.Phys.13, 261 (1977)
25. R.A.Powell: J.Vac.Sci.Technol.15, 1797 (1978)
26. Using the Moliere potential from the tables by M.T.Robinson: Rept.ORNL-4556 (Oak Ridge National Laboratory, 1970)
27. C.C.Lu, T.A.Carlson, F.B.Malik, T.C.Tucker, C.W.Nestor, Jr.: Atomic Data 3, 1 (1971)
28. U.Fano, W.Lichten: Phys.Rev.Lett.14, 627 (1965)
29. M.Barat, W.Lichten: Phys.Rev.A6, 211 (1972)
30. V.Pejčev, T.W.Ottley, D.Rassi, K.J.Ross: J.Phys.B10, 2389 (1977)
31. P.Dahl, M.Rødbro, G.Hermann, B.Fastrup, M.E.Rudd: J.Phys.B 9, 1581 (1976)
32. D.L.Walters, C.P.Bhalla: Phys.Rev.A4, 2164 (1971)

Interaction Between Metastable Rare Gas Atoms and Surfaces

Claude Boiziau

Service de Physique Atomique, Section d'Etudes des Interactions
Gaz-Solides - CEN.SACLAY, B.P. No.2
F-91190 Gif-sur-Yvette, France

1. Introduction

Surface physics is at present undergoing major development. More and more powerful techniques are being devised, each having its own limits, and new techniques are being worked out at the same time in order to have complementary information. In particular, there are often difficulties in accurately appreciating the respective parts of the signal issuing from the bulk and from the surface proper.

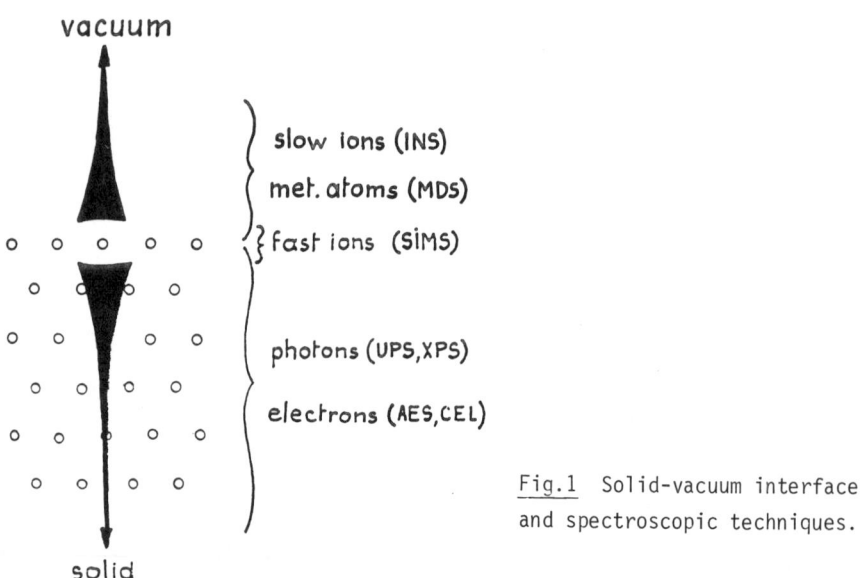

Fig.1 Solid-vacuum interface and spectroscopic techniques.

It is necessary moreover to define what is meant by the "surface". For instance, is it the first atomic layer which is seen when using Secondary Ion Mass spectroscopy? Is it the first atomic layers which are seen with incident photons or electrons, the depth of the surface then being determined by the escape probability of the secondary electrons ? Is it only the solid part which can interact with foreign atoms or molecules, and which is studied with incident particles which cannot penetrate into the solid, e.g. slow ions or slow atoms ?

These three definitions of the surface are evidently complementary : each of the three only gives a partial picture, and a complete description of the solid-vacuum interface requires that the three aspects of the surface be studied.

Study of the interaction between metastable rare gas atoms and solid surfaces is evidently a contribution to this objective :
- it is a non-destructive method if the kinetic energy of the metastable atoms is small;
- the chemical state of the surface is not changed since the incident particle is a rare gas atom;
- moreover, it will be shown that the information issues from a precisely located part of the surface, namely the exterior of the solid.

There is thus a question : why is the Metastable atom De-excitation Spectroscopy, which we call M D S, not undergoing major development, in the same way as U P S, A E S, S I M S, E S D, ... ?

Firstly, the physics of the interaction is not yet well known, and some work is necessary in order to allow complete explanation of the secondary electron spectra.

Secondly, experimental techniques are complicated, and the signal is always very small.

Thirdly, we will see later that there are some mathematical difficulties in extracting information from the experimental spectra.

It can thus easily be explained why only a few laboratories started with this spectroscopic method in spite of very promising results.

2. Interaction of a Metastable Rare Gas Atom with a Surface

Despite the small number of laboratories working on the interaction between metastable rare gas atoms and a solid surface, this field has a very long history. The first theoretical predictions were given by OLIPHANT and MOON (1) : they described the resonant transition of an electron between a metal and the electronic level of an atom. Then SHEKTER (2) studied Auger neutralization of an ion and Cobas and Lamb(3) worked on Auger de-excitation of an excited atom. But HAGSTRUM(4) developed the only global analysis of metastable atom-surface interaction, which remains the theoretical basis for interpreting experimental results.

Three interaction processes are possible when a metastable atom arrives in front of a solid surface :

- an elastic reflection (5, 6), which is dependent on the chemical state of the surface. But our recent work (7) shows that this process has a very low probability of interaction between metastable rare gas atoms and metallic surfaces,
- de-excitation with emission of a photon (8) is the metastable level is broadened enough to allow radiative transition to the ground level. This second possibility is also very weak with metallic surfaces,
- de-excitation with emission of an electron. But, in this last case, two de-excitation processes are possible.

2.1 Either the excited electron can undergo a resonant transition by tunneling to metal free levels (Fig. 2A), which is the case when the metal work-function $e\varphi$ is larger than the ionization potential of the metastable atom. The resulting ion is then neutralized according to an Auger process :

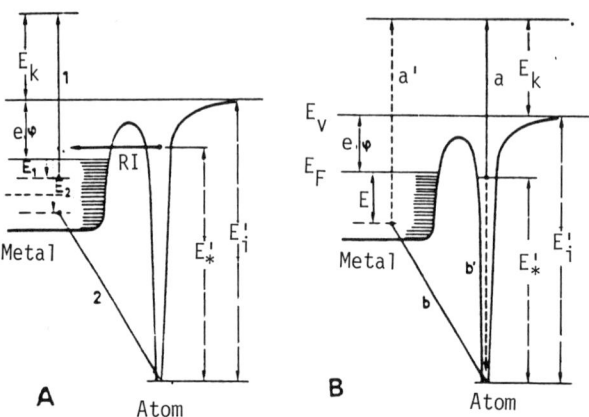

Fig.2 Two de-excitation processes are possible whether tunneling of the metastable electron to the free levels of the metal is possible or not

one valence electron drops to the atom ground level while another one is ejected with the kinetic energy

$$E_k = E'_i - 2e\varphi - 2x$$

where E'_i is the atom ionization energy perturbed by the surface potential.

It is easy to show that the electron energy distribution is the self-folded transition density with a width :

$$E = E'_i - 2e\varphi ; \qquad (1)$$

we call this mechanism a two-electron process, or RI + AN (Resonant Ionization + Auger Neutralization).

2.2 Or de-excitation follows a direct Auger process : (Fig. 2B) one valence electron drops to the ground level of the atom while its excited electron is ejected with the kinetic energy

$$E_k = E_* - e\varphi - x .$$

This process may be observed in particular when the metastable level is either in front of occupied levels or in front of a forbidden band. The energy distribution is as the single density transition with a width :

$$E = E_* - e\varphi . \qquad (2)$$

This second mechanism is called a one-electron process or AD (Auger De-excitation).

2.3 The situation is more complicated when the metal is covered either with adsorbed gases or with a thin layer having forbidden levels :

the metastable atom cannot approach the metal surface, and there is possible competition between the two mechanisms, as far as the excited electron tunneling remains possible.

In a recent paper (9), HAGSTRUM suggested that this competition is governed by the wave-function overlap at the position of the metastable atom, which has been confirmed by our recent results.

But, in each cases, we saw that the de-excitation processes giving electron emission follow Auger processes which are always short-range transitions. So only the electron located close to the metastable atom can be involved in the interaction.

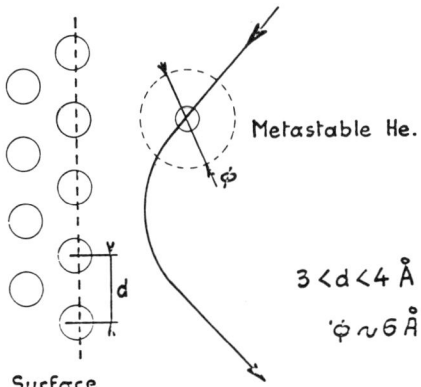

Fig. 3 Trajectory of metastable rare gas atom

Moreover, a metastable atom cannot penetrate into the solid network (Fig.3) and the de-excitation process takes place outside the surface.

Therefore, information contained in the secondary electron emission is characteristic of the external part of the surface, in fact, of the external extension of the surface electronic wave-functions.

3. Experimental Aspects

We saw before that experimental difficulties are probably a cause of the present limitation to the spread of the M D S.

A typical apparatus involves four essential parts, which are four sources of difficulties.
- the metastable atom source, and it is often impossible to obtain a high density of metastable atoms without many parasitic particles,
- the atom beam between the high pressure source and the very low pressure measurement chamber,
- the measurement chamber, which must contain diagnostic and measurement apparatus necessary for surface spectroscopy,
- electronic instrumentation, sufficient for a low level signal, and possibly, pulse techniques.

3.1 Metastable Atom Source

The classical method of producing metastable atoms is electronic bombardment of a gas,
- either in a high pressure chamber, and the beam being created by thermal effusion throughout a narrow diaphragm,
- or in an atom beam, either supersonic or obtained by thermal effusion throughout needles or a microchannel array.

But in each case, electronic bombardment of rare gas atoms - e.g. helium atoms, produces :
- metastable atoms, in two states
  the singlet one 2^1S with an energy of 20.6 eV
  the triplet one 2^3S with an energy of 19.8 eV
- Rydberg atoms, with principal quantum number between 20 and 100,
- photons, in particular 21.2 eV radiation,
- ions
- fast neutrals, obtained either by ion neutralization on a surface or by charge exchange between ion and neutral atoms.

All these particles may interact with the target, giving secondary electronic emission. Therefore, a good measurement method implies either elimination of impurities contained in the beam or separation of the individual effects of the beam constituents.

3.2 Cleaning of the beam

The first operation is the quenching of the 2^1S singlet state : the beam passes throughout a discharge lamp giving 2 radiation, and the singlet level is pumped to the radiative 2^1P level.

The separation of the two state effects is carried out by two measurements : with and without discharge lamp.

Secondly, a 2 kV/cm electrostatic field is enough to destroy the Rydberg atoms by Stark effect and extract the resulting ions from the beam. Figure 4 shows the decreasing of the He^+ signal extracted from the target as a function of an increasing

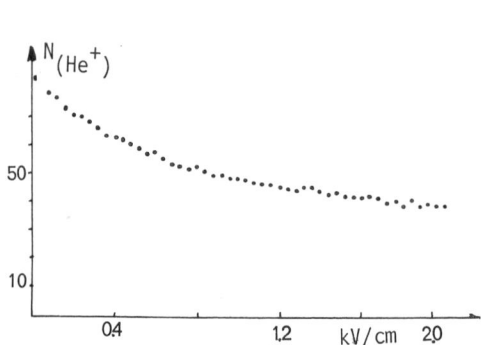

Fig.4 Effect of an electrostatic field applied to the beam. (target: Cu 110)

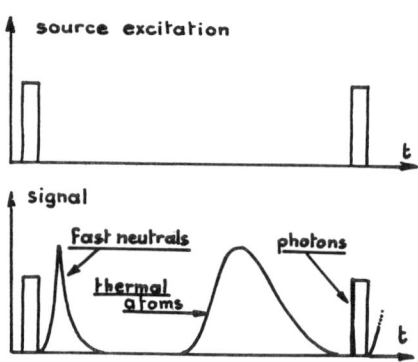

Fig.5 TOF method. The electronic discharge is pulsed and the distance between source and target is about 70 cm

electrostatic field applied to the beam. The residual signal, which remains constant up to 50 KV/cm, is due to ionization of the metastable atoms near the surface and the reflected component of the ions created. We may emphasize that, under our experimental conditions, this part may have the same order of magnitude as the reflection probability of slow ions measured by HAGSTRUM (10).

Finally, the only way to separate completely the effects of the metastable atoms, photons and fast neutrals is a time of flight method.

In our laboratory, the gas is bombarded in a pulsed electronic discharge, and the electronic equipment is so designed that the signals characteristics of each incident particle may be recorded simultaneously, giving, in particular, simultaneous UPS and MDS measurements. It may also be emphasized that the configuration and the operating conditions of the source allow large variations in the beam composition.

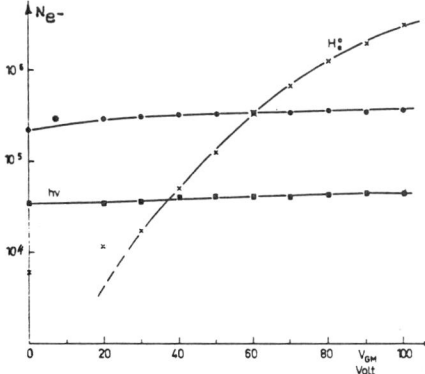

Fig.6 Increasing of the signal due to the fast neutrals as a function of their kinetic energy

In particular, we see (Fig.6) that the potentials in the source allow a large variation of the signal due to the fast neutrals, but we have verified that by suppressing metal parts in the source in direct view of the target, and including an electrostatic configuration in the source to avoid ions running in the target direction, it is possible in practice to eliminate fast neutral production (11). In this case, the time of flight method is only used to separate the photon and metastable atom effects.

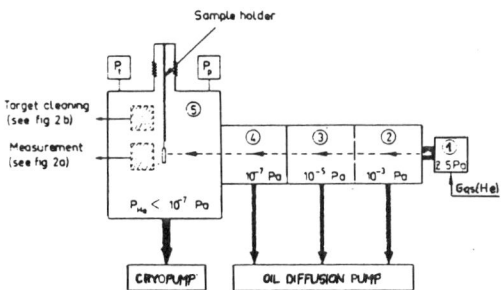

Fig.7 Experimental apparatus

3.3 Measurement Chamber

The measurement chamber is very classical and contains the typical apparatus for cleaning and diagnostics used in surface physics.

The electronic emission from the target may be :

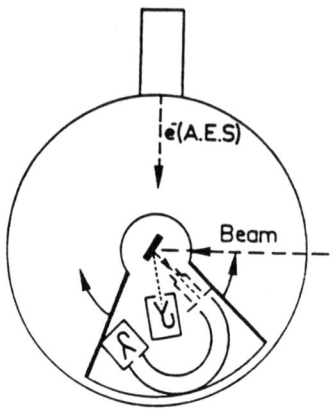

Fig.8 Electronic emission measurement

- either all collected by the channeltron multiplier A (Fig.8), which allows yield variation measurements
- or analysed by an electrostatic analyser which, in our case, is hemispherical (12).

In the latter case, two measurement modes are possible with or without an electrostatic field between the target and analyser.

With one, the electrons are concentrated along the direction normal to the target. We obtain a relatively high signal level, but the low energy electrons are favoured so that the spectra are deformed.

Without one, we operate according to the angle resolved mode, but the signal level is weak so that the treatment of the curves is more difficult.

3.4 Electronic Instrumentation

As mentioned above, it is always difficult to obtain a large number of metastable atoms, so that, with an angle resolved electrostatic analyser, a few tens - in some cases a few hundreds- of electrons per second may be counted. It was then necessary to develop very low noise counting techniques.

The electronic equipment also has two essential functions :

- to synchronize the source excitation pulses with the device directing the signal according to whether it is produced by fast or slow particles : four lines are possible.

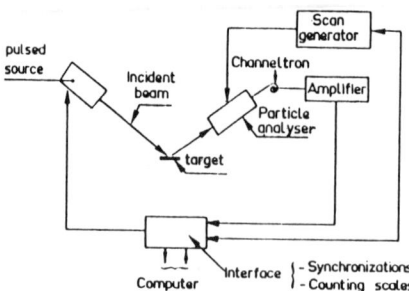

Fig.9 Electronic equipment for simultaneous U P and M D spectroscopies

- and to synchronize the voltage sweep of the electron analyser with the advance channel of the memory blocks corresponding to each type of emitting particle.

3.5 Fast Neutrals as a Surface Cleanliness Probe

It is obviously possible for a mass spectrometer to take the place of the electron analyser in order to study the nature of the secondary particles and obtain secondary ion mass spectra.

The T O F technique allows the separation of the secondary ions according to the incident particle types and shows that the fast neutrals, when they are present in the beam, are responsible for the largest part for the secondary ion emission.

Figure 10 presents an illustration of the possibilities that SIMS offers when operating with fast neutrals, giving a picture of the target composition.

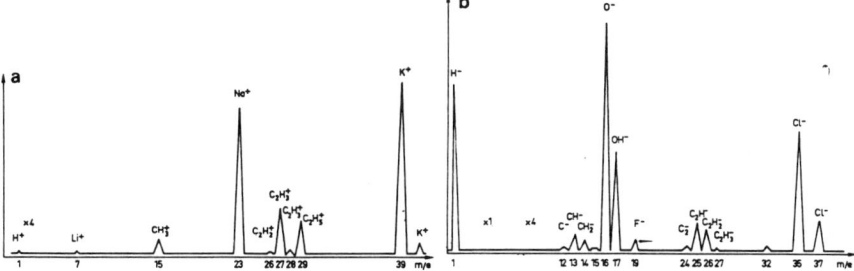

Fig. 10   SIMS operated with incident fast neutrals (E_K = 100 eV)

We may also emphasize the high sensitivity of the secondary electron signal as a cleanliness test : the curve of Figure 11 shows the result obtained in this way with a nickel (111) target, after oxygen adsorption . It is possible to compare this curve with those obtained by AES under the same conditions (13).

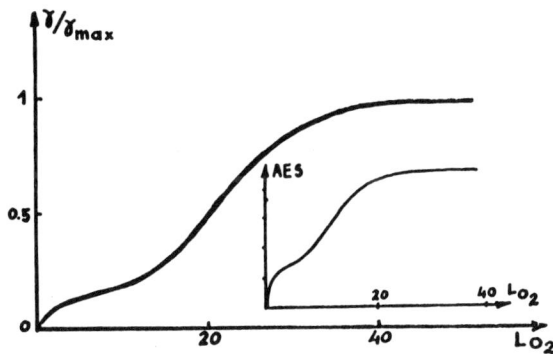

Fig.11 Variation of the electronic emission due to fast neutrals as a function of the exposure to oxygen

3.6 Electron Emission Spectra Obtained with Fast Neutrals

The fast neutrals may thus practically be eliminated. But when present, their effects may be a source of very interesting information about the chemical state of the surface studied. Moreover, study of their secondary electron emission shows that the essential part of the total electron number is found in the 0-5 eV kinetic energy range, which in UPS is that of the scattered electrons.

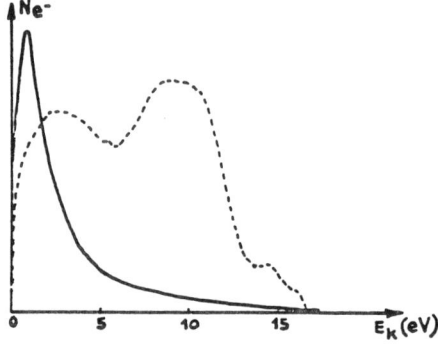

Fig.12 Energy distribution of electron ejected by fast neutral collision. The dotted curve shows the U P spectrum obtained simultaneously

It is then possible, with our electronically run T O F technique, to prolong the excitation in the source in order to separate only fast and slow particles and increase the signal in the same proportion.

The UPS spectra will contain electrons due to the residual fast neutrals, but these electrons cause contamination that is practically negligible and located in the less interesting part of the UPS data.

4. Recent Results in MDS

So, with some experimental precautions, study of the interaction between metastable rare gas atoms and surfaces is possible. To our knowledge, four laboratories in the world are working on MDS : in Tokyo, in Munich and in our laboratory in Saclay; the fourth laboratory, in Great Britain, is publishing electron spectra which look rather strange, like spectra induced by fast neutrals or slow ions.

It is then possible to give a short review of recently published work without excessively prolonging this paper. Moreover, it will allow us to understand the actual problems of the interaction physics and of the experimental techniques.

4.1 Interaction of Metastable Atoms with Crystal Surfaces of Condensed Aromatics.

The first work at the Japanese laboratory started with a scrupulous study of the experimental conditions before publication (14) : the beam contains only a few per cent of photons and the signal due to the fast neutrals is negligible insofar as the electronic bombardment energy in the source is less than 50 eV (15). So the measurements may be made without T O F operation and the electron analysis is performed by means of the retarding potential on a spherical collector surrounding the target.

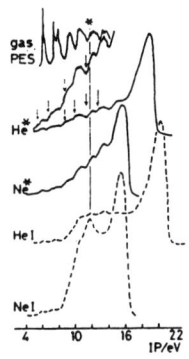

Fig. 13 Energy distribution curves for Coronene

Figure 13 shows a comparison between spectra obtained with MDS, UPS and gas phase photo-electron spectroscopy (16).

Two conclusions are possible : firstly, the good agreement between the MDS peaks with the corresponding photo-electron ones shows that the metastable atom de-excitation follows the direct Auger de-excitation process; secondly, the peak corresponding to the scattered electrons is smaller in MDS than in UPS. Since thermal metastable atoms do not penetrate into the solid, about half of the Penning electrons are emitted directly to the vacuum without interaction with other valence band electrons.

4.2. Interaction of Metastable Atoms with Surfaces of Alkali Halides (15)

The same laboratory studied surfaces of alkali halides.

Figure 14 gives an example of their results. The same conclusions as above are possible. Moreover, the peak positions of the valence band structure observed in MDS are shifted from the corresponding peaks in UPS, which may be explained by the bending of the valence band at the surface if MDS is - as expected - more sensitive than UPS to the outermost atomic layer of the solid.

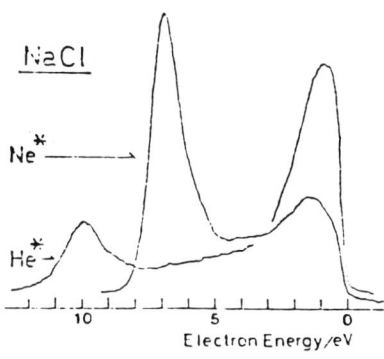

Fig.14 Energy distribution curves for NaCl

4.3. Interaction of Metastable Atom with Clean and CO-covered Pd(111).

This was the first work from the German laboratory (17). There are two essential differences between their instrumentation and that used in the Japanese laboratory :

- measurements of electron energy distributions are made with a two-stage 127° electrostatic analyser, with an electrostatic field between the target and the analyser entrance, in order to concentrate electrons along the normal direction and to have more signal,
- the singlet state atoms may be almost completely quenched and it is possible to distinguish between the singlet and triplet states effects.

They start with a comparison between UPS and MDS obtained with clean palladium (111) (Fig.15).

There is a complet difference in the shapes and in the extent of the spectra, which may easily be explained if a two-steps process (RI + AN) is assumed for the metastable atom de-excitation.

The difference observed between the MDS curve and the INS one obtained with a similar surface is probably due to the electrostatic field effect.

The MDS spectrum may thus be interpreted as the self-folded density of states of the clean palladium (111).

The results are quite different after adsorption of a half-monolayer of CO on the palladium (111). Experiments performed with singlet and triplet atoms (Fig.16) exhibit quite similar spectra. Some observations may be derived from these curves :

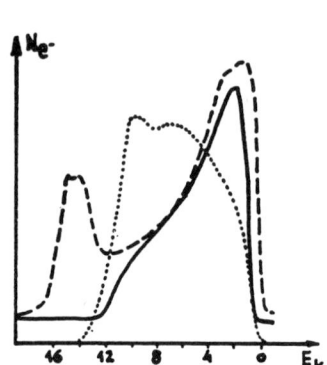

Fig.15 UP spectrum in dashed line and MD spectrum solid line with Pd (111). The dotted curve is an IN spectrum obtained by HAGSTRUM (18) with clean Ni(111).

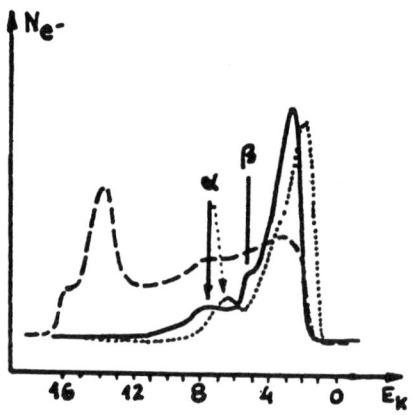

Fig.16 Study of CO-covered Pd(111) UPS : dashed line; MDS (2^1S) : solid line; MDS (2^3S) : dotted line.

- firstly, the peak corresponding to the palladium d band, very pronounced in the UPS spectrum, does not appear in the MDS spectra, showing the sensitivity of this last spectroscopic technique to the outermost layer,

- secondly, the shape variation of the MD spectra after CO adsorption suggests a large change in the interaction parameters,

-thirdly, the peaks marked α and β on the MD spectra may easily be attributed to electron emission from the CO ($5\sigma + 1\pi$) orbitals and the CO 4σ orbital respectively, in the picture of a direct Auger de-excitation process,

- fourthly, the validity of this interpretation is shown by the difference between the peak positions on the 2^1S and 2^3S spectra, which is just the difference in the excitation energy of the two species.

Nevertheless, these results suggest some questions :
- what is the origin of the shift observed between the two spectra for the low-energy cut-off ?

- how does one explain the large number of electrons which are not typically Penning electrons ? Are they secondary and backscattered electrons ? In this case, their number is much larger than that of the primary electrons, which would be surprising for spectroscopy only sensitive to the outermost layer, and it would be contradictory to the observations made by the Japanese laboratory.

4.4 Interaction of Metastable Atoms with Cs-Covered Pd (111).

This was the second experimental work presented by CONRAD et al. (18) and it is perfectly compatibile with HAGSTRUM's results obtained with a similar system : Ni (111) + K (9) : the electron emission takes place via the direct Auger de-excitation process.

5. Our Results in MDS (19 to 28)

5.1 Comparison of UPS, MDS, and AES with clean Mo(110).

A first question must be answered : what place can MDS take among the other spectroscopies, and what information is it possible to obtain with it ? We saw above, and we will see later, that MDS may be a tool well suited to studying molecules in a condensed or adsorbed state. We now want to investigate the possibilities of MDS with clean metal, in particular with molybdenum (110), by comparison of our results in UPS, AES and MDS.

The theoretical model for the de-excitation process when a metastable helium atom interacts with clean Mo(110) is the two-step one : resonant ionization + Auger neutralization. We may thus write that the three spectra expected are of the form :

$$F_{UPS}(E) = (U \cdot P_{UPS}) * Q_{UPS} * I$$

$$F_{MDS}(E) = ((V * V) \cdot P_{MDS}) * Q_{MDS} * I$$

$$F_{AES}(E) = ((W * W) \cdot P_{AES}) * Q_{AES} * I * G$$

where . is a single product
and $*$ is a folding product

In these equations, U, V, W are the transition densities in each mechanism, P is the escape probability for electrons, Q is the broadening function due to the energy losses, I is the broadening function due to the spectrometer (the same one for the three measurements) and G is the broadening function due to the electron gun in AES.

In a first approximation, we may neglect the effects of the I and G functions; if the P and Q functions are different, they are certainly not the essential reason for the dissimilarity of the experimental spectra. The main reason must be sought in the fundamental mechanisms of electron emission

- a one-electron process in UPS,
- a two-electron one in AES and MDS, but with different parameters which may be found in the matrix elements of the transition.

Experimentally, we may compare either the MDS and AES spectra with the self-folded UPS one, or the UPS spectrum with the curves obtained by unfolding of the MDS and AES data. This second possibility allows comparison between the calculated density of states and the experimental curves, and gives total confirmation of the realisticness of the models (Fig.17).

Nevertheless, scrupulous examination of the curves shows differences which may be attributed to the initial approximations, but above all, to the elementary parameters of the interactions : location of the emitting atom and correlation effects.

In a few words, these results show that MDS is actually a good surface spectroscopy and may be, in conjunction with UPS and AES, used to give a thorough knowledge of the electronic properties of a surface.

5.2 Effect of an Electrostatic Field between the Target and an Electrostatic Analyser. Trial with Clean Mo (110).

We saw above that it is sometimes necessary to place an electrostatic field between the target and the analyser entrance in order to increase the signal. Figure 18 shows the effects of this field, which, as expected, has a larger influence on the slow electrons than on the fast ones.

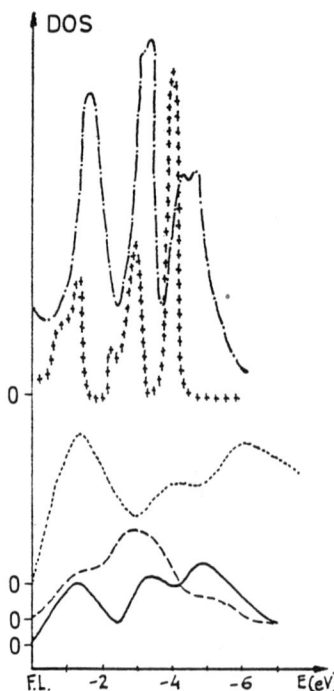

Fig.17 Study of clean Mo (110)
unfolded MDS ———
unfolded AES -------
UPS ....
calculated DOS (29) —.—.—
calculated surface states (30) ++++

In particular for MDS, we see that the scattered and secondly electrons form only a very weak part of the total electron emission : only the electrons emitted in the vacuum direction are collected and their probability of interaction with other electrons is very low. It can be neglected in further work.

5.3 Study of a Metallic Oxide

We shall now study the evolution of the MDS signal when the density of states varies at the surface. For this purpose, we choose a system where the metallic characteristics of the surface are not changed with adsorption, e.g. when oxygen is adsorbed on the Mo(110) surface.

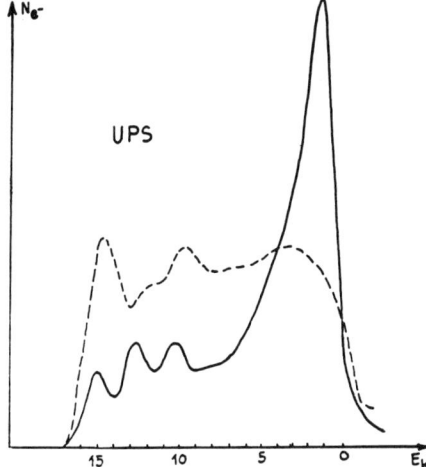

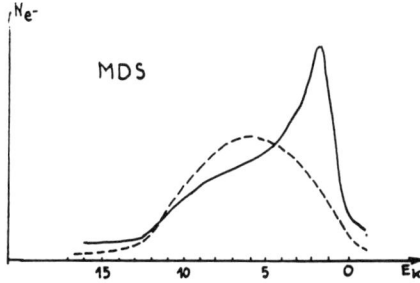

Fig.18 UPS and MDS measurements with an electrostatic field (full lines) and without (dotted lines).

Measurements are made with an electrostatic field between target and analyser.

Comparison between UPS and MDS shows :

- firstly, the spectra shapes remain different all along the adsorption kinetics, suggesting that the interaction processes remain quite different,

- secondly, MDS appears more sensitive to oxygen adsorption than UPS, with, in particular, a fast decrease of the electron number situated in the high energy range,

- thirdly, the shift observed in the low energy range, corresponding to the work function variation, is also seen, for MDS, in the high energy range, which is another confirmation of the validity of the RI + AN process.

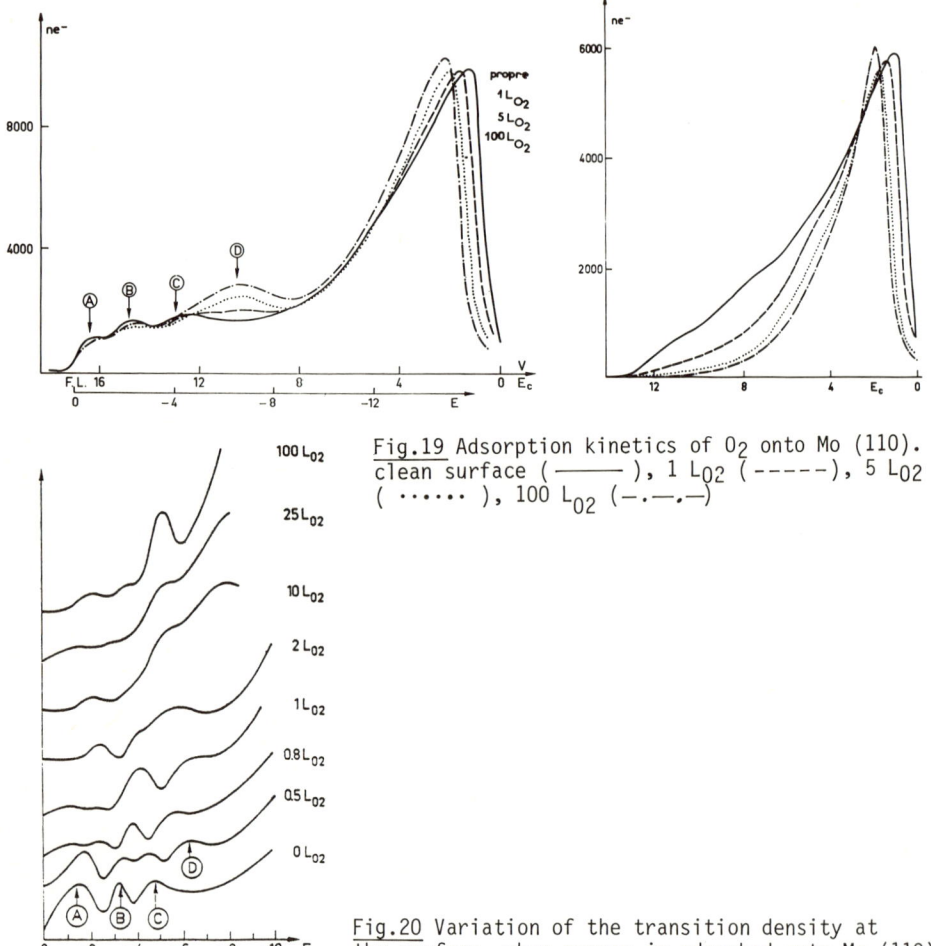

Fig.19 Adsorption kinetics of O_2 onto Mo (110). clean surface (———), 1 L_{O2} (-----), 5 L_{O2} (······), 100 L_{O2} (−·—·—)

Fig.20 Variation of the transition density at the surface, when oxygen is adsorbed onto Mo (110)

It is thus reasonable to unfold the MDS spectra (Fig.20) and obtain the variation of the transition density at the surface. We see, in particular, the increasing of a peak at about 6 eV below the Fermi level, corresponding to the peak seen with UPS and attributable to the 2p level, of adsorbed oxygen. Simultaneously, the peaks corresponding to the d band of Molybdenum vanish : the substrate is not seen by the metastable atom; only the outermost layer is involved in the interaction. It then becomes possible, with MDS, to follow adsorption kinetics with great sensitivity.

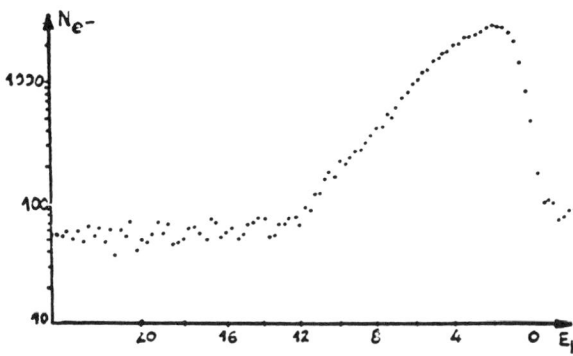

Fig.21 ARMDS with oxygen saturated Mo(110) surface

We now examine a spectrum obtained in angle resolved MDS, i.e. without electrostatic measurement perturbation. In order to guarantee oxygen saturation of the target, it is exposed to a few hundred langmuirs and maintained during the measurement to an oxygen pressure of 10^{-7} torr. We want to answer the question : is the RI+AN process the only mechanism available with such a surface ? Careful examination of the whole results obtained under these conditions shows a possible contribution of the direct Auger de-excitation process. (Fig.21).

- a first feature, with an energy of about 2 eV below the Fermi level, may be attributed to the d band wave-functions of the Mo(110) substrate,

- the second one, with an energy of about 6 eV below the Fermi level, which corresponds to the 2p level of oxygen.

But, we easily see that the largest part of the spectrum must be interpreted in terms of the RI + AN process. In fact, the possible AD process contribution, if it could be one, has the same order of magnitude as the statistical uncertainty of our data and may be neglected, at least in a first approximation.

5.4 Study of Adsorbed Molecules

Oxygen is thus adsorbed on Mo (110) in the dissociated form. Taking the same substracte, it will be interesting to study the electronic levels of a surface constituted of adsorbed molecules, e.g. CO.

Following the same procedure as for oxygen, we start with the CO adsorption kinetics (Fig.22).

The same type of comments may be made, leading to unfolding of the MDS spectra in order to obtain clear information.

Two stages can be distinguished (Fig.23) : for small exposures, we observe the disappearance of peak A, characteristic of the upper d band of molybdenum and the replacement of peak C by two peaks easily attributable to the 2p levels of atomic carbon and oxygen. This suggests a first phase in the adsorption where the CO molecule is dissociated.

However, already at this step, a slight bulge appears at the right of the diagram and its meaning becomes clear as exposure to CO is prolonged : a very broad peak with a maximum at about 8 eV develops as CO adsorption continues. This peak is characteristic of the CO (5σ and 1π) molecular orbitals.

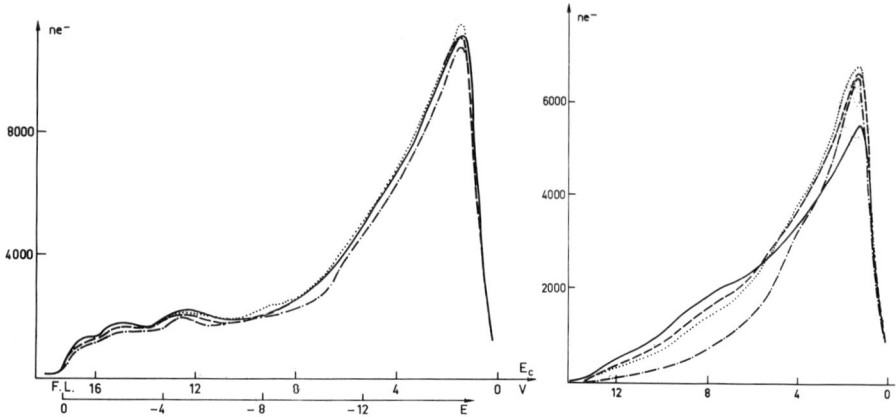

Fig.22 UPS and MDS during CO adsorption onto Mo(110)

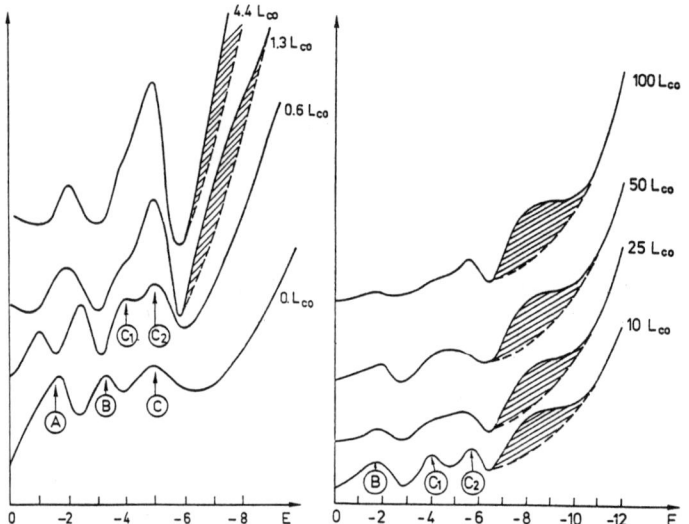

Fig.23 Transition density variation at the surface, when CO is adsorbed onto Mo(110)

The last curve, for 100 langmuirs, gives a good picture of the surface electronic levels and affords a schematic diagram (fig.24):

- the molybdenum d band is completely shielded;
- the dissociated from appears with the two atomic levels of oxygen and carbon (γ and β);
- the molecular form appears with the two peaks α and δ, the bonding being carried out by electronic transfer from the 5σ and to the $2\pi^*$ orbitals.

We now consider the saturated surface, with a CO pressure of 10^{-7} torr (Fig.25).

64

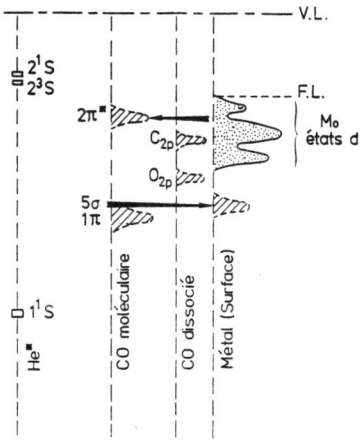

Fig.24 The energy diagram of the system He*-CO-Mo(110)

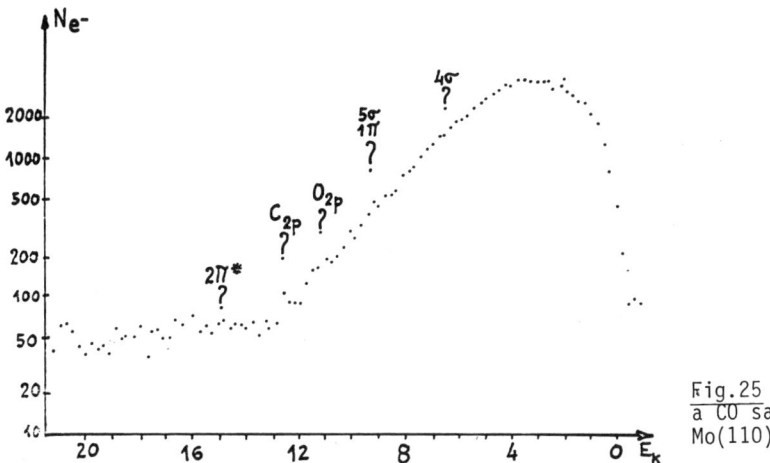

Fig.25 ARMDS of a CO saturated Mo(110) surface

In the same way as for oxygen, angle resolved MDS shows that if it could be one, the contribution of the AD process is practially negligible: interpretation of MDS according to the single RI + AN process is available, at least with a molybdenum substrate.

5.5. Study of Pure Molecular Adsorption

We know that CO adsorption on Ni (111) is purely molecular, without dissociative step. Moreover, nickel has a band structure very different to that of molybdenum, with, in particular, a high and narrow peak just below the Fermi level. We will see that MDS takes these peculiarities into account.

First, the UPS and MDS spectra obtained with clean nickel show that an RI + AN process is appropriate. The shapes of the spectra are different and the width of the MDS one is perfectly compatible with formula (1).

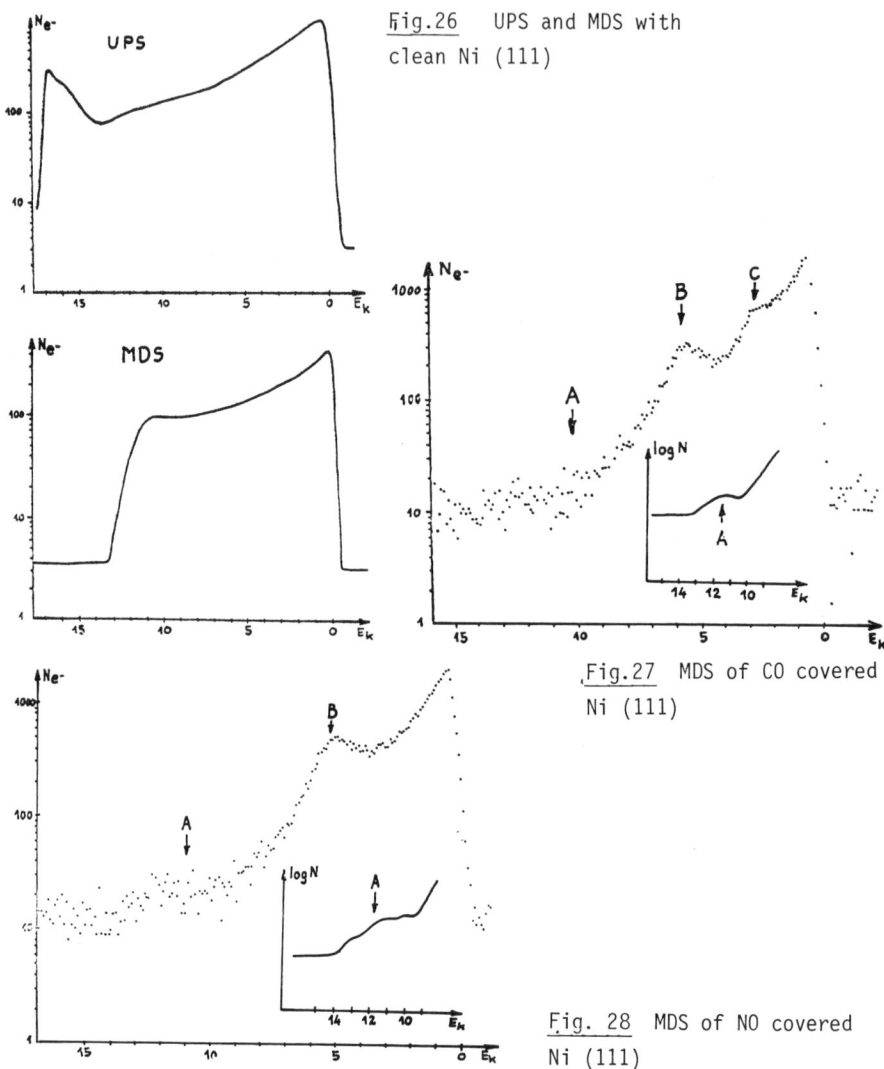

Fig.26 UPS and MDS with clean Ni (111)

Fig.27 MDS of CO covered Ni (111)

Fig. 28 MDS of NO covered Ni (111)

The situation is more complicated with adsorbed CO. Being in semi-logarithmic coordinates, the Figure 27 shows that the apparent threshold, in the high energy range, and corresponding to a RI + AN process, is not the true threshold : another one, very slight has the good energy value for the AD process. A special measurement in the corresponding energy range confirms this appreciation. Moreover, two peaks appear which have the good positions for the molecular orbitals levels of the adsorbed CO.

We verified that these features are not instrumental effects with measurement obtained when NO is adsorbed on the same substrate : the orbital levels are deeper in NO than in CO and the $2\pi^*$ is partially occupied in the free state.

In fact, with NO, we see (Fig.28) :
- a more noticeable threshold in the high energy range;
- a peak B shifted towards the low kinetic energy,
- the disappearing of the peak C, shifted on the other side of the emission threshold.

It is then possible to assign the peaks to the molecular orbitals, interacting directly with the metastable atom according to the AD process.

But a question remains : where do the electrons come from which are not involved in this description ? Are they only the result of the spectrum distortion due to the electrostatic field ? A preliminary calculation shows that they may be due - at least partially - to the contribution of the RI + AN process.

5.6 Study with a Surface Offering a Forbidden Band

If a surface presents a forbidden band in front of the metastable level, theoretical models point out that tunnel transition becomes impossible and de-excitation may only occur according to the AD process.

On the other hand , it is known that a few tens of oxygen langmuirs on Ni(111) produces about 2NiO layers, this oxide having a forbidden band with a width of around 4 eV (31-32).

It is then attractive to study the oxygen adsorption kinetics.

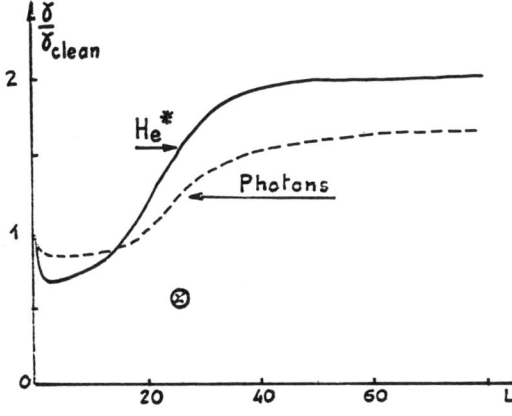

Fig.29 Electronic yield variation following oxygen adsorption onto Ni (111)

The two curves (Fig.29), respectively obtained by photoemission and meta emission , clearly show three regions which may be described using previous works :

The first one is the oxygen adsorption step, up to a few langmuirs, and giving a P(2x2) structure ; the second one is oxide layer formation, the layer developing by islands; finally, the oxide layer covers the surface.

The encircled point is deduced from HAGSTRUM's publication (33) and seems to be a good extrapolation of the first region of the meta-emission curve. Naturally, the interpretation of these curves will be made easier by examination of the spectra obtained for some characteristics exposures (Fig.30).

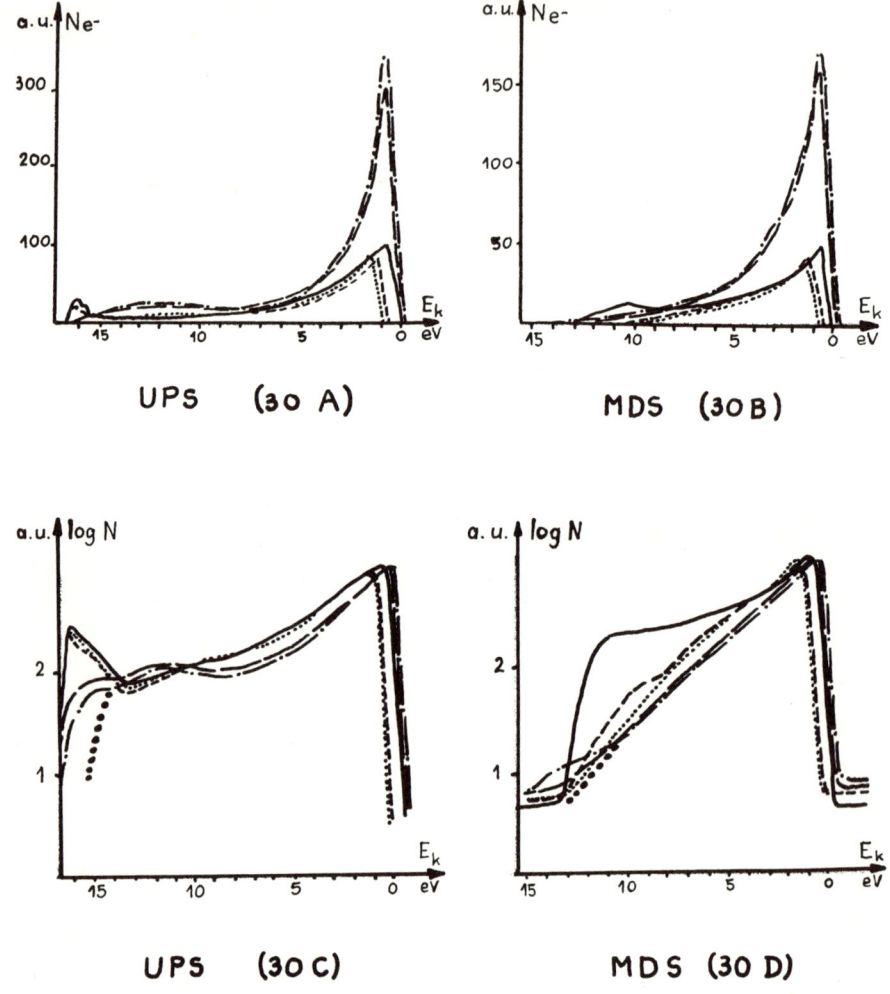

Fig.30 (A-D): UPS and MDS from Ni(111): clean surface: (———); oxygen exposure 5L: (-----); 20L: (······); 120L: (— — —); 500 L: (—.—.—)

First we consider, UPS and MDS data in a linear form (respectively Fig.30-A and 30-B). The adsorption region, below 5 langmuirs, produces a decreasing of the electron number extracted from the nickel d band, and another decreasing in the low energy range, owing to the work-function increasing.

The intermediate region is characterized by an increasing of the electron number in the low energy range, and, at least in the UPS spectra, by the development of a bulge in the 5-6 eV range below the Fermi level, which may be attributed to the oxygen 2p level.

The saturation region, which is the nickel oxide one, corresponds essentially to a large collection of low energy electrons.

Secondly, we consider the same data, but in a semi-logarithmic form, which reveals some shape particularities. The UPS data (Fig.30-C) allow good measurement of the apparent work-function, and the MDS ones (Fig.30-D) give the width of the spectra.

We used the terme apparent work-function : in fact, the photo-electrons have a mean free path sufficient to be extracted both from the oxide layer and from the metal substrate. The examination of UPS spectra suggests that for large coverages, we may find a second possible value for $e\varphi$ which is about 6.5 eV, obtained by extrapolation of the spectra.

The problem is more complicated with MDS : on the one hand, two de-excitation processes may be considered giving :

$$E = E'_i - 2e\varphi$$
and
$$E = E_* - e\varphi$$

On the other hand, electrons can issue either from the oxide layer or from the outer region of the metal electronic wave-functions, if it is.

Moreover, the spectra obtained for large coverages clearly exhibit a discontinuity in the high energy range ; extrapolated; this discontinuity gives a width of about 11.5 eV.

We may thus compile the following table E'_i is calculated using the

	clean	5L	20L	120L	500L
$e\varphi$	5.4	6.2	6.2	5.4 6.5	5.2 6.5
ΔE	11.6	11.1	10.4	13.2 11.5	13.2 11.5
E'_i	22.4	23.5	22.8	22.3	21.9
E_*	-	-	-	19.7	19.7

lowest ΔE value and the $e\varphi$ one corresponding to the metallic part of the target.

E_* is calculated using the largest ΔE value and the $e\varphi$ one corresponding to the outermost layer, and is given only when the result is compatible with metastable atom energy, here the triplet 2^3S one.

How does one explain these results ?

If the first step, corresponding to the adsorption phase, may easily be interpreted according to the RI + AN process, it is not the case with the second and third steps.

In particular, what is the origin of the large increasing of the electronic yield when oxide coverage takes place. We see that it is essentially the low energy range which undergoes a fast development, and this is more important with MDS (outer electrons) than with UPS (outer and bulk electrons). Low energy electrons are due to secondary emission and inelastic backscattering of primary electrons. Is this development proof that the nickel oxide layer reflects the electrons emitted towards the solid, which may then be collected after interaction with other electrons ?

And, if the true width of the 500 L spectrum is really 13.2 eV, suggesting an AD process, why are the peaks corresponding to the oxygen levels, at ~ 6 eV and ~ 11 eV below the Fermi level (values usually given for NiO), not more apparent ?

In fact, the feature giving the 13.2 width and those attributable to the oxygen levels are very weak; is the contribution of the AS process of the same order of magnitude ? Nevertheless, if the metastable de-excitation emission is due to the RI+AN process, it implies that the electron tunneling must be done throughout the oxide layer and it suggests a large extension of the substrate wave-functions throughout this layer.

In conclusion, there is probably a superposition of these phenomena, and further work is necessary to explain these results.

5.7 Unfolding of the Spectra Resulting from an RI+AN Process

We saw above that the spectra obtained when the de-excitation follows an RI+AN process must be unfolded in order to extract the transition density. But we ought to take some problems into account.

5.7.1. The statistical uncertainty of the experimental spectra obviously gives a much larger uncertainty for the unfolded curves.

5.7.2 The possible contribution of the AD process and that of the electrostatic field between the target and analyser (when it is) are difficult to evaluate.

5.7.3 The refraction effects for large measurement angles give inhomogeneous spectra : electrons are deviated and their deviation is a function of the surface barrier value, the kinetic energy of the electrons and their initial direction.

5.7.4. At least, when the spectrum width is smaller than the self-folded transition density one, experimental data are not complete, and the truncation effects are increased by those of the escape probability, being lower for the slow electrons than for the fast ones.

All the results must then be thoroughly studied and discussed, and their limitations evaluated.

5.8 Summary of our Results

Following our results, some conclusions are possible :

- MDS always gives information about the external extension of the electronic wave-functions, and the process is governed, as predicted by HAGSTRUM (9), by the overlap of the wave-functions at the metastable atom position.

- A clean metal surface with large work-function produces de-excitation according to the RI+AN process. The spectrum is then a picture of the self-folded transition density under the interaction conditions. The reflection probability of the electrons emitted towards the solid is very low; back-scattered electrons and secondary emission may be neglected.

- A metallic oxide, keeping a metallic structure, essentially follows the same rules. When the oxide has a non-metallic structure, tunneling of the metastable electron throughout the oxide layer seems possible, at least for thin layer. Spectra may be interpreted as a competition of the two de-excitation processes, and the probability of slow electron reflection by the oxide layer seems large.

- A metallic surface, covered with gases in the molecular form, produces spectra which exhibit a competition of the two de-excitation processes : there are peaks corresponding to the electronic molecular orbitals inter-acting with the metastable atom according to the AD process, protruding from a background which must be attributed to the RI+AN process.

6. Prospects for MDS

The physics of interaction between metastable atoms and surfaces must thus be developed before MDS becomes a trivial spectroscopic tool.

Secondary emission, diffraction, refraction and wave-function overlap will be studied in order to obtain a theoretical model taking all the parameters into account.

At the same time, MDS may make a large contribution to the fundamental work on Auger transitions. But experimental improvements are necessary; in particular the incident beam density is sometimes too weak to perform measurements under good conditions. Nevertheless, even as it is nowadays, MDS appears as a very attractive method to study clean metal surfaces, adsorbed gases and adsorption kinetics, ultra-thin films, non-metallic compounds and organic chemistry.

7. Acknowledgements

This paper is the result of a collective effort, I am very grateful to Catherine GAROT, Emilien LABOIS, Françoise MUCCHIELLI, Robert NUVOLONE, Cécile REYNAUD and Joseph ROUSSEL for their contribution to the measurements and discussions of the results, and to Roger BARRE, Christian JURET and Jean LE GRAND for their participation in the apparatus design.

REFERENCES

[1] M.L.E. Oliphant, P.B. Moon, Proc. R. Soc. (London), A127, 388 (1930).

[2] S.S. Shekter, J. Exp. Theor. Phys. (USSR) 7, 750 (1937).

[3] A. Cobas, W.E. Lamb, Phys. Rev. 65, 327 (1944).

[4] H.D. Hagstrum, Phys. Rev. 96, 336 (1954).

[5] J.H. Craig, J.T. Dickinson, J. Vac. Sci. Technol. 10, 1, 319 (1973).

[6] C. Boiziau, R. Nuvolone, J. Roussel, Rev. Phys. Appl. 13, 571 (1978).

[7] in preparation.

[8] J. Fricke, private communication.

[9] H.D. Hagstrum, Phys. Rev. Lett. 43, 1050 (1979).

[10] H.D. Hagstrum, Phys. Rev. 123, 758 (1961).

[11] J. Roussel, E. Labois, Surf. Sci. 92, 2, 561 (1980).

[12] Y. Ballu, J. Micros. Spectros. Electr. 2, 231 (1977).

[13] P.H. Holloway, J.B. Hudson, Surf. Sci. 43, 141 (1974).

[14] T. Shibata, T. Hirooka, K. Kuchitsu, Chem. Phys. Lett. 30, 241 (1975).

[15] T. Munakata, T. Hirooka, K. Kuchitsu, J. Electron. Spectrosc Relat. Phenom., 13, 219 (1978).

[16] T. Munakata, T. Hirooka, K. Kuchitsu, J. Electron. Spectros. Relat. Phenom., 18, 51 (1980).

[17] H. Conrad, G. Ertl, J. Küppers, S.W. Wang, K. Gérard, H. Haberland, Phys. Rev. Lett. 42, 16, 1082 (1979).

[18] H. Conrad, G. Ertl, J. Küppers, H. Haberland, Symposium on Atomic and Surf. Physics, Maria Alm/SBG, Austria (Feb.1980).

[19] H. C. Boiziau, V. Dose, J. Roussel, Surf. Sci., 77, 412, (1976).

[20] J. Roussel, C. Boiziau, J. Physique, 38, 757, (1977).

[21] C. Boiziau, C. Garot, R. Nuvolone, J. Roussel, J. Physique Lettres, 39, L339, (1978).

[22] C. Boiziau, C. Garot, R. Nuvolone, J. Roussel, 4th Colloq. on Solid Surface Physics and Chemistry, Antibes (1978).

[23] C. Garot, C. Boiziau, Le Vide, 96, 17, (1979).

[24] F. Mucchielli, C. Boiziau, R. Nuvolone, J. Roussel, J. Phys. C, 13, 2441, (1980).

[25] C. Boiziau, C. Garot, R. Nuvolone, J. Roussel, Surf. Sci., 91, 313, (1980)

[26] C. Boiziau, F. Mucchielli, R. Nuvolone, J. Roussel, Symposium on Atomic and Surf. Physics, Maria Alm/SBG, Austria (Feb. 1980).

[27] C. Boiziau, F. Mucchielli, R. Nuvolone, C. Reynaud, J. Roussel, Sixth Solid-vacuum interface conference, Delft (May 1980).

[28] C. Boiziau, F. Mucchielli, R. Nuvolone, C. Reynaud, J. Roussel, Fourth Internat. Conf. on Solid Surfaces, Cannes (sept. 1980).

[29] I. Petroff, C.R. Viswanathan, Phys. Rev., B4, 799, (1971).

[30] C. Noguera, D. Spanjaard, D.W. Jepsen, Phys. Rev. B17, 607, (1978).

[31] D. Adler, J. Feinleib, Phys. Rev., B2, 3112, (1970).

[32] D.E. Eastman, J.L. Freeouf, Phys. Rev. Lett., 34, 395, (1974).

[33] G.E. Becker, H.G. Hagstrum, Surf. Sci., 30, 505, (1972).

Deexcitation of Metastable He-Atoms Interacting with Clean and Adsorbate Covered Metal Surfaces

H. Conrad, G. Ertl, J. Küppers and W. Sesselmann
Institut für Physikalische Chemie, Universität München
D-8000 München, Fed. Rep. of Germany and

H. Haberland
Fakultät für Physik, Universität Freiburg
D-7800 Freiburg, Fed. Rep. of Germany

Abstract

Metastable, electronically excited He-atoms impinging at clean or adsorbate covered metal surfaces are deexcited with a high probability causing electron emission. The deexcitation either proceeds via a combined resonance-ionization + Auger neutralization or a Auger-deexcitation (Penning-ionization) mechanism. It is shown that the latter mechanism dominates at geometrically shielded (adsorbate covered) or electronically shielded (low work function) surfaces.

1. Introduction

He atoms can be excited by electron impact into 2^1S and 2^3S states which cannot deexcite to the ground state optically because of the dipole selection rules. The excitation energies and lifetimes of these excited states are high enough (^1S : 20.61 eV, 17.9 msec ^3S : 19,82 eV, 10^4 sec) to provide useful probes for collision experiments.

It is known that in gas phase collision experiments the following two processes may occur:

(1) $He^* + M \rightarrow He^* + M$
(2) $He^* + M \rightarrow He + M^+ + e$

(M : target molecule or atom).
Angular distributions of elastically scattered He^*-atoms (process 1) can be used to evaluate the interaction potential $V_{He^*-M}(R)$ [1].

The ionization process (2), known as Penning-ionization, has been studied experimentally [2] as well as theoretically [3] at various target particles. Measured electron energy distributions resemble-apart from intensity variations - those determined with UV photoemission. Within the framework of the Born-Oppenheimer approximation these electron spectra allow the potential $V_{He-M}(R)$ and transition probabilities in addition to the ionization energies of the target particle M.

The use of He^*-atoms to investigate the electronic structure of surfaces is favourable as they do not penetrate the solid and are therefore only probing the outermost layer. Recent work in different laboratories has shown that electron spectra excited by He^*-atoms at various metal surfaces indeed reflect this surface sensitivity [4]. Due to the fact that a metal surface exhibits a continuous spectrum of occupied and unoccupied states, there are complications involved which do not allow to apply the Penning-ionization model valid at gas particle targets directly to the surface problem.

2. Experimental

The experimental setup has been described earlier [5] and will not be dealt with in detail here. Briefly, it consists of a beam source adapted to a UHV chamber equipped with those techniques to handle and control solid surfaces such as eg. LEED and AES. An electron spectrometer allows to measure electron distributions emitted within the scattering plane.

3. Results and Discussion

Fig. 1 shows the electron spectrum measured at a clean Pd(111) surface upon impact of a He^* (^1S) beam. The observed maximum kinetic energy of the emitted electrons at $E_k \approx 12$ eV is not in accordance to that value expected from a Penning-type deexcitation process which would demand: $E_{k,max} = E^* - \emptyset$, E^* being the excitation energy of the impinging atom, $\emptyset$ the target surface work function. With $E^* = 20.6$ and $\emptyset \approx 5$ eV one would expect $E_{k,max} \approx 15.6$ eV.

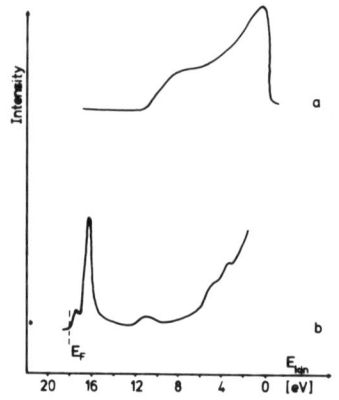

Fig. 1
a) Electron emission spectrum by impact of He^* (^1S) at clean Pd(111). The kinetic energy distribution is drawn with respect to the vacuum level above the surface, $\emptyset = 5$ eV.

b) As a) but Cs saturated surface, $\emptyset \approx 2.4$ eV. Only AD step operating.

The reason for this discrepancy is explained by the energy diagram (a) of fig. 2. If the approaching He^*-atom is resonance ionized (which is possible as empty states are present above the Fermi level) near the surface, the ion then can be Auger-neutralized in the same way as known from ion-neutralization spectroscopy (INS) developed by HAGSTRUM [6].

The $E_{k,max}$ value expected from the operation of sequential resonance-ionization (RI) and Auger-neutralization (AN) processes is then given by (see fig. 1a): $E_{k,max} = E_i - 2\emptyset$,

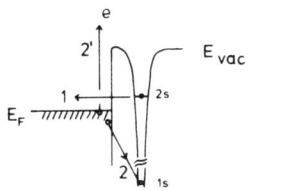

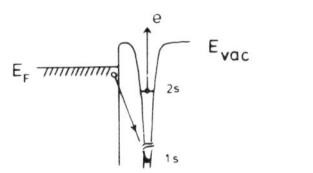

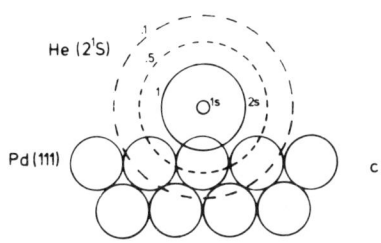

Fig. 2 a) Energy diagram representing electronic level positions for impact of He^*(^1S) at clean Pd(111).
1 : RI step, 2,2' : AN step

b) As a) but Cs saturated surface.

c) Geometric representation of the spatial expansion of the radial parts $r^2\psi\psi^*$ for ψ_{1s} and ψ_{2s} of He^*. The full lines indicate the maximum positions, the broken lines those corresponding to the indicated fractions. The distance He^*/surface is not scaled.

E_i being the ionization potential of He in front of the surface. This potential is about 2 eV lower than the gas phase value (24,8 eV) due to image force effects [7]. Inserting the respective numbers then yields $E_{k,max} \approx 12,8$ eV in agreement with the measured spectrum.

As suggested from the energy diagram, fig. 2a, the RI process will be suppressed, i.e. the excited 2s electron of the He^*-atom cannot tunnel into unoccupied states, by a sufficient decrease of the work function $\emptyset$. This situation is shown in fig. 2b. Since the ionization potential of He^* is about 4 eV, the work function has to be smaller than this value, which can be easily achieved by depositing alkali atoms at the surface.

The electron spectrum shown in fig. 1b has been measured with a Cs-saturated Pd(111) surface, $\emptyset$ = 2.4 eV. The onset of electron emission now occurs at E_k = 18.2 eV in accordance with the operation of the Penning-type Auger-deexcitation process (AD) depicted in fig. 2b.

As the electron spectrum now is to be interpreted by a one-electron excitation, rather than the two-electron mechanism present when RI + AN processes occur, the structures within fig. 1b can be used to identify electronic levels with respect to the Fermi level, E_F, which is fixed by the onset of electron emission.

The double peak structure right below E_F shows emission from Cs6s-derived levels, the peak at 6 eV below E_F originates from oxygen impurities within the Cs adlayer. Two peaks in the range of 11-13 eV below E_F are due to the spin orbit split $Cs5p_{1/2}$ and $5p_{3/2}$ electron levels.

A detailed study of the Cs-coverage-dependence of the He^*-excited electron spectra clearly demonstrates that the <u>local</u> work function rather than an average work function determines whether deexcitation proceeds via an AD or RI + AN steps [8].

This is very plausible when looking at fig. 2c which displays the spatial extension of the electron density of $He^{*1}S$ as well as a ball model of the target surface.

The probability of AD will to a first approximation be determined by the overlap of surface wave functions ψ_s with the empty 1s state ψ_{1s}. Vice versa the RI process depends on the overlap of ψ_{2s} and empty ψ_s states. Both quantities are only very locally different from zero, due to the character of the ψ_{1s}, ψ_{2s} wave functions.

It is suggested from fig. 2c that an impinging He^* atom can be resonance neutralized with high probability only at a small distance from the surface. A geometric shielding of the surface by an adsorption layer, e.g. by a compact CO adlayer (CO-diameter $\approx$3.5 Å), should prevent resonance ionization. Indeed, electron emission spectra of CO saturated Pd(111) surface excited by He^* impact substantiate this expectation. Moreover, in this case the spectral features resemble those expected from the operation of a AD process and can be used to identify the electronic energy levels of adsorbed CO, in accordance with the information obtained from UV photoemission experiments [9].

Up to which extent a geometric shielding of the surface is operating in the Cs/Pd(111) experiments is hard to decide. Judged from the ionic radius of Cs (1.7 Å) in comparison with the CO diameter (3.5 Å) it should be less effective. Experiments using Li (ionic radius 0.6 Å) may well help to solve this problem.

4. Conclusions

Interaction of metastable He*-atoms with clean and adsorbate covered metal surfaces leads to electron emission. The energy spectra of emitted electrons are caused either by the operation of an Auger-deexcitation (Penning-type) or resonance-ionization and subsequent Auger-neutralization process, depending on the actual surface condition. At surfaces with low work function ($\emptyset \leq 4$ eV) or a thick adlayer a Penning type deexcitation mechanism dominates which is in accordance with a simple model consideration.

References

1. B. Brutschy and H. Haberland, Phys. Rev. A19, 2232 (1979).
2. H. Hotop and A. Niehaus, Z. Physik 228, 452 (1970).
   A. Niehaus, Ber. Bunsenges. Phys. Chem. 77, 632 (1973).
3. W. H. Miller, J. Chem. Phys. 52, 3563 (1970).
   A.P. Hickman, A.D. Isaacson and W. H. Miller
   J. Chem. Phys. 66, 1483 (1977).
4. C. Boiziau, C. Garot, R. Nuvolone and J. Roussel
   J. Phys. 39, L 339 (1978).
   P. D. Johnson and T. A. Delchar
   Surface Science 77, 400 (1978).
5. H. Conrad, G. Ertl, J. Küppers, W. Sesselmann, H. Haberland and S. W. Wang, Proc. 8th Int. Vacuum Congr., Cannes 1980
   J. Küppers, Phys. Bl. 36, 212 (1980)
6. H. D. Hagstrum, in: Electron and Ion Spectroscopy of Solids, Ed.: L. Fiermans, J. Vennik and W. Dekeyser, Plenum, New York, 1978.
7. H. D. Hagstrum, Phys. Rev. 96, 336 (1954), 104, 672 (1956).
8. H. Conrad, G. Ertl, J. Küppers, W. Sesselmann and H. Haberland
   Surface Science, in press.
9. H. Conrad, G. Ertl, J. Küppers, S.W. Wang, K. Gerard and H. Haberland
   Phys. Rev. Lett. 42, 1082 (1979).

Electron and Photon Impact

The Use of Angle-Resolved Electron and Photon Stimulated Desorption for Surface Structural Studies

Theodore E. Madey
Surface Science Division, National Bureau of Standards
Washington, D.C. 20234, USA

ABSTRACT

We review recent experiments and models related to desorption processes induced by electrons and photons incident on surfaces. The utility of angle-resolved electron and photon stimulated desorption of ions for studies of molecular structure at surfaces is emphasized.

I. Introduction

A continuing need in studies of atoms and molecules on surfaces concerns the location of surface bonding sites and the geometrical structure of molecules and molecular fragments on surfaces. That is, where are adsorbed species bonded, what are the directions of the bonding orbitals between the atom (molecule) and surface, and what are the bonding directions of ligands in adsorbed molecular complexes? In a continuing series of experiments [1,2], we have established that the electron stimulated desorption ion angular distribution (ESDIAD) method has clear potential for providing <u>direct</u> information regarding the site location and geometrical structure of molecules adsorbed on surfaces.

In this method, a surface containing adsorbed molecules is bombarded by a focused low energy electron beam. Electronic excitation of the adsorbed species by electron bombardment can result in the desorption of atomic and molecular ions from the surface [3-6]. The ions desorb in discrete cones of emission in directions determined by the orientation of the surface molecular bonds which are "broken" by the excitation. The resultant ESDIAD patterns provide a visual display of the geometrical structure of surface molecules in the adsorbed layer.

In the present paper, we shall review the experimental and theoretical developments concerning the relationship between ESDIAD and surface structure. Recent measurements of Photon Stimulated Desorption [7-9] (PSD) and angle-resolved PSD [10] of ions from surfaces using synchrotron radiation will also be discussed. It will be seen that in angle-resolved PSD [10], the potential exists to "tune" the incident radiation to determine the orientation of specific surface bonds.

The paper is organized as follows: Section II provides a general introduction to ESD phenomena, and Section III is a discussion of experimental procedures. The application of ESDIAD to various molecular systems is given in Section IV, along with a discussion of the role of steps and defects in ESD of oxygen monolayers. Section V is concerned with theoretical concepts regarding ESDIAD, and Section VI outlines the principles and recent results of angle-resolved PSD.

II. Basic Experimental Observations in Electron Stimulated Desorption

When a low energy electron beam bombards a surface containing an adsorbed monolayer, electronic excitation in the adsorbed layer may result in the desorption of ion or neutral fragments (including metastables) [3-6]. Thresholds for these excitations are typically in the range 10-50 eV, and the physical mechanisms for ion production are considered in detail by FEIBELMAN in this volume [11]. Electron bombardment can also cause dissociation and/or polymerization in surface layers. For electrons in the energy range 10 to 1000 eV incident upon metal surfaces containing an adsorbed monolayer of atoms or molecules, the general observations are as follows [1-6, 12]:

(a) Most ESD processes have cross sections which are smaller than those for electron-induced dissociation and ionization of gaseous molecules. For 100 eV electrons, typical gas phase dissociative ionization cross sections for small molecules are 10^{-16} cm^2. Typical cross sections for ESD of adsorbed molecules lie in the range 10^{-18} to 10^{-23} cm^2, with both lower and higher values observed for certain systems. Also, the cross sections for neutral desorption are generally larger than the cross sections for desorption of ions. Although ions frequently comprise a small fraction of the desorbing species, the ease of experimental detection of ions means that their desorption characteristics are the most frequently studied ESD processes.

(b) Most ESD ions are atomic, with H^+, O^+, F^+ and Cl^+ being the most abundant. Negative ions of each of these four atoms have also been detected, [13, 14] but their yields are generally about 100 times smaller than for positive ions. The most common molecular ions reported from ESD of adsorbed monolayers are CO^+, OH^+, and OH^- [1-6]. ESD of multilayers can result in desorption of more complex molecular ions (i.e., $C_6H_{12}^+$ from a multilayer of C_6H_{12} [2], $H^+(H_2O)_n$ and $H^+(NH_3)_n$ from thick films of water and ammonia ice [15]).

(c) ESD cross sections are very sensitive to the mode of bonding. In general, cross sections for rupture of an internal bond in a weakly adsorbed molecule (the C...O bond in adsorbed CO, the H...C bond in adsorbed hydrocarbons, the H...O bond in adsorbed H_2O) are higher than cross sections for rupture of metal-atom bonds (low coverages of hydrogen or oxygen on a metal surface).

(d) The binding energies of chemisorbed molecules are sufficiently large that direct momentum transfer between electron and adsorbate does not provide sufficient energy to cause desorption of neutral species [5]. Power densities in most ESD studies are generally sufficiently low that thermal heating of the substrate does not induce desorption. These observations, coupled with the fact that ions are generally desorbed having most probable kinetic energies in the range 0 to 10 eV, indicates that ESD proceeds via an electronic excitation mechanism.

A one-dimensional FRANCK-CONDON excitation model of ESD has been developed independently by REDHEAD, and by MENZEL and GOMER [5], and is qualitatively consistent with the above observations. This model, along with the recent KNOTEK-FEIBELMAN Auger-Induced-Decay model of ESD have both been discussed extensively in the literature [1,3,6]. For a more extensive discussion of the mechanisms of ESD ion formation, the reader is referred to the paper by FEIBELMAN [11] in this volume.

The origin of angular effects in ESD will be discussed in V, and a summary of relevant experimental data relating ESDIAD to surface structure is discussed below.

III. Experimental Procedures

The ultrahigh vacuum apparatus used for most of the NBS ESDIAD studies has been described previously [1, 12] and is shown in Figure 1.

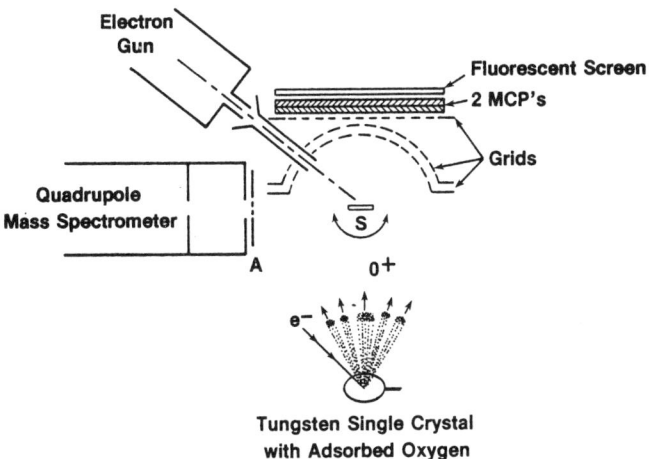

Figure 1. Schematic of ultrahigh vacuum ESDIAD apparatus. The sample S can be rotated about an axis normal to the plane of the drawing. ESD ions are mass analyzed in the quadrupole mass spectrometer, and ESDIAD patterns are displayed using the grid-microchannel plate (MCP) - fluorescent screen array. The lower drawing is a schematic of the ESDIAD process, in which O^+ ions are liberated in cones of emission during bombardment of the sample by a focused electron beam.

Briefly, a focused electron beam (50 to 1500 eV) bombards a crystal surface onto which gases have been deposited using a molecular beam doser. The ion beams which desorb from the crystal by electron stimulated desorption (ESD) pass through a hemispherical grid and are accelerated to a microchannel plate (MCP) assembly. The output signal from the MCP assembly is displayed visually on a fluorescent screen and photographed. By reversing the potential of the input of the MCP assembly, the elastic low energy electron diffraction (LEED) pattern from the sample can also be studied. Mass identification of ESD ions are made using a quadrupole mass spectrometer (QMS). In addition, the QMS may be used as a detector in thermal desorption studies from the adsorbed layers. The cleanliness of the sample crystal is verified using Auger Electron Spectroscopy.

A schematic illustration of the ESDIAD process is shown at the bottom of Figure 1. A focused electron beam (e^-) bombards a single crystal containing a monolayer of adsorbed oxygen. The ESD O^+ ions are liberated in cones of emission, in specific directions related to the bonding geometry. The beams are intercepted by the MCP detector assembly, and displayed visually.

NIEHUS [16] has employed a channeltron multiplier as a moveable ion detector for measuring ESDIAD. His data are in the form of computer-generated plots of ion intensity as a function of ion desorption angle; he determines the mass of ESD ions using a Time of Flight method.

IV. Experimental ESDIAD Results

A. CO Adsorbed on Transition Metal Surfaces in Different Binding Configurations

Two examples [17,18] will be given (CO on Ru (001) and Pd (210)) where ESDIAD has been used to complement and verify structural information predicted using other methods, such as angular resolved ultraviolet photoemission spectroscopy, low energy electron diffraction, and surface vibrational spectroscopy [19]. In a third example, the adsorption of CO on stepped surfaces vicinal to W(110), entirely new insights into CO bonding configurations have been provided by this work [20].

As will be discussed in detail below, we have observed that when molecular CO is adsorbed on the close packed Ru(001) and W(110) surfaces, the dominant mode of bonding is via the carbon atom with the CO molecular axis perpendicular to the plane of the surface. For CO on atomically rough Pd(210) and for CO adsorbed at step sites on surfaces vicinal to W(110), the axis of the molecule is tilted or inclined away from the normal.

a. CO on Ru(001)

Data previously reported using UPS (ultraviolet photoemission spectroscopy), EELS (electron energy loss spectroscopy), and reflection infrared spectroscopy have indicated that CO is terminally bonded to the Ru surface through the C atom, with the CO axis perpendicular to the surface. The ESDIAD results for CO confirm this orientation [17]: for all CO coverages in the temperature range 90K to 350K, the angular distributions of O^+ and CO^+ ESD ions are centered about the surface normal. The widths of the ion beams are temperature dependent; for both O^+ and CO^+, the corrected half widths at half maximum, α, of the ion cones are $\sim 14.5°$ at 300K and $\sim 11°$ at 90K. This temperature dependence, coupled with a simple model calculation, indicates that the dominant factors contributing to the width of ESD ion beams are initial state effects, i.e., CO surface bending vibrations of the type:

Since ion desorption times are short with respect to molecular vibration times, ESDIAD appears to sense the instantaneous statistical distribution of molecular orientations on the surface in this case. Thus, the data suggest that both the directions and widths of ESDIAD beams from adsorbed molecular species are determined largely by the structure and dynamics of the initial adsorbed state.

b. CO on Pd(210)

ESDIAD has been used to verify an unusual bonding configuration for CO on the (210) surface of fcc palladium [18]. In an infrared reflection-adsorption study [19] of CO on Pd(210) and (100), the measured values of the C-O stretching frequency indicated that at low coverage, the CO is bridge-bonded to two Pd atoms via the C atom. The Pd(210) is a rather open surface, and top layer atoms with the nearest neighbor distance of 2.73 Å do not exist (see Fig. 2); the shortest distance between top layer atoms is 3.88 Å in the [001] direction. Bridge-bonding is not known to occur in transition metal carbonyls for metal-metal spacings greater than ~ 2.78 Å. On Pd(210), it therefore appears that bridge-bonding can only occur on sites of the kind which exist between atoms in the first and second atom layers, so that the axes of adsorbed CO molecules are expected

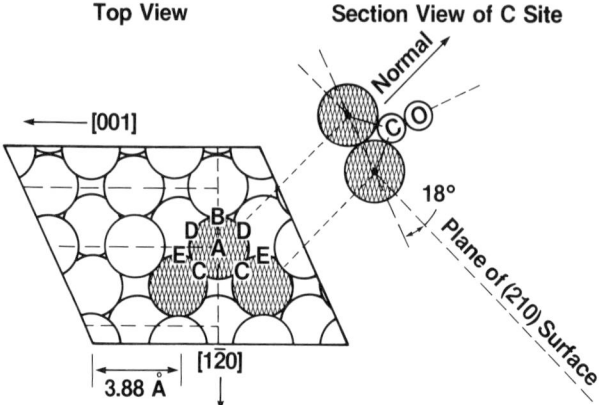

Figure 2. Model of Pd(210) surface with possible bonding sites for CO labeled. The section view illustrates how the CO molecule is "inclined" in the C sites (from Ref. 18, with permission).

to be inclined away from the normal by $\sim$ 18° as shown for the type C site in Figure 2.

The ESDIAD data are consistent with the infrared results, and for CO coverages Θ less than 0.5 monolayers, two-fold symmetric ion desorption patterns dominated by emission in directions away from the normal are observed. In addition, the ESDIAD patterns provide specific information about the desorption sites. CO populates the two equivalent type C sites at $\Theta<0.5$, and type B sites at higher coverages. Furthermore, the ESDIAD results indicate that the surface bending vibrational amplitudes for the bridge-bonded CO are different in orthogonal directions, in agreement with recent calculations [21]. At the saturation CO coverage ($\Theta=1$) at 90K, at least a fraction of the adsorbed CO appears to be bonded with the molecular axis normal to the Pd(210) surface.

c. CO on W(110) and Stepped Surfaces Vicinal to W(110)

The above results demonstrate that the ESDIAD method yields CO structures consistent with adsorbed molecular geometries deduced using other techniques. Based on these data, we have examined the role of surface steps in molecular adsorption in an ESDIAD study of CO on a multifaceted tungsten monocrystal [20]. The questions to be answered are: How do the CO adsorption geometries compare on flat surfaces and on stepped surfaces,

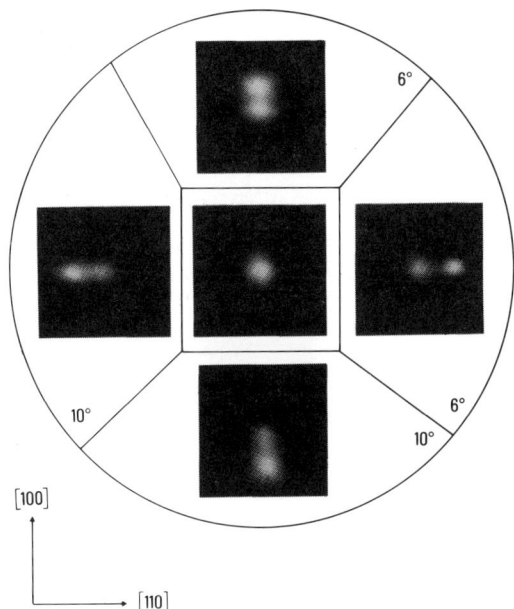

Figure 3. ESDIAD patterns for desorption of ions from CO adsorbed at 273K on the multifaceted W crystal described in the text. The central facet is oriented with its surface parallel to the (110) plane and the (110) terrace widths on the 6° and 10° stepped surfaces are 13Å and 22Å respectively. In all of the patterns, the spot in the center of the picture corresponds to a beam of O^+ and CO^+ ions desorbing normal to the surface. The off-normal beams from each of the faceted surfaces provide evidence for "inclined" CO. (From Ref. 20, with permission).

and are new structures seen on stepped surfaces? The experiments were performed on a 7mm diam. tungsten crystal cut to expose 5 separate facets. It was also used in an ESDIAD study of oxygen adsorption [22] which will be discussed in IV.B., and was similar to one used in studies of oxygen adsorption kinetics on stepped surfaces [23]. The central facet was oriented within 0.3° of the (110) plane, and the four surrounding facets were stepped surfaces of different step densities (6° and 10° off the (110) plane), and with step orientations parallel to [100] and [110] directions. The ESDIAD patterns seen for a monolayer of CO adsorbed on this multifaceted sample at 273K are shown in Figure 3.

The electron beam was scanned from facet to facet, and the patterns were photographed from the fluorescent screen. The central W(110) facet yields a single ESDIAD beam which desorbs perpendicular to the surface, giving a single spot in the center of the photograph. Each of the stepped surfaces also yields an ESD ion beam which desorbs perpendicular to the (110) terraces; the images of these normal beams appear in the center of the photos for each of the 4 outer facets in Figure 3. In addition, each stepped facet yields an extra ESD ion beam which desorbs in a down-step direction, along an azimuth perpendicular to the step edge. All beams consist of both O^+ and CO^+, in approximately equal intensities. The polar angle between the normal and off-normal beams on the right and left facets is $\sim 40°$.

We interpret these data as follows: At 273K, CO adsorbs in molecular form (the virgin state) along with some dissociated CO (the β states) [24]. The ESD signal is due primarily to the molecular CO. Ultraviolet photoemission data demonstrate that molecular virgin CO is bonded to tungsten through the carbon atom [25]. The single normal beam seen in the ESDIAD pattern for CO on W(110) indicates that the molecular CO is bonded perpendicular to the W(110) facet. On each of the stepped surfaces, a fraction of the molecular CO is also bonded with the molecular axis perpendicular to the W(110) terraces. In addition, the observation of the down-step ion beams indicates that the CO molecules are tilted away from the normal to the terraces by $\sim 40°$, and are probably adsorbed directly on the edges of the steps. A preliminary account of the temperature dependence of CO bonding configurations on the multifaceted W crystal has been published elsewhere [20].

Finally, we note that JAEGER and MENZEL [26] have also found evidence for "inclined" CO in an ESDIAD study of CO on W(100). They conclude that the CO is bound on different sites, including both symmetric and asymmetric bridges.

B. The Role of Steps in ESD of Adsorbed Atoms: Oxygen on Tungsten

In a large number of studies of oxygen on tungsten using ESD [27], it has been found that there is little or no O^+ ion yield at low oxygen coverages. Only where the surface coverage is greater than 0.5 to 0.75 monolayers is significant O^+ ion signal seen. Several models of this "induction period" in the appearance of O^+ during oxygen adsorption have been offered, including local oxide formation, the adsorption of a new molecular state, and bonding at special sites on the surface [27]. In order to systematically investigate to what extent sites such as atomic steps and defects might influence the ESD of O^+ from adsorbed oxygen, we have studied the adsorption of oxygen [22] on the same polyhedral W crystal discussed in IV. A.c. Upon adsorption at 300K, there is little or no ESD O^+ emission from the flat W(110) plane at any oxygen coverage. In contrast, adsorption of oxygen on stepped surfaces vicinal to W(110) yields intense O^+ emission normal to the terraces and in "downstep" directions, as seen using ESDIAD and shown in Figure 4.

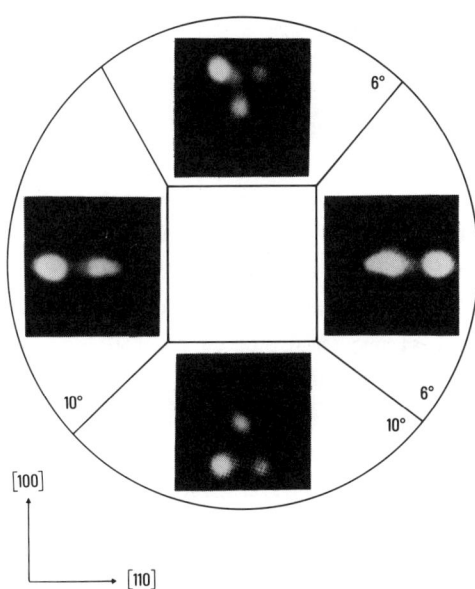

Figure 4. ESDIAD patterns for adsorption of oxygen on multi-faceted tungsten crystal described in the text. $T_{ads} \simeq$ 300K. The center of each pattern corresponds to the normal to the (110) terrace. The outer beams are due to ESD of O^+ in directions away from the normal. See Fig. 3 for more details. (From Ref. 22, with permission).

The data suggest that the "ESD-active" sites for O^+ desorption from oxygen on tungsten are low coordination sites which are absent on a perfect W(110) surface, but present on the stepped surface. It appears that the adsorption of oxygen atop substrate atoms at step edges is necessary for the ESD of O^+ from W surfaces containing (110) terraces; oxygen adsorbed on the (110) terraces does not yield a significant ESD ion signal. In contrast, oxygen adsorbed on the more open W(100) surface does yield an appreciable O^+ signal; even for this surface, however, adsorption at step sites leads to enhanced O^+ desorption [28].

Two possible explanations can be offered to explain this sensitivity of ESD to atomic adsorption at low coordination sites. First of all, on the basis of the one dimensional REDHEAD-MENZEL-GOMER [5] model of ESD, the probability of ionic desorption P_I is exponentially related to the neutralization rate for an ion formed at the surface by electron bombardment, viz.,

$$P_I = \exp - \left\{ \int_{x_0}^{\infty} \frac{R(x)}{v(x)} dx \right\}$$

where $R(x)$ is the neutralization rate ($\sec^{-1}$), $v(x)$ is the ionic velocity in the repulsive final state, x is the distance from the surface, and x_0 is the separation between atom and surface at the point of excitation.

R(x) is a measure of the rate of electron transfer (i.e., tunneling, Auger neutralization) from the substrate to the ion created by electron bombardment, and it is reasonable to assume that R(x) is different for adatoms bonded in different sites. Oxygen atoms adsorbed on W(110) flats are located in multiply-coordinated sites, bonded to 2 or 3 substrate W atoms [29]; oxygen atoms adsorbed at step edges may be in atop sites, bonded to single W atoms. An increase in coordination (i.e., the "number of bonds" to the substrate) should result in an increase in R(x) and a corresponding decrease in P_I at that site. Since P_I is exponentially dependent on R(x), even small changes in R(x) will have a strong influence on the ESD ion desorption probability P_I.

In accordance with the KNOTEK-FEIBELMAN [6] model of ion desorption, another interpretation of the enhanced O^+ emission from step edges may involve the formation of oxide-like complexes at step sites, with the enhanced yield due to the Auger decay mechanism. Careful measurements of the thresholds for ion desorption are not available for stepped W(110) surfaces, but O^+ desorption thresholds measured in recent PSD studies of oxygen on W(111) [10] and W(100) [9] indicate that the O^+ signal may indeed originate from oxide-like species on these surfaces. These data will be discussed in Section VI.

Thus, both for non-dissociative adsorption (in which mainly internal molecular bonds are ruptured by ESD, i.e., O^+ from CO/W) and for dissociative adsorption (in which metal-adsorbate bonds are ruptured by ESD, i.e., O^+ from O/W), the ESDIAD structures are sensitive to the presence of steps and defects on metal surfaces. The widespread occurrence of "off-normal" ESDIAD beams from stepped surfaces indicates that there are many instances of "inclined" molecular structures on surfaces. Finally, the ESD ion yield is rather high from molecular adsorbates (i.e., CO) on planar W(110) surface but very low for atomic oxygen.

C. Application of ESDIAD to Other Molecular Adsorbates.

ESDIAD has been applied to a number of molecular systems in addition to those already discussed. Particularly interesting observations have been made in the cases of H_2O and NH_3 on Ru(001) [2,30] and NH_3 on Ni(111) [31]; for low coverages of these adsorbates, the characteristic ESDIAD pattern in all 3 cases has the appearance of a "halo", i.e., a continuous band of emission on an off-normal direction with little emission normal to the surface. An example is shown in Fig. 5, corresponding to ~ 0.5 monolayers of NH_3 on Ni(111) at 160K. The structures suggested by the halo patterns are arrays of species such as

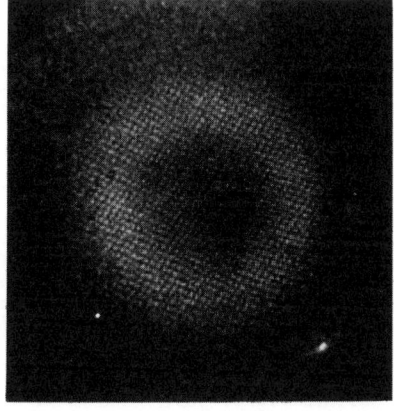

Figure 5. ESDIAD pattern for ESD of NH_3 on Ni(111); $\theta<1$, $T_{ads} \gtrsim 160K$. Total electron energy was 350 eV: a bias potential of 200 V was applied to the sample to "compress" the ion pattern.

Such structures are also consistent with UPS evidence, as well as with an EELS study of H_2O on Ru(001) [32]. Evidence for rotational (azimuthal) ordering at higher coverages ($\theta > 0.5$) was seen for H_2O and NH_3 on Ru(001), but not for NH_3 on Ni(111).

A fractional monolayer of C_6H_{12} adsorbed on Ru(001) yields a hexagonal H^+ ESDIAD pattern [2]. These results indicate the azimuthal orientation of the C-H bonds, and are consistent with a simple model of C_6H_{12} adsorption. The C_6H_{12} appears to be adsorbed directly over Ru substrate atoms in a "tilted chair" configuration. These data for H_2O, NH_3 and C_6H_{12} indicate several of the advantages of ESDIAD in determining the structure of molecules at surfaces, namely:

- ESDIAD reveals structures consistent with the molecular geometry predicted using other methods

- ESDIAD is particularly sensitive to the azimuthal orientation of hydrogen ligands in adsorbed molecules. In general, electron scattering from H in molecules is too weak to allow LEED to be useful.

- ESDIAD is sensitive to the <u>local</u> bonding geometry. It is not a diffraction method, so long range surface order is not necessary to produce a sharp ESDIAD pattern.

V. Origin of Angular Effects in ESD

Since the first experimental observations of the ESDIAD phenomenon, there have been only a few articles which have specifically addressed the theory of ion angular distribution in ESD. GERSTEN, JANOW and TZOAR [33] were the first to use dynamical arguments to link the observed angular distribution patterns to details of the bonding of adatoms on surfaces. They reported model calculations of angular distributions of O^+ ions desorbing from oxygen adsorbed at different sites on model W(100) and W(111) surfaces. A MAXWELL-BOLTZMANN distribution was used to approximate the distribution of atom positions about each site due to vibrations. The atoms were converted to ions by the excitation process, and the positive ion-solid potential was modeled assuming atomic wave functions and an unrelaxed lattice. Asymptotic ion trajectories were computed classically and plotted graphically. Details of the patterns were found to vary sensitively with changes in surface bonding geometry. Beams desorbing in off-normal directions were invariably due to oxygen atoms located in non-high-symmetry sites; beams desorbing normal to the surface were due to oxygen bonded in sites atop substrate atoms.

CLINTON [34] has formulated a quantum scattering theory of ESD in which he assumes that the final state potential experienced by the ion following excitation is a sum of central potentials. In this model, the initial direction of ion desorption occurs along a chemical bond direction (ignoring anisotropies in reneutralization). Thus, he concludes that ESDIAD processes are dominated by the initial state (ground state) structures of atoms and molecules on surfaces. He also suggests that the widths of ion beams are due, in large part, to bending vibrations of the adsorbed species.

In the KNOTEK-FEIBELMAN Auger-induced-decay model [6], ion desorption from maximal valency transition metal oxides involves COULOMB repulsion between a multiply charged cation (W^{6+} in WO_3, Ti^{4+} in TiO_2, etc.) and an anion which has acquired a positive charge (e.g., O^+) due to an interatomic Auger transition. Since the dominant force is the Coulomb repulsion along the line of centers between cation and anion, the initial impulse experienced by the desorbing ion (O^+, in this case) will be directly determined by the original bond direction. Of course, a complete description of the desorbing ion's trajectory requires a knowledge of the electrostatic potential outside the surface of the ionic crystal.

A consideration of the timescales in ESD also leads to the conclusion that in most cases, the initial ion desorption direction should be related to the initial state bond angle. Molecular vibration and rotation times are 10^{-12} to 10^{-13} sec, much slower than typical ion desorption times (10^{-14} to 10^{-15} s). For example, an H^+ ion desorbing with 2 eV kinetic energy will travel 1Å in 5 x 10^{-15}s, and an 8 eV O^+ ion will travel 1Å in 1x10^{-14}s. After an ion has moved ~ 1Å from the equilibrium bond distance, the probability of recapture by neutralization is small. Ion desorption times are sufficiently rapid with respect to vibration times that significant molecular rearrangements are unlikely to occur prior to desorption. Thus the ion desorption angle should be related to the initial bond angle.

Now, we consider to what extent final state effects perturb the ion trajectories. Several final state factors which have been suggested previously [17,26] include: anisotropy in the reneutralization rate (structure in the imaginary part of the final state potential), "defocussing" due to structure or curvature in the real part of the final state potential, and deflection of the ion trajectory due to the image force acting on the desorbing ion. In the absence of detailed knowledge of the final state potentials, we have no basis for estimating the first two factors. We can, however, estimate the influence of the image potential. CLINTON [34] has shown that an ion desorbing with an initial angle α_i with respect to the surface normal will arrive at the detector with an apparent desorption angle α_o given by

$$\cos \alpha_o = \cos \alpha_i \left[\frac{1 + V_I/[(E_K-V_I) \cos^2 \alpha_i]}{1+V_I/(E_K-V_I)} \right]^{1/2} \quad (1)$$

Here, V_I is the (screened) image potential at the initial ion-surface separation Z_o [35a] (using Gadzuk's procedure [35b] to locate the image plane), and E_K is the final (measured) kinetic energy of the desorbed ion. Note that V_I is a negative quantity so that $|V_I/(E_K-V_I)|$ is ≤ 1 and $\alpha_o > \alpha_i$. The straightforward derivation of eq. (1) assumes a step-like "hard wall" repulsive final state potential.

From eq. (1), it is clear that the image potential acts to systematically increase the measured desorption angle α_o over the initial desorption angle α_i <u>in all cases</u>. The magnitude of the correction is greater for large values of α, and low values of E_K, and vice versa. For example, if we insert in eq. (1) values appropriate to the desorption of O^+ from CO on Ru(001) (V_I = 1.52 eV, E_K = 7 eV, Z_o = 1.9 Å) eq. (1) predicts that α_i = 14.5° when α_o = 16° and α_i = 10.8° when α_o = 12°, corrections of the order of 10% in the polar desorption angle. A much larger effect is seen for lower energy ions desorbing with a large value of α_i. For V_I=-2.62 eV, E_K=4 eV, and Z_o=1.04 Å (appropriate to a "bent" NH species adsorbed on Ni(111)), α_o = 80° when α_i = 50°, a substantial correction.

Implicit also in eq. (1) is the existence of a cut-off angle for ion desorption. Specifically, for

$$\left| \frac{V_I}{(E_K-V_I) \cos^2 \alpha_i} \right| > 1 \qquad (2)$$

there will be no escape of the desorbing ion. For values of α_i slightly greater than the cut-off value, the ions will follow shallow trajectories and strike the surface at some distance from the point of excitation. An interesting consequence may be that the bombardment of surface molecules by low energy ESD ions following shallow trajectories can induce chemical changes in the adlayer!

Thus, the image potential can have a major influence on the <u>polar</u> component of the ion desorption angle. The question now arises: What about the azimuthal component of the ion desorption angle, i.e., the desorption angle projected into the plane of the surface? For desorption from a perfectly plane surface, or along an azimuth of symmetry, there are no "torques" expected to influence the azimuthal angle. For desorption from a site of lower symmetry, azimuthal anisotropy in either the final state repulsive potential or the reneutralization probability could cause deviation of the ion trajectories in the azimuthal direction. From an experimental point of view, the frequent occurrence of ESDIAD beams having nearly circular cross sections suggests that azimuthal deflections of the desorbing ions are not a general problem.

It thus appears that, in general, the directions of ion desorption are determined largely by the structure of the adsorbed complex in its ground (initial) state. The only consistently predictable perturbation of the ion trajectories is due to the image potential; this invariably results in a deflection of the polar angle to larger values. There do not appear to be systematic effects which influence the azimuthal component of the ion desorption angle.

Finally, we note that measurements of the angular distributions of neutral species released in ESD would be useful in avoiding those electrostatic effects which perturb the trajectories of ions. Desorption angles in neutral ESD should bear a much more direct relation to bond angles than in ESDIAD! These (difficult) experiments are being planned in several laboratories.

VI. Photon Stimulated Desorption

A. Angle-Resolved PSD Using Synchrotron Radiation

For a number of years, it was thought that the photon stimulated desorption (PSD) of species from surfaces was a relatively inefficient and unimportant process [36]. A new impetus was given to PSD studies when KNOTEK and FEIBELMAN [6,11] proposed their core hole Auger decay mechanism, which predicts that ion desorption from ionically bonded species at surfaces is initiated by the formation of shallow core holes in surface atoms. An essential feature of the model is that ion desorption occurs independent of the manner of production of the core hole, whether it is excited by electrons or photons. The first demonstration of PSD of ions via core hole excitation using synchrotron radiation was made by KNOTEK, JONES and REHN [7], who observed ion desorption from adsorbed species on TiO_2. The PSD of ions from a metal surface (O^+ from oxygen on W(100)) was observed by WOODRUFF, TRAUM et al. [9], who also showed the essential equivalence between ESD and PSD threshold energies. The PSD of ions by X-Ray photons was seen by FRANCHY and MENZEL [8], and will be discussed in VI (B) below.

To date, there are only two clearcut examples of the use of angle-resolved PSD for studying surface geometrical structures: VAN DER VEEN et al. [37] measured the energy and angular distributions of PSD ions from a cleaved V_2O_5-(010) surface and MADEY et al. [10] have measured ion angular distributions for O^+ desorption from a W(111) crystal, as well as photon excitation spectra for O^+ ion desorption. In both cases, good agreement was found with ESDIAD results for the same systems. However, as shall be seen below, the sharpness of the PSD threshold energies indicate that angle resolved PSD has clear potential for determining the bonding structures of adsorbed atoms and molecules, by selective excitation of surface species having different energy thresholds for ion desorption.

Using a unique display-type analyzer, VAN DER VEEN et al. [37] studied the PSD of O^+ from a V_2O_5(010) surface. The ellipsoidal mirror analyzer with a microchannel plate detector array was designed by EASTMAN et al. [38] primarily for angle-resolved UPS studies, and was adapted specifically for this study. In addition to determining ion energy distributions, the authors were able to mass-analyze the desorbing ions using time-of-flight gating techniques. The angular distribution of desorbed O^+ ions was found to be strongly peaked in the direction of the surface normal. The strongly directional desorption pattern reflects the local bonding geometry of the initial state surface oxygen atoms, and is consistent with a previously proposed structural model of the V_2O_5 (010) surface [39]. In this model, the outermost surface oxygen atoms ($\sim 5 \times 10^{14}$ atoms/cm^2) occupy sites directly atop vanadium atoms, with a bond direction parallel to the surface normal. The observed photoexcitation spectrum of the O^+ ion yield was seen to be consistent with the core hole Auger decay model [6,11], and showed convincingly that the desorbed oxygen had been originally bonded to substrate V atoms. Thus, both the bonding site and the surface bond angle were identified experimentally.

In a search for strong angular anisotropies in PSD from a well-characterized monolayer adsorbed on a metal surface, a joint NBS-IBM group [10] used the EASTMAN [38] analyzer to study oxygen adsorbed on a W(111) surface. The primary objective was to determine whether or not the angular distribution of O^+ ions observed in ESDIAD [40, 41], i.e., discrete off-normal O^+ beams, would also be seen in PSD ion angular distributions.

Figure 6 is a PSD angular distribution pattern for an oxidized W(111) surface [10], and corresponds to O^+ ion desorption excited by photons of energy hv = 45eV. The three O^+ beams are symmetrically disposed about the surface normal, each having a polar angle α of 41 ± 2° with respect to the normal. The value of α depends on coverage and temperature, and is 27° ± 3° in the similar PSD pattern observed for monolayer oxygen at 300K. These experiments indicate that the symmetry, azimuthal orientation and angular separations of the PSD patterns are identical to those of the ESDIAD patterns excited by 500 eV electrons as well as those reported previously [40, 41].

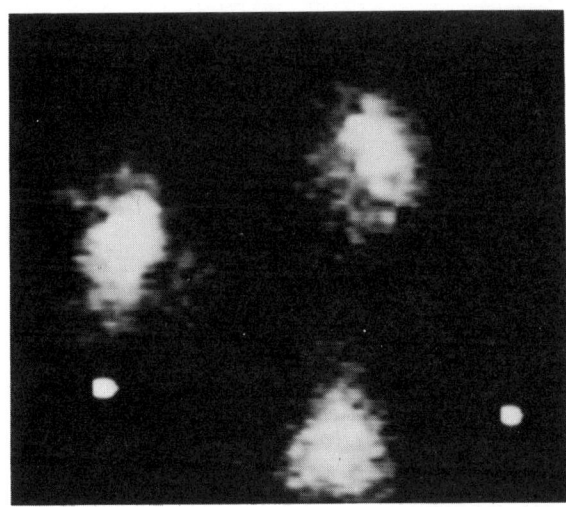

Figure 6. PSD ion angular distribution patterns for O^+ desorption from the oxidized W(111) surface for a photon energy of 35.4 eV. The two small white dots are markers on the detector screen (From Ref. 10, with permission).

Figure 7 is taken from Ref. [10], and contains plots of the O^+ ion yield, corrected for photon flux, as a function of photon energy. Fig. 7b corresponds to the oxide surface ($\gtrsim$ 3 monolayers of oxide) 7c is the yield from an oxygen monolayer and 7d is the yield from $\sim$ 0.5 monolayer of oxygen. For each of these curves, the ions were determined to be O^+ using a time-of-flight method. The distinct "breaks" or onsets in the yield curves correspond roughly to the core-hole binding energies for tungsten atoms in solid W and WO_3 [42] indicated on the figure. We note that curve 7c is similar to PSD data for an oxygen monolayer on W(100) [9]. Of particular interest is the similarity between the ion yield curves and the secondary electron yield curve of Fig. 7a measured for the surface with the monolayer of oxygen. Such a constant-final state ($\sim$ 3 eV kinetic energy electrons) plot of the secondary electron cascade has been shown to be directly proportional to the soft x-ray absorption coefficient.

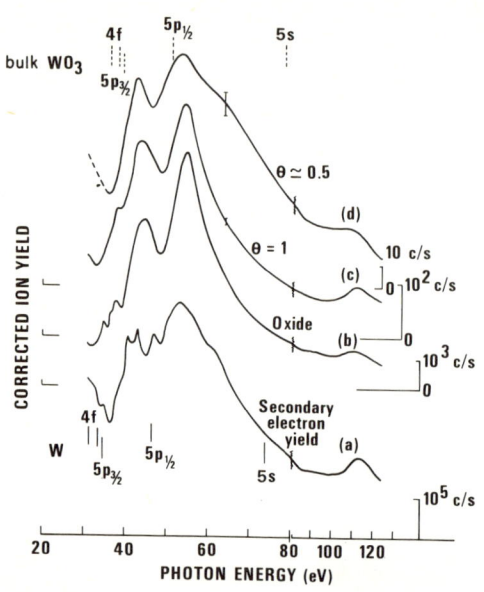

Figure 7. Electron and ion yields for W(111) as a function of photon energy, corrected for monochromator transmission and second-order contributions. Ion yields are normalized to the same incident flux. Curve a, secondary electron yield at constant final state (E_{kin} = 3eV) for an oxygen monolayer. Curve b, O^+ ion yield from an oxide layer. Curve c, O^+ yield from an oxygen monolayer. Curve d, O^+ yield from an 0.5 monolayer coverage. Binding energies for W core levels in pure W (solid lines) and WO_3 (dashed lines) are shown. The energy scale changes at 80 eV (From Ref. 10, with permission).

Although most of the structure in Fig. 7a is due to such inelastic processes, the sharp peaks at 40 and 43 eV are due to direct emission via W 4f levels.

The overall agreement between ion yields and secondary electron yields indicates that photo-induced excitations of substrate W atoms plays a major role in the desorption of ions, consistent with the KNOTEK-FEIBELMAN model. The differences in detail are likely due to the fact that PSD ions originate only from the top layer of surface atoms for which the local density of states is different from that in the bulk, whereas the secondary yield curve results largely from pure W metal.

All three of the ion yield curves in Fig. 7 are dominated by peaks at 45 and 55 eV. This suggests that the O^+ - yielding species for $\theta < 1$, $\theta \sim 1$ and the oxide layer have similar electronic configurations. The simplest formulation of the Auger decay model of ion desorption requires maximal valence for the cationic species (e.g., W^{6+} as in WO_3); reduced forms of the oxide result in little or no PSD ion yield due to the increased valence electron density on the cation, as in W^{4+}. The data thus indicate that maximal valence species are present even in monolayers and fractional monolayers of oxygen. Such species could be WO_3 - like molecular species, or they could be oxygen atoms bonded to special sites (steps, defects) at which the valence charge density on the W is lowered due to its reduced coordination to the substrate. As indicated by the relative intensities of curves b-d in Fig. 7, such species must have a low concentration at coverages < 1 monolayer, and hence ESD and PSD of O^+ from W is due to a "minority species." As discussed in Section IV B and Ref. [22], the O^+ yielding sites on stepped surfaces vicinal to W(110) were shown to be minority sites, i.e., steps and defects.

The question logically arises: since angle resolved PSD requires a synchrotron, isn't it easier to use ESDIAD for characterizing the structures of molecules at surfaces? In many cases, the answer is yes. However, the sharp thresholds for PSD of ions (in contrast to the rather broad excitation spectra seen in electron bombardment, due to inelastic effects) [6] means that selective excitation and desorption of specific adsorbed atoms or molecules is possible. In a complex overlayer, PSD using a tuneable source offers the promise of selective desorption and structural determination of specific adsorbed species bonded to single substrate elements. In addition, the ion current above a core hole threshold is proportional to the core ionization cross section; its energy dependence in PSD will exhibit extended x-ray absorption fine structure (EXAFS) [7, 8, 43]. If the PSD ion current is used as an EXAFS monitor, the distance between an adsorbate atom and its neighbors can be determined, in principle.

B. Deep Core Excitations by Electrons and Photons

FRANCHY and MENZEL [8] have shown that soft-X-ray induced ion desorption from a covalently bound adsorption layer on a metal surface (CO on W(100)) is caused by an intrinsic photoprocess identified as adsorbate core hole ionization, followed by Auger decay. This process is the surface analog of gas phase molecular decomposition processes following deep core ionization [44]. They further demonstrated a substantial enhancement of the electron stimulated desorption cross sections for CO^+ and O^+ desorption at primary energies greater than the C1s and O1s core hole ionization energies, respectively. HOUSTON and MADEY [45] have also found that the cross section for O^+ desorption for virgin CO on W(110) increases sharply for electron energies greater than the O1s binding energy; very recently, similar observations were made for CO and NO on Ni(111) [46].

The first explanation of desorption processes initiated by adsorbate core ionization [8] is the effect known as "Coulombic Explosion" in molecules [44]: Auger decay, including Auger cascades, of the primary core hole results in accumulation of positive charge on the originally core-ionized atom and its bonded neighbors, which fly apart by Coulomb repulsion. It has been recently suggested [47-49] that the desorption mechanism hinges on final state hole localization (such as two holes in the 1π bonding orbital of molecular CO) which can lead to destabilization of bonding, and dissociative ionization. The effect of multiple electron excitations in ESD and PSD of molecular adsorbates is an active area of experimental [50] and theoretical [47-49] interest and we are on the threshold of new insights into the mechanisms of these important processes.

VI. Epilogue

Angle resolved ESD and PSD studies are providing direct and useful insights into the geometrical and electronic properties of surfaces with adsorbed layers. It is clear that these two tools are unique additions to the ever-growing arsenal of methods for probing surfaces.

VII. Acknowledgements

The author acknowledges with pleasure a host of collaborators who have contributed substantially to the research and ideas discussed herein:
J. T. Yates, Jr. and R. Stockbauer (NBS), J. E. Houston (Sandia), S. C. Dahlberg (Bell), J. F. van der Veen and D. E. Eastman (IBM), A. M. Bradshaw and F. M. Hoffmann (Fritz Haber Inst.) and C. Seabury and T. N. Rhodin, (Cornell). Valuable conversations were held with D. Menzel, P. Feibelman, and W. L. Clinton. This work was supported in part by the U. S. Office of Naval Research.

REFERENCES

1) T. E. Madey and J. T. Yates, Jr., Surface Sci. <u>63</u>, 203 (1977).

2) T. E. Madey and J. T. Yates, Jr., Chem. Phys. Letters <u>51</u>, 77 (1977); Surface Science <u>76</u>, 397 (1978).

3) D. Menzel, in Interactions on Metal Surfaces, ed. by R. Gomer, Topics in Applied Physics, Vol. 4 (Springer Berlin, Heidelberg, New York 1975) p. 101; and Surface Sci. <u>47</u>, 370 (1975).

4) T. E. Madey and J. T. Yates, Jr., J. Vac. Sci. Technol. <u>8</u>, 525 (1971).

5) D. Menzel and R. Gomer, J. Chem. Phys. 41, 3311 (1964); P. A. Redhead, Can. J. Phys. 42, 886 (1964).

6) M. L. Knotek and P. J. Feibelman, Phys. Rev. Lett. 40, 964 (1978); P. J. Feibelman and M. L. Knotek, Phys. Rev. B 18, 6531 (1978).

7) M. L. Knotek, V. O. Jones, and V. Rehn, Phys. Rev. Lett. 43, 300 (1979).

8) R. Franchy and D. Menzel, Phys. Rev. Lett. 43, 865 (1979).

9) D. P. Woodruff, M. M. Traum, H. H. Farrell, N. V. Smith, P. D. Johnson, D. A. King, R. L. Benbow, and Z. Hurych, Phys. Rev. B 21, 5642 (1980).

10) T. E. Madey, R. L. Stockbauer, J. F. van der Veen and D. E. Eastman, Phys. Rev. Lett. 45, 187 (1980).

11) P. J. Feibelman, this volume.

12) T. E. Madey, in Methods of Experimental Physics, Vol. , "Methods of Experimental Surface Science," R. L. Park, ed., (Academic Press, N.Y.) to be published.

13) A. Kh. Ayukhanov and E. Turmashev, Soviet Physics - Tech. Phys. 22, 1289 (1977); J. L. Hock and D. Lichtman, Surface Sci. 77, L184 (1978).

14) M. L. Yu, Surface Sci. 84, L493 (1979).

15) R. H. Prince and G. R. Floyd, Chem. Phys. Lett. 43, 326 (1976).

16) H. Niehus, Surface Sci. 78, 667 (1978); Surface Sci. 80, 245 (1979).

17) T. E. Madey, Surface Science 79, 575 (1979).

18) T. E. Madey, J. T. Yates, Jr., A. M. Bradshaw, and F. M. Hoffmann, Surface Sci. 89, 370 (1979).

19) A. M. Bradshaw and F. M. Hoffmann, Surface Sci. 72, 513 (1978).

20) T. E. Madey, J. E. Houston and S. C. Dahlberg, in D. A. Degras and M. Costa, eds. Proceedings of the 4th International Conference on Solid Surfaces and the 3rd European Conference on Surface Science, Cannes, France (Supplement a la Revue "Le Vide, les Couches Minces" no. 201) p. 205.

21) N. V. Richardson and A. M. Bradshaw, Surface Science 88, 255 (1979).

22) T. E. Madey, Surface Science 94, 483 (1980).

23) K. Besocke and S. Berger, in: Proc. 7th Intern. Vac. Congr. and 3rd Intern. Conf. on Solid Surfaces, Vienna, 1977, Eds. R. Dobrozemsky et al. (Berger, Vienna, 1977) p. 893.

24) R. Gomer, Japan J. Appl. Phys., Suppl. 2, Pt. 2, 213 (1974); Ch. Steinbruchel and R. Gomer, Surface Science 67, 21 (1977).

25) E. W. Plummer, B. J. Waclawski, T. V. Vorburger and C. E. Kuyatt, Prog. in Surface Sci. $\underline{7}$, 149 (1976).

26) R. Jaeger and D. Menzel, Surface Sci. $\underline{93}$, 71 (1980).

27) See, for example, (a) V. N. Ageev and N. I. Ionov in: Progress in Surface Sci. Vol. 5, Ed. S. G. Davison (Pergamon, New York, 1975) p. 1, (b) T. E. Madey, Surface Sci. $\underline{33}$, 355 (1972), (c) D. A. King, T. E. Madey and J. T. Yates, Jr., J. Chem. Soc., Faraday Transactions I, $\underline{68}$, 1347 (1972).

28) B. Krahl-Urban and H. Niehus, Surface Science $\underline{88}$, L19 (1979).

29) M. A. van Hove and S. Y. Tong, Phys. Rev. Letters $\underline{35}$, 1092 (1975).

30) T. E. Madey and J. T. Yates, Jr. in: Proc. 7th Intern. Vac. Congr. and 3rd Intern. Conf. on Solid Surfaces, Berger, Vienna, 1977) p. 1183.

31) T. E. Madey, J. E. Houston, C. Seabury and T. N. Rhodin, J. Vac. Sci. Technol., to be published.

32) P. A. Thiel, F. M. Hoffmann and W. H. Weinberg, in D. A. Degras, Proceedings of the 4th International Conference on Solid Surfaces and the 3rd European Conference on Surface Science, Cannes, France (Supplement a la Revue "Le Vide, les Couches Minces" no. 201) p. 307.

33) J. Gersten, R. Janow and N. Tzoar, Phys. Rev. Lett. $\underline{36}$, 610 (1976).

34) W. L. Clinton, Phys. Rev. Lett. $\underline{39}$, 965 (1977). W. L. Clinton, to be published.

35) a) J. W. Gadzuk, Surface Sci. $\underline{67}$, 77 (1977).
    b) J. W. Gadzuk, Phys. Rev. $\underline{B14}$, 2267 (1976).

36) D. Lichtman and Y. Shapira, Crit. Rev. Solid State Sci. $\underline{7}$, 167 (1978).

37) J. F. van der Veen, F. J. Himpsel, D. E. Eastman and P. Heimann, Solid State Communications, $\underline{36}$, 99 (1980).

38) D. E. Eastman, J. J. Donelon, N. C. Hien and F. J. Himpsel, J. Nucl. Inst. Methods, $\underline{172}$, 327 (1980).

39) L. Fiermans and J. Vennik, Surf. Sci. $\underline{9}$, 187 (1968).

40) T. E. Madey, J. J. Czyzewski and J. T. Yates, Jr., Surface Sci. $\underline{57}$, 580 (1976).

41) H. Niehus, Surf. Sci., $\underline{87}$, 561 (1979).

42) M. Cardona and M. Loy, Photoemission in Solids I, Topics in Applied Physics, Vol. 26 (Springer Berlin, Heidelberg, New York 1978)

43) R. Jaeger, J. Feldhaus, J. Haase, J. Stohr, Z. Hussain, D. Menzel, and D. Norman, Phys. Rev. Lett. $\underline{45}$, 1870 (1980).

44) T. A. Carlson and M. O. Krause, J. Chem. Phys. $\underline{56}$, 3206 (1972).

45) J. E. Houston and T. E. Madey, to be published.

46) R. Treichler and D. Menzel, to be published.

47) D. Jennison and J. Kelber, to be published.

48) D. A. Ramaker, C. T. White and J. Murday, to be published.

49) D. Menzel, in D. A. Degras and M. Costa, eds. Proceedings of the 4th International Conference on Solid Surfaces and the 3rd European Conference on Surface Science, Cannes, France (Supplement a la Revue "Le Vide, les Couches Minces" no. 201) p. 1259.

50) T. E. Madey, R. Stockbauer, S. A. Flodstrom, J. F. van der Veen, F. J. Himpsel and D. E. Eastman, to be published.

Auger-Initiated Desorption from Surfaces: Review + Prospects[1]

Peter J. Feibelman[2]
Sandia Laboratories[3], Albuquerque, NM 87185, USA

Abstract

A brief introduction is given to recent work showing that Auger decay of low-lying core-holes is a fundamental mechanism leading to desorption of ions from surfaces. I introduce the important concepts, show how they are verified experimentally and how they lead to laws governing the resistance of ionic surfaces to damage by ionizing radiation. Problem areas for future research are noted, particularly the importance of establishing the connection between valence hole localization and ion emission.

1. Introduction

In considering the effects of ionizing radiation on surfaces, the general questions are:
 1) How will a given surface be altered by a given beam ?
 2) What product distribution will emerge from a given surface under bombardment of a specific kind ?
These questions can be answered only if
 i) we can develop surface structure analysis tools adequate to tell us the nature of pre- and post-bombardment surfaces, and
 ii) we can gain insight into the mechanisms which lead to radiation damage.

In this regard, electron and photon stimulated desorption (ESD and PSD) are particularly important phenomena to study because they enable us to explore the effects of electronic transitions, in radiation damage, separately from the effects of direct kinetic energy transfer. Recent work has revealed, moreover, that for a rather general class of surfaces ESD [1,2] and PSD [3] are initiated by the creation of a core hole followed by an Auger decay. As we shall see, this discovery has far-reaching implications, for example:
 1) that ESD and PSD can straightforwardly be used to answer surface structure questions such as what atom is bonded to which, and what are surface atom valences,
 2) that one can state laws governing the resistance of various materials to radiation damage, and
 3) that one can explain one aspect of preferential sputtering - why anions should be more readily sputtered than cations.

---

[1]Work sponsored by the U.S. Dept. of Energy, under contract DE-AC04-76-DP 00789
[2]Address till 1 June 1981: IFF der KFA, D-5170 Jülich, Fed. Rep. of Germany
[3]A.U.S. D.O.E. Facility

While most of what is presented here concerns the initial excitation in ESD and PSD and the laws of radiation damage which follow from the understanding that it is an Auger decay, the question of what follows the initial event is also, obviously, important in deciding whether or not that event actually results in a desorbing ion. This question has only begun to be explored and will surely be the subject of future research. A general discussion of prospects for further research into the theory of ESD and PSD is given at the end of the article.

2. Evidence for Auger Initiated Description

In order to understand the mechanism of ESD from a TiO_2 surface, M.L. KNOTEK [1,4] carried out electron energy loss spectroscopy (ELS) and ESD in the same UHV system. Typically the experiments were carried out within moments of each other. Structure in ELS data (number of reflected electrons vs. the energy they lost) corresponds to surface elementary excitations. Thus the coincidence of an ELS structure with an ESD onset would be strong evidence that a particular elementary excitation is the initial excitation in desorption.

2.1 Menzel, Gomer, Redhead (MGR) Model

According to a Franck-Condon model proposed by MENZEL and GOMER [5] and REDHEAD [6], it is clear what one should see in ESD from TiO_2. Because electrons are $10^{-4} \sim 10^{-5}$ lower in mass than ions, only electronic transitions can lead to desorption for low energy ($\lesssim$ a few x 10^2 eV) beams. MGR's model (see Fig. 1) is that the incident beam causes an electron to be transferred from a bonding to an anti- or nonbonding level. If such a transition causes an atom that is originally bound to find itself on a repulsive potential surface, then the atom will desorb. When this picture is correct then

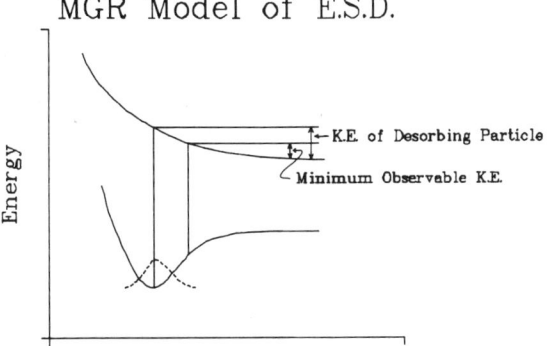

Fig.1 (After Ref. [2]) Franck-Condon diagram illustrating the usual picture of how ESD works. The equilibrium position of the desorbate species is represented by the (dashed) Gaussian distribution in the lower potential. An incident electron causes an electronic transition to a state in which the desorbate feels the upper, purely repulsive potential. Note that when the minimum in the equilibrium well is at sufficiently small R internuclear separation, the probability of desorption with zero kinetic energy is quite small.

1) ESD thresholds correspond to valence-electron bonding to antibonding transitions, which typically occur at energies $\lesssim$ 20 eV, and

2) Desorption of ions produced via a large charge transfer is unlikely, since an incident electron will not cause a multielectron transfer with appreciable probability.

2.2 Knotek's Data and Their Implications

What Knotek observed (Fig.2) agrees with neither of these predictions. Both H^+ and O^+ desorption from TiO_2 were seen. The O^+ desorption threshold is at the Ti(3p) ionization threshold. H^+ has ESD onsets at the O(2s) and Ti(3p) ionization energies. Neither species is found to desorb in the energy range of *observed* bonding-antibonding transitions. TiO_2 is an ionic material in which the O is close to O^{2-}. For example, no d-electrons are seen in Ti Auger spectra from TiO_2. Thus the appearance $at\ all$ of O^+ ions in desorption is remarkable. It implies that the incident beam is causing the transfer of 2 to 3 electrons from negative O ions into the conduction band (or the vacuum). To summarize, Knotek's data pose several important questions:

1) How are O^+ ions produced at all ?
2) Why do O^+ and H^+ show desorption onsets at the Ti(3p) core ionization potential (IP) ?
3) Why does O^+ not start desorbing at the O(2s) IP as H^+ does ?
4) Why is there no desorption at the valence excitation energies ?

2.3 Auger Initiated Desorption

A hypothesis which provides an answer to questions 1), 2), and 4) is that the initial event leading to desorption is an Auger decay. Such a decay follows the creation of a core hole and results in the removal of two or more [1,7] electrons from an atom. The answer to question 3) is provided by energy conservation; since the H, even if it exists on TiO_2 as an H^-, only needs to lose two electrons to become an H^+ while O^{2-} needs to lose three, it can easily be shown that an O(2s) hole has enough energy to produce an H^+ but barely enough to produce an O^+ [1].

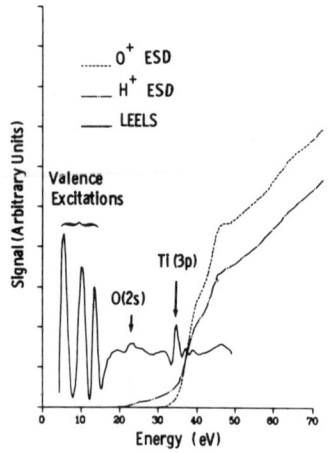

Fig.2 (After Ref. [14]) Comparison of O^+ and H^+ desorption yields vs. energy, with (2nd derivative) electron loss spectrum from TiO_2. Note that no desorption occurs at the large valence electron transitions at low energies. The H^+ yield shows onsets at the O(2s) and Ti(3p) levels, while the O^+ yield is essentially zero up to the Ti(3p) ionization potential.

Let us now pursue this Auger picture further in order to arrive at a test of its validity. How is it that the Auger decay of a *Ti* hole leads to emission of an O^+? The answer is that in a *maximal valence* solid such as TiO_2 no valence electrons reside on Ti's which are nominally Ti^{4+} ions, and thus *intra*-atomic Auger processes cannot quench a Ti(3p) hole. *Inter*-atomic Auger decay is necessary for the 3p-hole to decay which removes charge from O^{2-} ions. What pushes the O^+ ion out of the crystal is then the Madelung force, which is highly attractive to an O^{2-} and conversely very repulsive to an O^+.

Consider in contrast the case of Ti_2O_3. Here each Ti atom has a 3d electron while the O's remain O^{2-}. Now if a Ti(3p) hole is created one of the electrons involved in the Auger decay will almost certainly be a Ti(3d) since intra-atomic wave-function overlap is much greater than inter-atomic. Thus the probability of removing 3 electrons from an O^{2-} in Ti_2O_3 is negligible and Auger induced desorption of O^+ unlikely. This prediction is verified by comparing O^+ ESD from sputtered TiO_2 (which has a Ti_2O_3-like surface) with an annealed surface, and O^+ ESD from V_2O_3 with that from V_2O_5 [1].

Consider next the case of NiO. Here, if a hole is produced in the Ni(3p) shell, the Auger decay will with very high probability be completely *intra*-atomic, since Ni^{2+} has roughly 8 3d-electrons. Thus the Auger decay will result in a positive Ni ion's being made still more positive. Accordingly, there are no force to push the Ni out of the crystal and neither Ni nor O^+ desorption should be seen. This prediction is also verified [1].

3. Consequences of the Auger Model

Once one is convinced of the validity of the Auger decay picture, one can see immediate important consequences of it.

3.1 ESD as a Surface Analysis Tool

Whenever inter-atomic Auger decay leads to desorption, the desorption onset energy tells, since it is a core ionization potential, what the atom neighboring the desorbate species was. Thus the fact that there are H^+ thresholds both at the O(2s) and Ti(3p) energies in Fig. 3 tells us that H is bonded *both* to surface Ti's and O's on TiO_2 [4]. What we do not know, in the absence of absolute cross-sections for desorption, is the extent to which the desorbing ions represent minority surface species. Nevertheless, one can think of few other ways to learn directly that atom A has atom B as its neighbor on a surface.

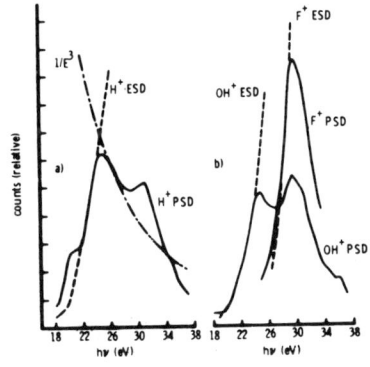

Fig.3 (After Ref. [3]) Spectral dependences for the PSD of a) H^+, b) OH^+ and F^+ compared to their respective ESD curves at threshold. The high-energy sides of the PSD curves are dominated by the E^{-3} dependence typical of photoionization cross-sections, giving rise to peaked yield curves. This is illustrated in a).

The identification of ESD as a core hole process suggests many other techniques as well. For example, one might use an electron below the core IP of atom A of a surface but above that of atom B to remove adsorbed atoms from the surface B's while leaving them on surface A's. This sort of selective desorption could be useful in tracing reaction paths. As will be discussed later, PSD provides a particularly interesting way to do surface EXAFS.

3.2 Principles of Radiation Damage

From the Auger model of ESD, one obtains some clear cut rules for when the creation of a core-hole will lead to desorption. The core-hole must be deep enough that energy is available to remove the requisite number of electrons from the species which is to desorb. Maximal valence is required for interatomic Auger decay to occur. These rules immediately explain why NiO or Cr_2O_3 do not reduce under electron bombardment while TiO_2, V_2O_5, WO_3, BeO, ... do. The fact that Auger decay *removes* electrons from a species means that the sign of the charge of negative ions can be changed but not that of positive ions. This implies that electron bombardment will cause the anions of a solid to desorb in preference to the cations.

To the extent that the core-hole creation occurs in ion-surface scattering, these "principles of radiation damage" will be important there too. A much more complete discussion of this aspect of our ESD results has recently been published [8].

4. Universality of Auger-Initiated Desorption

Until now, most of the discussion has concerned *electron* stimulated desorption from surfaces of *ionic* crystals. It is of course of great interest to know whether the Auger picture is more general than that. Thus let us consider other systems and other beams.

4.1 Desorption of Adsorbates from Metals

Most of the early desorption literature concerned specifically this case, notably the papers of MGR [5,6]. A glance at the ESD yield curves for typical cases, O/Mo [6] and O/W [9], shows the most important ESD onsets to lie at the Mo(4p) and W(5p) and W(4f) thresholds [2]. These results strongly suggest an inter-atomic Auger mechanism, but lead one to wonder how that can be possible without maximal valence coordination (i.e. without having surface MoO_3 or WO_3). Here we take note of the fact that for both Mo [6] and W [10] desorption of O^+ does not occur with appreciable probability for the tightly bound initially deposited O, but only after considerable exposure to O. Thus it seems likely that surface MoO_3 or WO_3 *is* formed before O^+ desorption occurs [2]. (This picture has been experimentally confirmed to some extent, by MADEY et al.'s recent PSD work, as mentioned below [11].)

In the cases of Cl^+ and F^+ desorption from W, which have been studied in ESD and PSD by WOODRUFF et al. [12], thresholds are not seen at W core levels. This is a satisfying result in that WCl_6 or WF_6 coordination should lead to the evolution of these species as gases. In the case of Cl/W the threshold corresponds to the IP of the Cl(3s) electron. For F/W the threshold lies several volts below the 31 eV that one expects for the F(2s) IP. One presumes that this corresponds to a large excitonic effect, but the explanation of the low F^+ threshold for F^+ off W remains an interesting question.

In general, then, the conclusion is that when ionic surface bonding can occur in chemisorption systems, the Auger model appears to explain the corresponding ESD [2] (and PSD). For covalently bonded systems (e.g. CO/W) the situation is unclear, cf. below.

4.2 Incident Photons and Ions

If one believes that a given electronic excitation, i.e. a core-hole, is what leads to desorption, then *a fortiori* anything which can produce such an excitation must give rise to desorption. Specifically photon and ion incident beams should result in desorption onsets at core thresholds and the product distributions from the core-initiated desorption should be identical to that found in ESD (apart, perhaps, from the effects of incomplete relaxation or, in other words, coherence of the initial hole production and the final state).

The case of ion induced desorption is more difficult to analyze because electronic effects leading to desorption must be separated from those of direct kinetic energy transfer. Nonetheless desorption yields have been found which are proportional to core-hole production cross-sections [13]. In the case of photon beams, the lore had it that PSD of ions would not occur [14], or that if it did it would be via the production of photoelectrons which would then cause ESD [15]. On the other hand, photon cross-sections for core-hole production are not appreciably smaller than electron cross-sections [16], so it was of considerable interest to show that the lore was wrong. This was first accomplished by KNOTEK, JONES and REHN [3], using time-of-flight techniques, at the Stanford Synchrotron. They showed (Fig.3) that ESD and PSD onset energies for H^+, OH^+ and F^+ were identical from a TiO_2 surface. Moreover, the yield of ions is much too large to be explained as photoemission followed by ESD. These results are just what one would expect. On the other hand, PSD has some advantages over ESD as a surface analytical tool, as noted in Ref. [3]. PSD yield curves show much finer detail than ESD because the incident beam suffers no losses before the core-hole is produced, and also because the $1/E^3$ falloff of photon cross-sections above threshold reduces the overlap of signals from different core hole onsets. Also PSD offers the possibility of site specific surface EXAFS via the measurement of ion desorption above a particular core threshold, at a given exit angle, as a function of photon energy. This EXAFS experiment is particularly clean compared to electron detection methods, because the ion background is negligible, and because the signal comes uniquely from the surface [17].

More recent PSD work by WOODRUFF et al. [12] has demonstrated the equality of ESD and PSD thresholds for Cl^+ from Cl/W, and shown the close correspondence of total photoemission yield and O^+ desorption for O/W. The yield measurement shows the W core level IP's clearly and the correlation of O^+ desorption onsets with these IP's then shows that PSD proceeds via the same mechanism as ESD [12]. WOODRUFF et al. also show that the O^+ energy and angle distributions in desorption from O/W are virtually the same in ESD and PSD.

Finally a recent paper by MADEY et al. [11] shows yield curves for PSD of O^+ from O/W as a function of O coverage, from about 1/2 monolayer through formation of the oxide. The yield curves change very little as the O coverage increases, in agreement with the idea that only O^+ is desorbed from regions of the surface where coordination is WO_3-like.

5. Where Do We Go From Here ?

In conclusion, I wish to point out a number of problem areas all relating to the further understanding of the mechanism for desorption.

5.1 How Ionic Is Ionic Enough ?

As the electronegativity difference between the constituents of a compound is reduced, the idea that *inter-atomic* Auger decay is necessary to quench a cation core-hole becomes less obvious. Thus one should pursue trend studies, for example, looking at ESD from not only TiO_2 but also TiS_2 and $TiSe_2$ to see if the core-hole induced desorption mechanism persists.

5.2 Can Atom Specificity Break Down ?

Imagine an OH simply coordinated to a metal atom in the configuration M-O-H. It is not clear that H^+ could not be desorbed when a *metal* atom core hole decays, by removing a number of electrons from the O, and thereby disrupting the OH bond. This situation occurs most plausibly for singly coordinated species.

5.3 What Forces Give Rise To Desorption Of Neutrals And Negatives ?

Core hole thresholds for H^-, OH^-, and O^- have been seen [17] but it is far from clear that these ions are produced by the Auger decay mechanism followed by electron transfer. YU's isotope effect data for O^- from O/Mo in fact suggest a direct O^- production mechanism [19]. His result is that desorption of $^{18}O^-$ is less probable than of $^{16}O^-$, indicating that the slower ^{18}O ions have a greater chance of losing their negative charge before escaping. Thus the direction of charge flow in his experiment is *off* the O^-'s, indicating that O^-'s were produced directly.

Little is known about desorption of neutrals. Since these species are thought to be the bulk of desorbed particles their direct observation would be highly desirable.

5.4 What Happens Between An Initial Auger Decay and Desorption ?

Most of the early theoretical efforts on ESD concerned the question of reneutralization and recapture of an ion after its initial production [5,6]. This question is somewhat less serious in the case of a maximal valence oxide, where the nearest neighbors of an anion have no available valence electrons for reneutralization. Nevertheless in the general case it is of interest to ask questions like:

  i) when is reneutralization important ?
  ii) do all interatomic Auger decays lead to desorption or only some ?
  iii) how can we make use of energy and angle distributions of ions to learn something about surfaces ?

All these questions, unfortunately, lead back to the same theoretical problem, that there is no "operator" for desorption, whose matrix elements we would use in a Fermi's golden rule calculation of a desorption cross-section. Some recent progress has been made in recognizing that the reneutralization issue can be restated as a question of hole-localization [20]. First one notes that electron transfer onto an ion is identical to hole-transfer off it. Thus any wisdom one has concerning hole motion has immediate implications for the question of reneutralization. For example, it is known that strong hole-hole repulsion between holes on a single site causes the holes to be bound to that site [21]. Thus the hole transfer time following an Auger decay in which *two* holes are produced on a single atom can be much slower than the resonance tunneling time of a single hole (which is of the order of an inverse band width). Also hole hopping times are increased because of the orbital shrinkage attendant on the creation of a multi-hole state. Thus consider an Auger decay in which a surface Cl^- becomes a Cl^+. The orbital into which an electron must hop to neutralize the Cl^+ has the radius of a Cl^0 while the Cl^+ is still located at a distance from the surface appropriate to a Cl^-. The hopping matrix element is therefore considerably smaller than it would be for conversion of a Cl^0 to a Cl^-. Both these arguments show that the Auger production of a 2-hole state leads to considerably longer (to $\sim 10^{-13}$ to 10^{-14} sec [20]) reneutralization times for surface ions, which are comparable to desorption times. The question of how localization occurs in more general situations, and specifically, whether it plays a role in desorption for covalent systems, is obviously of high interest.

Acknowledgement

I have benefitted from many conversations with M.L. Knotek, J.E. Houston, D.R. Jennison, and G. Loubriel. I would like to thank the KFA Jülich, where the manuscript was prepared, and the Alexander von Humboldt Foundation for their hospitality and support.

References

1. M.L. Knotek and P.J. Feibelman, Phys. Rev. Letters 40, 964 (1978).
2. P.J. Feibelman and M.L. Knotek, Phys. Rev. B18, 6531 (1978).
3. M.L. Knotek, V.O. Jones and V. Rehn, Phys. Rev. Lett. 43, 300 (1979).
4. M.L. Knotek, Surf. Sci. 91, L17 (1980).
5. D. Menzel and R. Gomer, J. Chem. Phys. 42, 886 (1969).
6. P.A. Redhead, Can. J. Phys. 42, 886 (1964).
7. T.A. Carlson and M.O. Krause, Phys. Rev. Lett. 14, 390 (1965).
8. M.L. Knotek and P.J. Feibelman, Surf. Sci. 90, 78 (1980).
9. M. Nishijima and F.M. Propst, Phys. Rev. B2, 2368 (1970).
10. D.A. King, T.E. Madey and J.T. Yates, Jr., J. Chem. Soc., Faraday Trans. I 68, 1347 (1972).
11. T.E. Madey, R. Stockbauer, G.F. van der Veen and D.E. Eastman, Phys. Rev. Lett. 45, 187 (1980).
12. D.P. Woodruff, M.M. Traum, H.H. Farrell, N.V. Smith, P.D. Johnson, D.A. King, R.L. Benbow and Z. Hurych, Phys. Rev. B21, 5642 (1980).
13. K. Wittmaack, Phys. Rev. Lett. 43, 872 (1979); P. Williams, Proc. 40th Phys. Electronics Conf., Ithaca, N.Y. 1980 (unpublished).
14. D. Lichtman, Surf. Sci. 90, 579 (1979); D. Lichtman and Y. Shapira, CRC Crit. Rev. Sol. Sc. Mat. Sci. 8, 93 (1978).
15. R. Franchy and D. Menzel, Proc. 7th Intl. Vacuum Cong. and 3rd Intl. Conf. on Sol. Surf., Vienna 1977, R. Dobrozemsky et al., ed. (F. Berger & Söhne, Vienna, 1977) p. 1209.
16. E.J. McGuire, Phys. Rev. A16, 73 (1977), Sandia Laboratories Report No. SC-RR-70-721 (unpublished).
17. M.L. Knotek, V.O. Jones and V. Rehn, Surf. Sci. (in press).
18. M.L. Knotek, unpublished.
19. M.L. Yu, Phys. Rev. B19, 5995 (1979).
20. P.J. Feibelman, Surf. Sci. (in press).
21. M. Cini, Sol. State Commun. 24, 681 (1977); 20, 605 (1976); G.A. Sawatzky, Phys. Rev. Lett. 39, 504 (1977).

Optical Radiation from Electron Sputtering of Alkali Halides

N.H. Tolk, L.C. Feldman, J.S. Kraus, R.J. Morris*, T.R. Pian, M.M. Traum and J.C. Tully
Bell Laboratories, Murray Hill
Murray Hill, NJ 07974, USA

1. Introduction

Electron-surface collisions have been observed to result in the emission of optical radiation from excited atoms and molecules desorbed from alkali halide single crystal surfaces. The detected radiation included Na and Li resonance lines, hydrogen Balmer emission, and OH molecular radiation. Much previous experimental and theoretical effort has been devoted to electron stimulated desorption (ESD) of ground state neutrals and ions [1], and in particular from alkali halides [2,3,4]. The few studies concerned with excited particles have dealt primarily with ejected metastable neutrals [5]. In one case involving electron bombardment fluorescence of ice, a tentative identification has been made of OH molecular radiation [6]. The results reported here include a) the first definitive work on optical radiation from electron bombardment induced emission of excited free substrate particles [7], and b) the first observations of characteristic radiation from previously adsorbed free atoms and molecules.

2. Experimental Results

The apparatus for these experiments (Fig. 1) included an electron gun (100 eV to

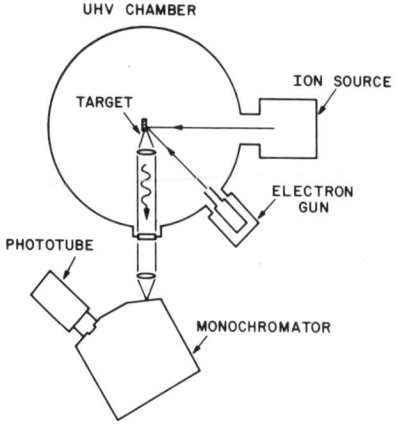

Fig.1 Schematic of experimental apparatus.

---

* Present address: Columbia University, New York, NY 10027

1 keV), an ion source (100 eV to 10 keV) and a target chamber maintained under UHV conditions (2×10^{-10} torr background) [8]. These studies used a 1 keV electron beam with $10\mu A$ ($1 mA/cm^2$) bombarding current. Single photon signals were observed at a variety of angles to the surface normal, which enabled measurements to be made of radiation arising from the front of the surface and of the surface itself. Counting rates as high as 7×10^4 counts/sec were observed from the alkali resonance lines. NaCl, NaF and LiF were the targets used in this study. The monochromator was set for a 2.4 nm resolution.

When viewing both the bulk and near-surface region, we observed radiation from desorbed free atoms and molecules plus bulk luminescence. Fig.2 shows the spectra of electron bombarded NaCl, LiF and NaF under these conditions, with prominent resonance alkali emission and continuum radiation. The intensity of discrete radiation could be emphasized by manipulating the target to view only the near-surface vacuum region. Fig.3 presents spectra for this case for each sample. For NaCl, Fig.3a, we see the Na D doublet (589.0-589.6 nm), unidentified "molecular-like" structure in the 240-310 nm region, and the H_α Balmer line at 656.3 nm arising from bulk or adsorbed H. We estimate the Na D radiation rate to be $3.2(\pm 1.5)\times10^{-5}$ photons/incident electron. Electron bombardment of NaF (Fig.3b) produced only the doublet. For LiF (Fig. 3c), electron bombardment produced the Li resonance line (670.7 nm), the H_α 656.3 nm line, and faint higher F excited states around 690.0 nm. The resonance halogen lines lie in the ultraviolet outside the spectrometer range.

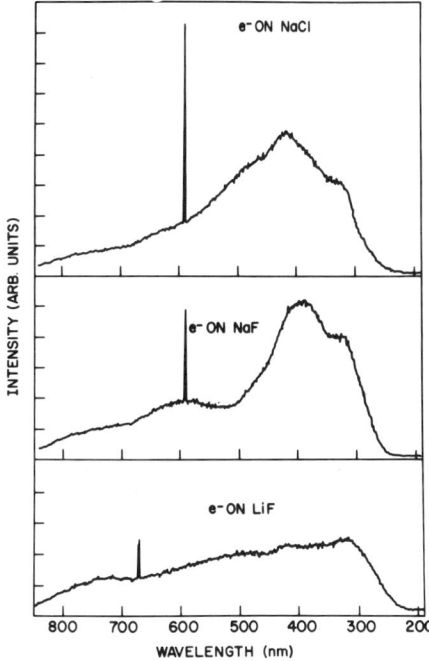

Fig.2 Spectra of radiation produced by 1 keV electron impact on NaCl, NaF, and LiF. Emission from surface and near-surface vacuum region viewed 45° to the surface normal, with a 2.4 nm monochromator resolution.

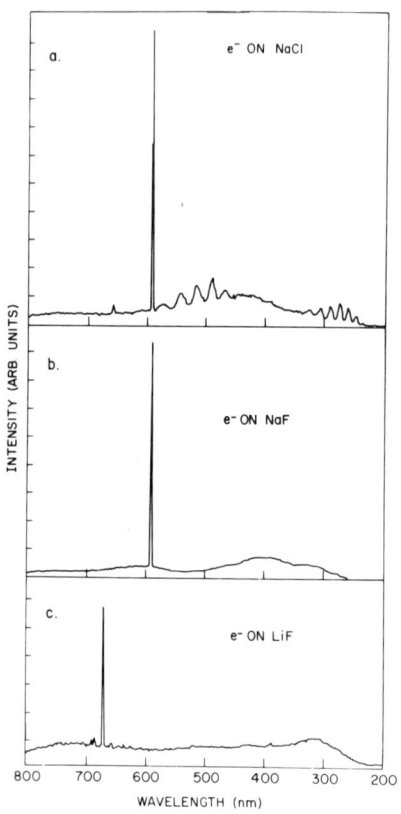

Fig.3 Spectra of radiation produced by 1 keV electron impact on NaCl, NaF and LiF. Emission from near-surface vacuum region only viewed perpendicular to the surface normal, with a 2.4 nm monochromator resolution.

Concurrent measurements were also performed of optical radiation from a 6 keV Ar$^+$ beam in the same chamber to compare spectra [10]. Such an ion bombardment spectra of NaCl (Fig.4) exhibits additional Na lines in contrast to our ESD experiments where only Na and Li resonance lines are measured (Fig.3a and 3b). This indicates an as yet undetermined selection process associated with the ESD excitation mechanism.

To enhance ESD excitation of H, we introduced a small partial pressure of H_2 gas into the chamber. We show electron bombardment photon spectra arising from NaCl and NaF targets with $p_{H_2} = 2-4 \times 10^{-6}$ torr (Fig.5). Results for an LiF target are shown in Fig.6. The H_2 appears to reduce the Na and Li resonance line intensity. The striking feature is the increased H Balmer emission; at least four series terms are readily observed. An additional feature due to OH radical gas phase emission (0-0, $\tilde{A}^2\Sigma^+ \rightarrow \tilde{X}^2\Pi$) can be observed around 306 nm in Fig.5. Removing the target from the field of view caused the signals to disappear, indicating no contribution from electron gas collisions. Finally, after removing the H_2 gas and waiting around 20 seconds, the photon spectra reverted to approximately that shown in Fig.3.

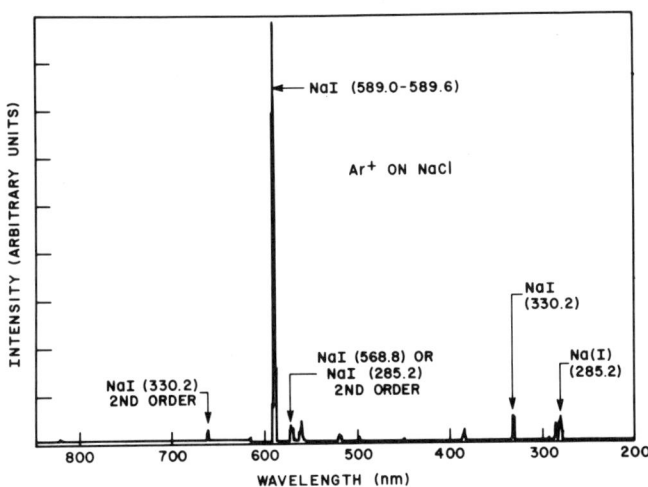

Fig.4 6 keV Ar$^+$ bombardment induced photon emission of NaCl, with prominent Na peaks identified.

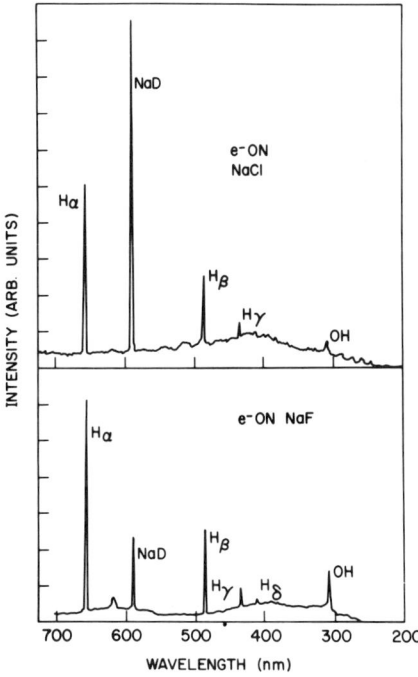

Fig.5 Spectra of radiation produced by 1 keV electron impact on NaCl and NaF, with $2-4 \times 10^{-6}$ torr H$_2$ partial pressure background. Viewed as in Fig.3 perpendicular to the surface normal.

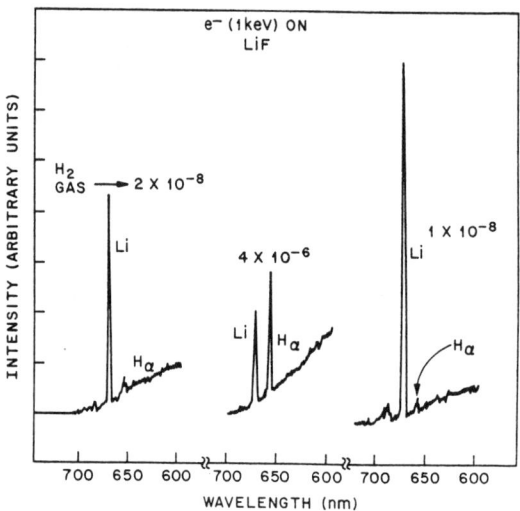

Fig.6 Photon spectra of 1 keV electron bombardment on LiF at indicated partial pressures (torr) of H_2.

3. Discussion

Our measurements support the view that the line and molecular-like emission originates in front of the surface from ESD-excited particles for the following reasons:

a. The optical lines and bands are not shifted or broadened as they would be if the radiation originated from the surface.

b. Tilting and translating the target effectively separates the radiation originating from the vacuum outside the surface from the radiation emitted from the surface itself.

c. Preliminary incident electron energy-dependent photon emission measurements on the alkali halides indicated a threshold for discrete radiation lower than that for the bulk continuum [9]. This shows that the discrete and continuum emission arise from different mechanisms

d. We rule out electron-alkali vapor collisions as the source of the observed alkali resonance excitation for the following reasons: 1) the calculated sample temperature rise during electron bombardment yields a vapor pressure too low to account for such a mechanism; 2) for electron impact excitation, we should see higher excited state photon emission given the magnitude of the cross-sections; 3) the photon signals completely disappeared by translating the sample 3mm away from the viewing region, too sudden a drop for the density of evaporating alkali target atoms; 4) as shown in Fig.7 the Na D doublet signal is linear with electron current, indicating that free excited particle creation is an isolated event which cannot arise from electron-gas collision excitation, which is a second order process.

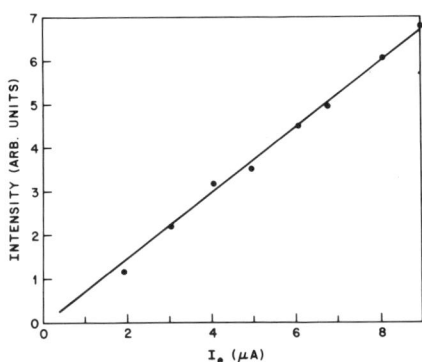

Fig.7 Intensity of Na D doublet (589.0-589.6 nm) radiation as a function of incident electron current.

From these results we are not presently able to specify mechanisms responsible for the observed alkali, hydrogen, and hydroxide radiation. However in accounting for alkali halide electron sputtering two mechanisms have been invoked [2,3]. An H-center produced by radiationless decay of an electron-impact induced exciton may diffuse to the surface by a focussed collision sequence causing directional ejection of a halogen with sputtering. Electron impact may also create a V_k center which diffuses to the surface and results in emitting a thermal halogen, "thermal" sputtering. The resulting non-stoichiometry caused by the halogen sputtering somehow leads to excess metal ejection. We do not know the ESD yield of metal ejection compared to that of thermal evaporation; the comments in paragraph d) above are pertinent.

In addition, alkali and surface contaminant atoms may be desorbed by ESD due to either outer [11,12] or inner [13] shell electronic excitation processes. The recent Auger decay ESD theory [13] may be particularly relevant to our studies since our alkali halide samples are ionically bonded. Careful incident energy dependent optical radiation measurements may contribute to the understanding of this question [9].

The intensity of the optical radiation from the OH molecule monotonically increases with H_2 partial pressure. This is particularly interesting because it appears to represent electron impact induced synthesis of molecules created from absorbed hydrogen and an oxygen containing surface contaminant which did not exist in the system prior to bombardment. We propose that this may have special relevance to the synthesis of molecules in planetary bodies and interstellar dust grains subject to electron bombardment in space. Similar suggestions have been made involving ion bombardment of interstellar surface [14] and recent experiments have given support to this contention [15].

In future experiments we plan to measure incident energy photon emission thresholds, monitor the sample temperature dependence, and follow radiation polarization to define possible directional ejection and excitation mechanisms. SIMS analysis of desorbing species may well clarify the role of various electron exchange processes involved in desorption. Finally, we note the utility of ESD excitation in detection of H at surface which makes it a potentially useful addition to surface science diagnostics.

References

1. See for example, D. Menzel, Surf. Sci. *47*, 370 (1975); T. E. Madey and J. T. Yates, Jr., Surf. Sci. *63*, 203 (1977); P. J. Feibelman and M. L. Knotek, Phys. Rev. B*18*, 6531 (1978); D. P. Woodruff, M. M. Traum, H. H. Farrell, N. V. Smith, P. D. Johnson, D. A. King, R. L. Benbow and Z. Hurych, Phys. Rev. B*21*, 5642 (1980).
2. P. D. Townsend et. al. Radiation Effects *30*, 55 (1976).
3. H. Overeijnder, M. Szymonski, A. Haring and A. E. deVries, Radiation Effects, *36*, 21 (1978).
4. R. Kelly, Surface Science *90*, 280 (1979).
5. I. G. Newsham and D. R. Sandstrom, J. Vac. Sci. Technol. *10*, 39 (1973).
6. R. H. Prince, G. N. Sears, and F. J. Morgan, J. Chem. Phys. *64*, 3978 (1976).
7. In the course of their studies of electron bombardment induced exciton luminescence from NaCl, Townsend et al. mentioned observations of sodium D lines emitted from neutral excited sodium atoms which had evolved from the surface bulk, P. D. Townsend, A. Mahjoobi, A. J. Michael and M. Saidoh, J. Phys. C: Solid St. Phys. *9*, 4203 (1976).
8. N. H. Tolk, L. C. Feldman, J. Kraus, R. J. Morris, M. M. Traum and J. C. Tully, submitted to Phys. Rev. Letters.
9. Absolute threshold potentials were difficult to determine due to uncertainties in alkali halide surface potentials. Measuring such thresholds would enable us to correlate our results with the recent Auger decay ESD theory, see Refs. 11-13.
10. For further references on SCANIIR, see for example: C. W. White and N. H. Tolk, Phys. Rev. Lett. *26*, 486 (1971); C. W. White, D. L. Simms, and N. H. Tolk, Science, *177*, 481 (1972); N. H. Tolk, I.S.T. Tsong and C. W. White, Anal. Chem. *49*, 16A (1977).
11. D. Menzel and R. Gomer, J. Chem. Phys. *41*, 3311 (1964).
12. P. A. Redhead, Can. J. Phys. *42*, 886 (1964).
13. M. L. Knotek and P. J. Feibelman, Phys. Rev. Lett. *40*, 964 (1978).
14. M. Maurette, Nucl. Instr. Meth. *132*, 579 (1976).
15. R. Kelly, S. Dzioba, N. H. Tolk and J. C. Tully, To be published in Surface Science.

Electron Transfer

Electron Capture and Loss to Continuum States in Gases and Solids*

I.A. Sellin
University of Tennessee, Knoxville, TN 37916, and
Oak Ridge National Laboratory, Oak Ridge, TN 37830, USA, and

R. Laubert
East Carolina University, Greenville, NC 27834, USA

I. Introduction

Among the many processes that occur when fast particles interact with target atoms there is the interesting case where the projectile captures target electrons, not necessarily into well-defined bound states, but into slightly unbound continuum states [1]. The net result is that one or more target electrons leave the target with speeds and directions that are nearly coincident with that of the bombarding projectile, but are not bound to the projectile. Present theories inadequately describe single and multiple capture processes to either bound or unbound states in the medium-velocity range.

Cognizance of the continuum electron capture phenomenon is important in experiments measuring pure bound state capture cross sections as well. Recent theory and experiment indicate that in some cases this channel can account for more than 50% of the total ionization cross section [2], and in others carried out in our laboratory, one or more bound state capture events are found to accompany continuum capture $\gtrsim$ 20% of the time. Hence experiments that separate bound state capture from continuum capture must be compared to theories that explicitly include the simultaneous continuum capture contribution. For solid targets the situation is even more complicated. Until very recently it has not even been known whether the origin of the forward continuum electrons observed in ion-solid collisions is associated with the surface or the bulk of the material.

The characteristic feature of electrons captured into the continuum (ECC) is that they possess a velocity, in magnitude *and* direction, close to that of the projectile ion. Hence the projectile ion emerges from the interaction region, which can be a gaseous or solid target, accompanied by electrons which move with the same velocity. Measurements by Crooks and Rudd [3] of the electron velocity distribution resulting from proton bombardment of He revealed a cusp-shaped peak in the forward ejected electron velocity distribution with the peak velocity equaling the ion velocity. Independently, Harrison and Lucas [4] observed a similar peak in the velocity distribution of electrons accompanying protons emerging from solid carbon targets. These authors have continued such work with other collaborators in a series of investigations [4], [5], [6], [7] for various H, He projectiles, mostly in thin solid and occasionally in gaseous targets, at projectile energies up to 1.2 MeV/u. Extensive work for both solid and gaseous targets and H, He projec-

---

* This work was partially supported by the National Science Foundation; the Office of Naval Research; the Fundamental Interactions Branch, Division of Chemical Sciences, Office of Basic Energy Sciences, U. S. Department of Energy, under contract W-7405-eng-26; and the East Carolina Research Council.

tiles has also been undertaken by Menendez, Duncan and co-workers [8], and for solid targets by Meckbach et al [9]. The latter authors have written an excellent summary article, whose title "Do present 'charge transfer to the continuum' theories correctly describe the production of $v_e = v_i$ electrons in ion beam-foil collisions?" emphasizes several controversial disagreements between the various data and theoretical predictions. Chiu, McGowan, and Mitchell [10] have extended the work of Meckbach et al. to gaseous targets, and find support for the electron capture to continuum (ECC) description in gaseous targets, in opposition to deficiencies they find in such a description for solid targets. However, Steckelmacher et al. [7] have criticized the interpretation given the data of Meckbach et al. on instrumental resolution and background treatment grounds, and argue for the validity of the ECC model for the solid-target case as well, and against the rival, solid-state "wake-riding" description [9], [11]. The latter model considers the possibility that electrons may be trapped in an oscillatory electron density fluctuation polarization potential extending behind and moving with the projectile, finally being liberated at the exit surface.

An excellent review of the theoretical aspects of the ECC process in ion-atom collisions has recently been provided by Shakeshaft and Spruch [12]. A number of experimental findings complementary to these discussed here have been covered by Meckbach et al. [9].

II. Experimental Procedure

A key feature of our experimental procedure is the easy interchange of short gaseous and thin solid targets at the same physical position, with all apparatus aperture sizes, dimensions, positions, and other experimental details unaltered. It has therefore been possible to cancel most systematic apparatus effects in comparing gaseous and solid target results. By using single ion-atom collision techniques, by using bare and few-electron ions of appreciably higher charge than heretofore, by extending the velocity range of measurement appreciably above that of earlier experiments, and by studying charge-state variation over an appreciably wider range than used previously, we have been able to test experimentally features of continuum electron-capture and -loss theories which have been inaccessible in previous experiments. These somewhat orthogonal tests complement rather than duplicate earlier experimental tests, but, like them, raise new questions about still other conspicuous disagreements.

Schematic diagrams of suitable apparatus are displayed in Figs. 1 and 2. Projectile ions from the Oak Ridge National Laboratory Tandem, Brookhaven National Laboratory Tandem, Lawrence Berkeley Laboratory Super-HILAC, or other accelerator traverse a thin gaseous or interchangeable solid target over a velocity range corresponding to 0.7-8.5 MeV/u. Beams are typically collimated to 1/3 mm diameter and $\pm$ 0.025° angular spread, and then traverse a 4 mm thick target cell terminated by $\sim$ 2 mm apertures (or a foil target centered at the same location). In the case of Fig. 1, the beam and accompanying electrons then enter along the central ray of a 180° spherical-sector analyzer of mean radius 3.8 cm, whose $\Delta E/E$ of 1.4% full width at half maximum is set by a 0.71 mm analyzer exit aperture, and the source dimensions at the center of the cell. Typical gas pressures of $\sim$ 15 mTorr are established by a standard feedback-controlled capacitance-manometer system. Single-collision conditions and negligible charge changing are established by extremely linear plots of integrated cusp cross sections versus pressure for the various incident ions and energies. The plot intercepts typically coincide precisely with very small residual cusp cross sections measured at zero pressure.

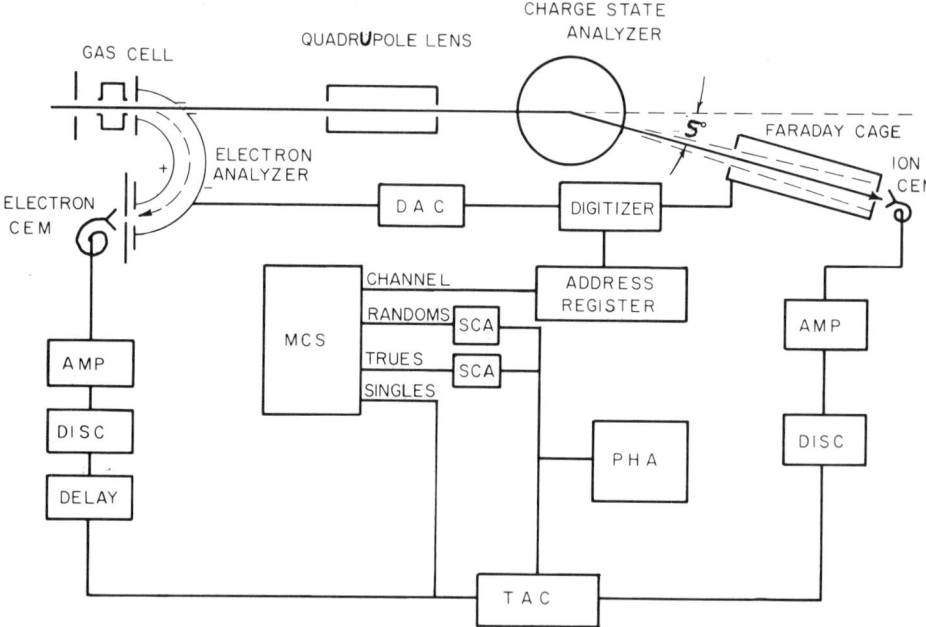

Fig. 1 Schematic diagram of electrostatic version of the experimental apparatus, showing coincidence arrangement used.

Spurious signals from electron loss by contaminant beams can be ruled out, as deliberate changes in the accelerator vacuum by a decade in pressure produced no noticeable effect on pressure linearity. The final collimation guarantees that the beam never encounters any aperture or surface in the gas cell or spectrometer regions, as verified by very low spurious signals obtained at zero pressure. Magnetic fields are reduced to $< 3 \times 10^{-6}$ T over the spectrometer volume by three orthogonal external coils. Changes in coil current of $\lesssim 20\%$ produce negligible changes in cusp shape and integrated production cross sections.

Upon emerging from the target, the ion beam passes through an opening in the larger radius sector of the electron energy analyzer and is subsequently focused by a magnetic quadrupole doublet, dispersed horizontally according to its charge state by a perpendicular magnetic field, and finally collected by a channel electron multiplier located approximately 6 m from the target. This arrangement allowed us to determine the charge state of the emergent ions and permitted beam normalization. Hence, multi-channel scaling techniques could be employed in collecting the data; standard coincidence electronics and techniques were also used. The start signal for a time-to-amplitude converter (TAC) was generated by the detection of an energy-analyzed electron, while the stop signal was generated by the detection of a charge-analyzed ion. The resultant TAC distribution has a width ~ 6 ns at full width at half maximum, and indicates that a coincidence event between the detected electron and a charge-analyzed particle has occurred. In a similar manner the number of accidental events is counted. The total particle flux was adjusted so that the true-to-accidental ratio is typically 100 to 1 but is always constrained to exceed 5 to 1 by beam flux adjustment. This experimental arrangement allowed us to measure the number and energy distribution of all the emitted electrons (i.e., the "singles" spectrum), the number and energy distribution of electrons (corrected for accidental events) in coincidence with a particular charge state of the emergent ion (coincidence spectrum), the total number of ions of a certain charge state, and the total number of ions.

Figure 2 shows a magnetic variant appropriate to use of projectiles above the Coulomb barrier, where problematic nuclear radiation backgrounds from entrance collimations can be suppressed through use of "virtual" entrance slits produced by non-magnetic collimators located *within* the analyzer. Using such experimental arrangements, we have employed bare, one-, and multi-electron projectile ions on He, Ne, and Ar targets (under single collision conditions); thin self-supporting polycrystalline C, Al, Ag and Au targets; and monocrystalline Au targets oriented for axial channeling (<110> and <100>) and random directions. Since there are several distinct aspects to the results obtained, we discuss them separately.

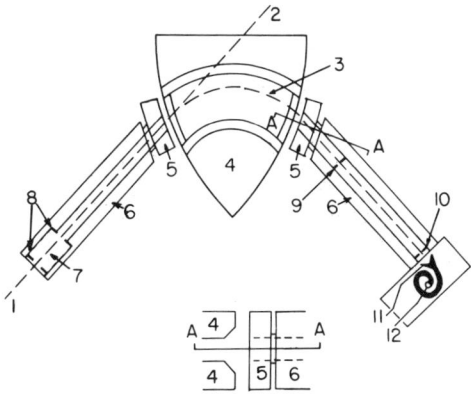

Fig. 2 Schematic diagram of magnetic sector spectrometer. Ion beam enters spectrometer (1) and exits to Faraday cup (or CEM) (2). Electrons follow curved path (3) through pole pieces (4). Mirror pieces (5) shape field edges for improved focusing. Shield pieces (6) minimize magnetic fields along electron path, house gas cell (7) defined by apertures (8), angle-defining aperture (9), and exit aperture (10). Channel electron multiplier (CEM) detector has negatively biased cone (11).

A. Electron Capture to Continuum in Gases by Bare Projectiles

When the incident projectile is bare and the target is a dilute gas, we find that the cusp-shaped ECC velocity distribution is skewed toward low velocities. It is a matter of current controversy whether the asymmetry discussed in the continuum electron velocity distribution is produced by contributions from the second Born amplitude (which are predicted to be asymmetric) added to those arising from the first Born term (which are symmetric). Hence experiments in this velocity domain offer unique tests for the validity of many approximations employed. The full width at half maximum (FWHM), Γ, of the velocity distribution can be represented approximately by the first Born theoretical result $\Gamma \approx (3/2) v_i \Theta_0$ (Θ_0 is half the analyzer polar acceptance angle and v_i the ion velocity). Regarding yields for Ne and Ar targets, we find that the cross section has a projectile charge dependence of Z_1^n ($n \sim$ 2.3) and energy E^{-m} ($m \sim 2.75$) dependence, dependences more closely resembling those found for bound state capture at comparable projectile velocities than the n and m dependences theoretically predicted for ECC at asymptotically high velocities using a hydrogenic target approximation.

While we have not yet been able to establish the relative importance of multiple continuum capture, we do find it possible to examine quantitatively the importance of simultaneous bound- and continuum-state capture measured in coincidence, relative to the processes of single-electron bound-state capture and of continuum-state capture alone. For example, for bare O^{8+} ions with velocities $\sim$ 9 au undergoing single collisions in Ar gaseous targets, we find that for every 100 events observed in which a continuum capture occurs with *no* additional simultaneous bound-state capture, additional simultaneous capture of one, two, and three electrons into bound states is observed in about thirty, about ten, and about one collision events, respectively. Since our results directly demonstrate that often one or more closely associated additional bound-state captures occur whenever a continuum-capture event is observed, it is natural to raise the question, in what fraction of the events do one or more continuum-state captures occur when a bound-state capture occurs? Though this question is not directly answered in our recent experiments because of counting-rate limitations in the ion-counting coincidence channel, we *can* answer this question indirectly by comparing the total cross section for coincident continuum and bound-electron capture with total cross sections measured by other authors for what has always tacitly been assumed by them to represent bound-state captures alone. We find that we can consistently account for a significant fraction of what have traditionally been termed one-, two-, and three-electron bound-state-capture cross sections in terms of two-, three-, and four-electron capture events instead, where an unobserved electron has been ignored. The immediate consequence is that such comparisons as are made with bound-state capture theories are thus being made with the wrong theories. Comparisons must instead be made with theories which explicitly include the correlated continuum-capture contributions.

Figure 3 presents representative ECC coincidence spectra for bare O^{8+} ions in Ar, with the coincident final ion charge labeling each curve. From comparative yields, ratios of 1:2:3:4 electron events (ECC + zero, one, two, or three bound-state captures) are extracted, where in each case the start signal was registration of a continuum-electron capture. While in principle the apparatus does not distinguish between single and multiple ECC leading to an electron start signal, multiple ECC is thought very likely to cause large scatter of the continuum electrons in phase space well away from the condition $\vec{v}_e \simeq \vec{v}_i$ as a result of their mutual Coulomb repulsion, a hypothesis consistent with an unsuccessful search we made to detect them.

In Fig. 4, data concerning the projectile Z, v_i, and scattered-ion charge-state (q_e) dependence of the cross section for zero, one, two, and three bound-state captures for O^{8+} and C^{6+} ions in Ar gas are presented along with lines to guide the eye. A number of interesting comparisons with the single and multiple bound-state-capture data (ECC tacitly ignored) obtained by Macdonald and Martin [13] for the same collision system at overlapping velocities can be drawn. In earlier, non-coincidence experiments, we had already noted many qualitative similarities concerning velocity and Z dependences for such total capture data [13], [14] with our ECC data. Several such dependences are mimicked.

Interesting properties of the data presented in Fig. 4 include the following:

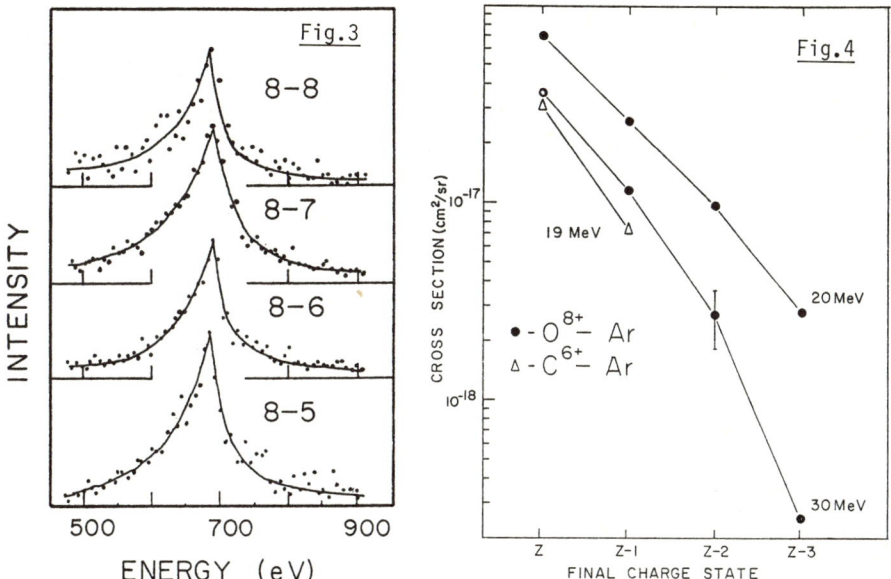

Fig. 3 Spectrum of continuum-capture electrons measured in coincidence with b = 0, 1, 2, 3 additional bound-state captures (scattered ion charge q' = 8, 7, 6, 5) for ∼ 9 au (30 MeV) O^{8+} in Ar.

Fig. 4 Differential cross section for production of continuum-capture electrons in coincidence with b = 0, 1, 2, 3 additional bound-state electrons for O^{8+} in Ar at 20 and 30 MeV, and for C^{6+} in Ar at 19 MeV, averaged over a cone of half angle Θ_0 = 1.6° centered on 0° and integrated over velocity intervals $v_e = v_i (1 \pm \alpha)$, α = 0.04.

(1) Given an ECC event for 20 MeV O^{8+} on Ar, cross sections for additional coincident captures of b = 0, 1, 2, 3 bound-state electrons scale approximately as 2.1:1:0.38:0.11, or ∼ $(1/3)^b$. This can be compared with corresponding ratios at 20 MeV read from [13] of $\sigma_{87}:\sigma_{86}:\sigma_{85}$ ∼ 1:0.25:0.036. The corresponding scaling for ECC + bound-capture events for C^{6+} in Ar at 19 MeV scales as ∼ $(1/4)^b$.

(2) In the range from 20 to 30 MeV, the trends with beam energy of ECC coincident with b = 0, 1, 2 are qualitatively similar to those observed in [13], and scale approximately as $E^{-1.8}:E^{-2.0}:E^{-2.8}$.

(3) If the O^{8+} ECC data on Ar at 20 and 30 MeV are interpolated to the same energy per nucleon as our 19-MeV C^{6+} data, the projectile Z scaling for ECC with b = 0 is ∼ $Z^{1.7}$, whereas that for ECC with b = 1 is ∼ $Z^{2.9}$. This observation is pertinent with respect to interpretation of the ∼ $Z^{2.3}$ scaling we had noted earlier [15] in non-coincident ECC data for C^{6+}, O^{8+}, and Si^{14+} on Ar at similar velocities. Now one sees that the exponent represents a statistical average over b = 0, 1, 2, 3, ... processes coincident with ECC.

Conversion of the $\sim$ 20%, $\sim$ 5%, and $\sim$ 1% likelihoods of b = 1, 2, and 3 bound-state captures accompanying 9 au O^{8+}-Ar ECC events into cross sections leads to the following results:

(1) Assuming a cusp shape of the form given by Dettmann and co-workers [4], and integrating the ECC cusp over intervals of $v_e = v_i (1 \pm \alpha)$, $\alpha = 0.5$, and over a cone of half-angle $\Theta_0 = 30°$ leads to cross section estimates of $\sim 9.3 \times 10^{-18}$, $\sim 3.0 \times 10^{-18}$, $\sim 9.9 \times 10^{-19}$, and $\sim 8.9 \times 10^{-20}$ cm^2 for b = 0, 1, 2, 3, respectively. Here use is made of a spectrometer solid angle and efficiency calibration based on absolute cross sections measured by Stolterfoht, Schneider, and Ziem [16] for L-Auger production by H+ on Ar and on absolute cross sections for continuum capture measured by Rødbro and Andersen [17] and by Cranage and Lucas [5].

(2) Alternative determinations of the ratio of integrated ECC yields to the projectile stripping yields we have measured for O^{7+} on Ar at the same energy, followed by normalization to the total-electron-loss cross sections obtained from [13], leads to independent coincident b = 0, 1, 2, 3 cross section estimates of $\sim 8.0 \times 10^{-18}$, $\sim 2.6 \times 10^{-18}$, $\sim 8.5 \times 10^{-19}$, and $\sim 7.7 \times 10^{-20}$ cm^2, respectively. Comparing these estimated cross sections with those of [13] leads to the conclusion that coincident ECC accompanies bound-state-capture events roughly 20% of the time for b = 1, 2, or 3. Hence we conclude that many measurements tacitly assumed to correspond to pure bound-state-capture processes - and indeed generally a large number of other capture cross-section measurements [13], [14] tacitly assumed to correspond to pure bound-state-capture processes - are actually compound, because of the neglect of an important coincident-electron-capture channel!

A word of caution about the dependence of the FWHM of ECC distributions on such dynamical variables as v_i, Θ_0 is in order. While the first Born result of Dettmann et al. [4] does yield $\Gamma \approx (3/2) v_i \Theta_0$, in approximate agreement with our experimental work and that of others for ECC in gases, in our results there is a strong left-right asymmetry in the dependence observed on v_i. While we find that Γ in most cases scales roughly as v, only the total FWHM does so. If divided into left-right (L, R) HWHM's corresponding to $v_e \leq v_i$, $v_e \geq v_i$, Γ_R turns out always to be of the order expected from consideration of analyzer resolution (typically $\gtrsim$ 1%), while Γ_L accounts for the bulk of the growth of Γ with v. Furthermore, there are significant shell effects. For Ne, Γ_L is seen to rise significantly just as the beam velocity is varied through the range $v_i \sim v_K$, where v_K represents the K-shell electron velocity. Similar K-shell effects, though less well resolved, are seen in Ar. Ar possesses three shells whose contributions overlap. Theory has yet to fold in the internal velocity distribution of ECC electrons in explicit quantitative calculations, as the approximation $v_i \gg Z_{TARG} v_B$ (v_B is the Bohr velocity) is normally made. The onset of the rise in Γ_L occurs just where the Bohr velocity-matching criterion for electron capture would predict.

B. Electron Loss to Continuum in Gases (ELC)

Drepper and Briggs [18] have put forward a binary collision theory concerning electron loss by fast projectile ions (ELC), in which the laboratory-frame distribution of the large forward ejected-electron peak displays a cusp in both energy and angle, centered in velocity near $\vec{v}_e = \vec{v}_i$. The theoretical cusp shape closely resembles that predicted in ECC theories [4], [19], [20] for bare ions.

Such cusps can be studied if multi-electron projectiles are used on gaseous targets. A typical result is shown for various O^{q+}, Si^{q+} ions incident on an Ar gaseous target in Figs. 5 and 6. We note a nearly symmetric cusp which is considerably (a factor of ~ 2) narrower than a typical ECC cusp. Superimposed on the cusp are very definite structures which are identified as projectile Auger lines resulting from the transition $(1s^2 2pn\ell) \rightarrow (1s^2 2s) + ke^-$, with n = 6, 7, 8, ... Transitions for $n \leq 5$ in oxygen are energetically forbidden, as then the 2s-2p energy difference is insufficient to liberate the Rydberg electron.

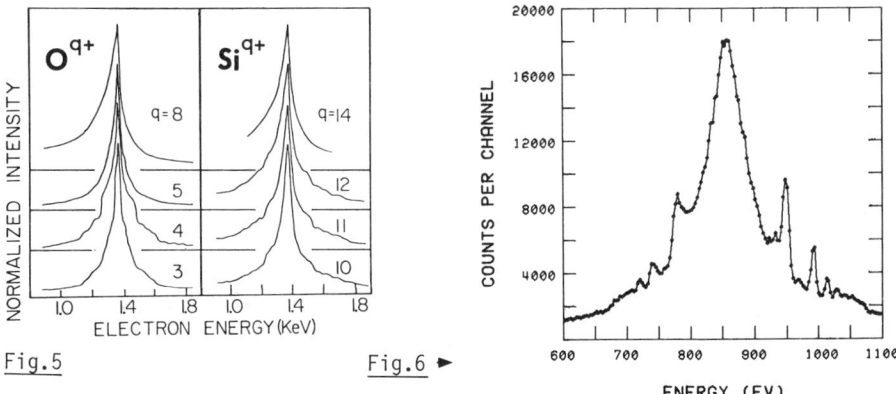

Fig. 5   Fig. 6

Fig. 5 Normalized spectra for electron capture and loss to continuum states in the forward direction for 2.5 MeV/A O^{q+} and Si^{q+} on Ar. Zero-count base lines are shown for each spectrum. Statistical errors are about 2% at the peaks.

Fig. 6 Energy spectrum of electrons emitted in the forward direction in collisions of 25 MeV O^{4+} on Ar. The spectrum shows a cusp-shaped peak produced mainly by projectile electron loss, centered at an energy corresponding to the projectile velocity. In both wings of this cusp a series of lines is visible, which originate from projectile autoionizing states of the following kind: $(1s^2 2p\, n\ell) \rightarrow (1s^2 2s) + ke^-$. For O^{4+}, $n \leq 5$ is energetically forbidden.

The wide range of charge states used for Si^{q+} on Ar (q = 5-14) gives rise to a monotonically decreasing ELC integrated cusp yield with q, producing a factor of ~ 10 drop in going from q = 5 to q = 11. These yields have been found to be weakly dependent on v_i, verifying a prediction of Briggs and Drepper that in the present velocity range, such electron-loss cross sections may tend to velocity independence. Significant departures (e.g., a factor of > 2 over the range 2.5-3.9 MeV/A) from this weak velocity dependence set in only for q = $\geq$ 11, just in the range where the ELC cross sections are dropping below the more strongly velocity-dependent ECC cross sections.

Coincidence experiments identifying the final projectile charge states allow allocation of the rich Auger structure observed among excited projectile states, which can thus be tagged according to parent ion charge state. The interpretation of these structures is presented elsewhere [21].

Generally speaking, there is much to be learned about basic collision dynamics from the coincidentally observed Auger spectra. For example, in experiments involving 18 MeV C^{4+} ions incident on Ar, Elston et al. [22] find it possible to intercompare absolute cross sections for single C K excitation, single C K ionization (sometimes accompanied by an excitation of a second electron), and double C K ionization (sometimes accompanied by an excitation of a third electron) from a unified set of spectra, all obtained under single-collision conditions.

Moreover, kinematics greatly aids the performance of such coincidence experiments. Owing to the high projectile velocity, the emitted electrons are folded into a forward cone, thereby aiding collection. Near zero deg, there is *zero* Doppler spread to first order, thereby reducing the need for very high resolution to overcome line blending problems. At the relatively high ion energies employed, the coincidence registration efficiency (including detection and collection efficiencies in the ion channel as well as electronic dead time and discriminator losses) can be made to exceed 85 to 90%. This high registration efficiency arises because the ion detection efficiency is high, and scattering angles are small. Coincidence data of useful statistical significance can be obtained with total ion currents in the 10^{-12} to 10^{-14} A range within a few hours.

As in the case of ECC, a word of caution about the FWHM of ELC distributions is in order. Though the ELC distributions are much more symmetric than the corresponding ECC ones, the predicted growth [18] in width $\propto v_i$ is *not* observed. Instead, for C^{q+}, O^{q+}, Si^{q+}, and Ar^{q+} traversing He, Ar at energies in the range 1-8.5 MeV/u, in numerous instances we find Γ to be only weakly dependent on v_i.

C. Convoy Electron Production in Solid Targets

"Convoy" electrons is the generic name [11], [23] that has been given to electrons that accompany ions as they emerge from solid targets. As remarked earlier, the ease of interchange of gaseous and solid targets in our apparatus while leaving all other geometrical and kinematic conditions fixed has permitted us to compare gaseous and solid target results under conditions where most systematic apparatus effects cancel to first order. A key parameter for comparison of theory and experiment is the full width at half maximum Γ_ℓ of the longitudinal velocity distribution of the electrons emerging in the forward direction. In the ECC theory of Dettmann et al. [4], Γ_ℓ can be expressed as $\Gamma_\ell = (3/2)v_i \Theta_0$, where v_i is the projectile velocity in au and Θ_0 is half of the acceptance angle, in radians, of the electron velocity analyzer. For ECC in gaseous targets, experimental results are only roughly consistent with this prediction (see the conclusion of Section A). For solid targets, some authors [9], [10], [24] observe Γ's that are essentially independent of the velocity of the projectile. These results are at odds with the ECC prediction and the assumption that the "last target layer" is the source of the continuum electrons. The properties of the distribution of continuum electrons coming from wake-riding (WR) states are predicted to be significantly different from those for ECC electrons. The predicted FWHM's of the longitudinal velocity distribution for wake-riding electrons are shown in Fig. 7; they depend on the target material and the atomic number of the projectile and are a decreasing function of the projectile energy. Results of our corresponding measurements are also shown in Fig. 7.

Results of our experiments with protons, oxygen, silicon, and nickel projectiles on polycrystalline C, Al, Ag, and Au targets all give a (somewhat

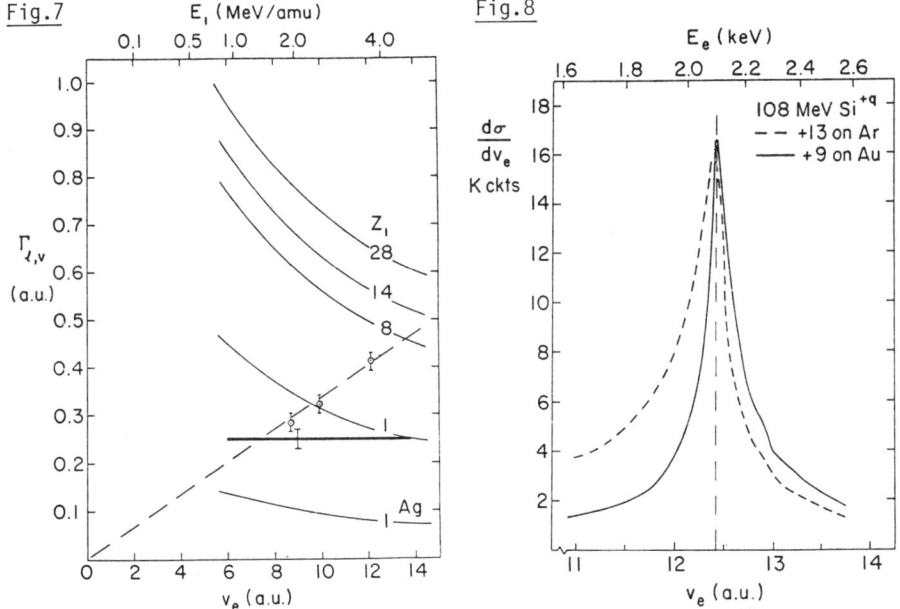

Fig. 7 The full width at half maximum of the longitudinal electron velocity distribution, Γ_ℓ, in au, for continuum electrons as a function of the electron velocity v_e, in au. The incident projectile energy, in MeV/u, appears at the top of the figure. The dashed line is the ECC prediction. The solid lines are the predictions of the WR theory for an Aℓ target with the indicated projectiles. The lowest solid curve marked Ag is the prediction of WR theory for protons incident on Ag. The experimental data for solids correspond to $\Gamma_\ell \simeq 0.25$ au and are represented by the heavy solid line. The open points represent the gaseous target (Ne and Ar) results for O^{8+} and Si^{14+}.

Fig. 8 The differential production cross section $d\sigma/dv_e$ as a function of the electron velocity, v_e, of continuum electrons emerging near 0° with respect to the ion beam, for 108 MeV Si^{13+} in Ar gas at 30 mTorr and for Si^{9+} on a 100 μg/cm² Au foil. The electron energy scale appears at the top of the figure. The mean emergent charge from the solid target is also 13.

geometry-dependent) FWHM of $\Gamma = 0.25$ au, but do *not* show the dependences on the variables (e.g., Z, v_i) discussed in present theories. The experimental results are clearly at variance with present theoretical considerations, although distinct differences in the cusp electron distribution are evident for solid compared to gaseous targets.

A typical velocity spectrum is shown in Fig. 8, for 108 MeV Si incident on Ar (gaseous) and Au (solid) targets. For solid targets the velocity distribution is narrower and exhibits a skew toward the high velocity side.

One shortcoming of the above convoy experiments (of about two years ago) is that the electron velocity distribution is averaged over all of the final projectile charge states. Guided by parallel experimental experience in our ECC studies, it became evident that a coincidence experiment specifying the

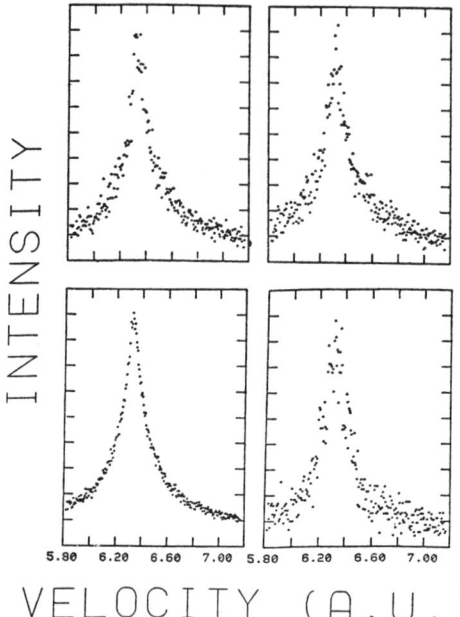

Fig. 9 The longitudinal velocity distribution of convoy electrons emerging from a solid when 12 MeV $^{12}_{6}C^{2+}$ is incident on a 40 μg/cm² Al target. The "singles" or non-coincident distribution is shown in the lower left-hand corner. The electron velocity distributions in coincidence with the final charge state of C^{6+}, C^{4+}, and C^{3+} are shown in clockwise arrangement starting in the upper left-hand corner. The vertical scale (intensity) is arbitrarily normalized for each spectrum.

relationship between the final charge state of the projectile and the electron velocity distribution and yield would clarify the situation. A typical experimental result is shown in Fig. 9, for 12 MeV C^{2+} ions incident on a 40 μg/cm² Al target. A total, or singles, spectrum is shown together with electron velocity spectra observed in coincidence with emergent C^{6+} ions (∼25% of the total), C^{4+} ions (∼20% of the total), and C^{3+} ions (∼1.2% of the total). The C^{5+} distribution accounted for 45% of the total and hence need not be shown. Lower final charge states were observed, but with intensities of less than 0.5% of the total. Hence reliable electron velocity distributions (i.e., distributions corresponding to a true-to-accidental ratio greater than 5) could not be obtained. Each particular coincident distribution required ∼ one to three hours of collection time. The sum of the electron spectra obtained in coincidence account for ≳ 92% of the total number of electrons observed, indicating a satisfactory ion collection geometry and a high detection efficiency for the electron multiplier used as a particle detector.

Inspection of Fig. 9 leads to the conclusion that no evident differences in the velocity distribution of the convoy electrons are observed as the final charge state of the projectile is changed. In fact, the FWHM is the same for all the final charge states of the projectile. Changing the incident energy of the projectile (from 12 MeV to 33 MeV carbon), the charge state of the projectile, the atomic number of the projectile (carbon, oxygen, and silicon [all at 1 MeV/amu]) or the target material (aluminum, carbon, and gold) does not alter the conclusion that to a good approximation the final convoy electron velocity distribution from solids is independent of the final projectile charge state. This statement also applies to crystalline targets. Recent theoretical considerations of Day [23] indicate a diffraction structure in the electron velocity distribution and a Z_1 dependence of the FWHM of approximately $Z_1^{1/4}$. No diffraction structure is observed in Fig. 9; however, it is unclear whether the present electron analyzer would be able to resolve the proposed structure.

The yield of convoy electrons may be estimated from an appropriate normalization and integration of the longitudinal velocity distribution. The measuring instrument normally has a solid angle of acceptance centered on zero deg, $\Delta\Omega$, and an energy resolution $\Delta E/E$ which fold the actual transverse and longitudinal velocity distribution characteristics. Hence if the width and shape of the longitudinal and transverse velocity distributions were known accurately, we could obtain the total electron yield from a measurement of the yield in the longitudinal direction. They are currently known only approximately; a corresponding Ansatz is needed to permit total yield estimation.

We represent the velocity distribution in the longitudinal ($||$) and transverse ($\perp$) directions by cusps of exponential shape, of half widths α and β, respectively. The total yield, in terms of the yield measured along the longitudinal direction, Y_R, is:

$$Y = Y_R \left\{ \frac{1}{1 - (1 + Cv_i)\exp(-Cv_i)} \right\}, \tag{1}$$

where v_i is the projectile velocity, $C = \frac{2\Theta_0 \ln 2}{\Gamma_t}$, and Θ_0 is the half-angle of the forward acceptance cone.

For solid targets we find $\Gamma_\ell = 0.25$ au, and infer $\Gamma_t = 1.8\,\Gamma_\ell = 0.45$ au, independent of the projectile velocity. (Here we have used the relationship $\Gamma_t \simeq 1.8\,\Gamma_\ell$, measured by Meckbach et al. in lower-energy ion-solid collision studies.)

For $\Theta_0 = 1$ deg, C takes the value 5.38×10^{-2}. (For solid targets we measure a higher fraction of the total electrons as the projectile velocity is increased than for ECC, where the corresponding cusps are wider.)

For completeness, the yield Y_R in the longitudinal direction can be expressed as:

$$Y_R = 1.6 \times 10^{-7} \frac{\bar{q}\,y}{Q} \frac{\Delta V}{k} \frac{E_i}{\Delta E_i} \frac{1}{E_i} \frac{1}{\Theta_0^2}, \tag{2}$$

where $\bar{q}$ is the mean charge of the collected ions, Q the charge in pC collected in the Faraday cup, E_i the energy of the convoy electrons in eV, $\Delta E_i/E_i$ the energy resolution of the analyzer, k the analyzer constant in V/eV, and ΔV the change in volts of the analyzer voltage for each step. For typical analyzer parameters $\Delta E_i/E_i = 1.4 \times 10^{-2}$, $k = 0.416$ V/eV and $\Theta_0 = 1.8°$, (2) reduces to:

$$Y_R = 8.5 \times 10^{-6} \frac{\bar{q}\Delta V y}{Q E_i} \frac{\text{electrons}}{\text{particle - deg}^2}. \tag{3}$$

Here y is the total number of electrons collected within the interval $(v_i - \beta)$ to $(v_i + \beta)$, where β was chosen as twice the full width at half maximum of the longitudinal electron energy distribution, Γ_ℓ. Numerically, $\beta = 0.5$ au, since Γ_ℓ is independent of the target material and of the charge, velocity, and atomic number of the projectile. The yield of convoy electrons,

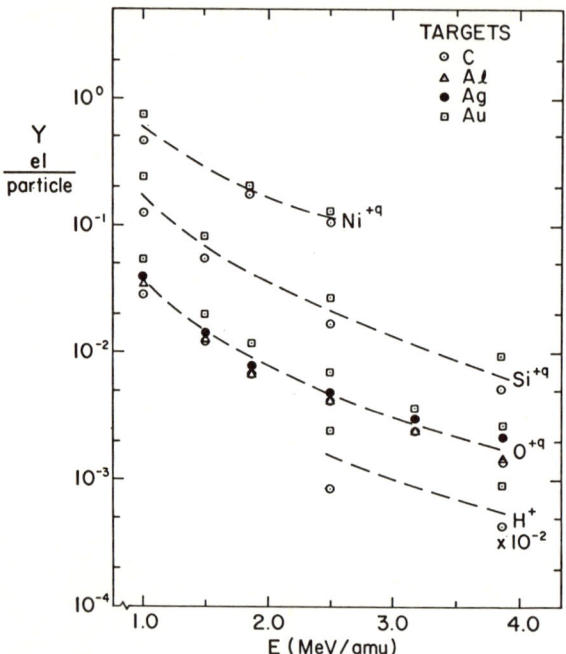

Fig. 10 The yield of convoy electrons prepared according to (1) and (4) for different target materials and different incident projectiles. The dashed lines are to guide the eye.

as given by (2) or (3), is independent of the incident charge state of the projectile and of the target thickness (the target thickness was changed by a factor of two), but does change for different target materials as well as with the energy and atomic number of the incident projectiles.

The data analyzed accordingly are shown in Fig. 10 for the various projectile-target combinations used as a function of the incident particle energy. In the interest of clarity, only the oxygen projectile data is shown for all of the targets. For the H^+, Si^{q+}, and Ni^{q+} particles, only the C and Au target data are shown. The near linearity and parallelism of the similar target curves for each projectile shown in Fig. 10 suggest a common velocity dependence for the various projectiles, and an approximate velocity-independent Z_1 dependence. Indeed, we find that we can summarize all of our results by the empirical equation

$$Y = 1 \times 10^{-4} \, C(Z_2) \, Z_1^m \, E_1^{-n}, \qquad (4)$$

where Z_1 is the atomic number of the incident particle of energy E_1 in MeV/amu; $m = 2.75 \pm 0.2$; $n = 2.25 \pm 0.1$; and $C(Z_2)$ is a constant depending on the target material: $C(Au) = 1.\overline{6}5$, $C(Ag) = 1.25$, $C(Al) = C(C) = 1.0$. All the values have estimated uncertainties ± 0.15.

To our knowledge, the only other measurement of the yield of convoy electrons from solids is by Dettmann et al. [4] for 225 keV protons on carbon. Eq. (4) predicts a result within 30% of their measurements. Meckbach et al. [9] report relative yield measurements for 200-500 keV protons on carbon and find an energy dependence of $n \simeq 3.1$, which is in reasonable agreement with our results.

D. Convoy Production Measured in Coincidence with Emergent Ion Charge

When we examine the yield of convoy electrons in coincidence with a particular final charge state of the projectile, we find the curious result [24] that the convoy electron yield, per coincidently registered projectile, is independent of the final projectile charge state. That is, the fraction of electrons coincident with a particular final charge state is just mirrored by the fraction of projectiles having that particular charge state. If the atomic number and energy of the incident projectile or the target material is changed, we find that each coincident electron yield varies in the same proportions as the non-coincident convoy electron yield.

These results strongly mitigate against the surface production of convoy electrons and suggest the bulk of the material as the origin of these electrons. To further explore this possibility, we utilized O^{6+}, O^{7+}, and O^{8+} beams traversing a Au single-crystal target and employed channeling techniques, where it is known that incident ion charges often remain intact (frozen) throughout their entire passage (e.g., $\sim 80\%$ of 2.5 MeV/u O^{8+} ions traversing the <110> direction will remain 8+ throughout a 300 $\mu g/cm^2$ gold crystal). As mentioned previously, the shapes of the velocity distribution for channeled ions obtained in coincidence with the final projectile charge state are indistinguishable from those obtained with polycrystalline targets or from non-coincident (singles) distributions. However, in the channeling data, the yields which were found to be independent of the final projectile charge state for polycrystalline targets exhibit instead a strong diminution, dependent on incident as well as exit charge state. Table I presents in matrix form the yields per emergent ion for various incident, q, and exit, q_e, charge state pairs, together with the corresponding measured charge state fractions and coincident fraction values. It is immediately evident that convoy production for well-channeled ions is much suppressed, with the greatest suppression arising in the most open channel <110>. It is also evident that the highest yields occur when distances of approach to lattice sites is likely to be closest. The yield entries in Table I can be surprisingly well reproduced by postulating two classes of channeled ions, those that approach close enough to a lattice site (effective distance ~ 0.65 Å) to produce a convoy with high probability (equal to the random or polycrystalline rate) and those which remain at larger separations, producing convoys with negligibly low probability. Two immediate tests of this extreme assumption are well verified. First, the number 0.65 Å explains both the <110> and the <100> yield figures. Second, it happens that for Au atoms the 6s and 5d electrons have kinetic energies < 10 eV. But the 5p and 4f electrons - which have binding energies of ~ 250 to 460 eV and are therefore far more efficient at contributing to capture according to the Bohr $v_e \sim v_i$ matching criterion - have mean radii ranging from 0.60 to 0.28 Å. Therefore, the "magic" distance of 0.65 Å can be assigned a most plausible physical interpretation.

Table I Convoy electron yield (%) per emergent ion, for O^{8+} incident at 2.4 MeV/u on Au in the <110>, <100>, and random directions. The yield is normalized to the measured random yield of $\sim 3.8 \times 10^{-4}$ electrons/ion. The number in parentheses is the fraction (%) of emergent ions in state q_e.

q in	q_e out	8+		7+		6+	
8+	Y<110>	21	(68)	39	(28)	82	(4)
	Y<100>	37	(59)	58	(35)	79	(6)
	Y<Rand>	100	(26)	100	(59)	100	(15)
7+	Y<110>	29	(42)	24	(51)	58	(7)
	Y<100>	37	(52)	47	(42)	71	(6)
	Y<Rand>	100	(25)	100	(60)	100	(15)
6+	Y<110>	37	(31)	29	(42)	21	(27)
	Y<100>	39	(49)	45	(42)	47	(9)
	Y<Rand>	100	(27)	100	(57)	100	(16)

E. Model for Convoy Electron Production in Solids

The experimental evidence to date allows us to construct the following model for convoy electron production in solids. Convoy production is initiated in a close collision, most probably a single or multiple-electron capture event ($\sim 10^{-17}$ cm² capture cross section [13]), which, according to the work of Brown et al. [25], is dominated by electron capture to excited states ($\gtrsim 90\%$ of the time). The excited states most copiously populated [25] have n = 2 and 3, with populations of higher states possible at the $\gtrsim 10\%$ level. Because binary encounter theory predicts that e⁻ loss cross sections scale approximately as n², and because of the peak width and symmetry considerations noted above, we suggest that convoy production is initiated by capture(s) followed *immediately* by electron loss to the continuum. If the mean free path for such dual event convoy production were, say, $\gtrsim 250$ Å, then several convoys per projectile would be produced in a solid a few thousand Å thick. Subsequent electron-electron scattering (elastic and inelastic) leads mainly to scattering into a wide range of angles, effectively extinguishing the convoy population, though it is possible that secondary elastic scattering would cause some unknown degree of repopulation. In any event, since the escape depth of ~ 1 keV electrons is ~ 10-20 Å, the net production of several convoy electrons per emergent ion is depleted by electron scattering to $\sim 10^{-2}$ to 10^{-4} observable electrons per ion.

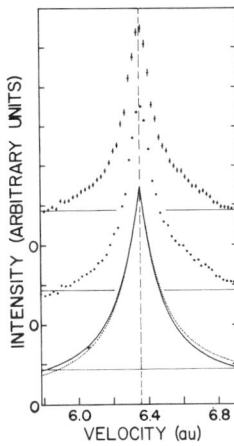

Fig. 11 Spectrum of convoy electrons emergent near 0 deg from 16 MeV O^{3+} ions traversing a 30 μg/cm^2 C foil. The upper data points are obtained from the raw spectrum (lower points), through a correction factor ($v_e^{-1.6}$) to account for the electron escape depth velocity dependence estimated. The lower curves represent respective fitted cusp shapes, which better display the degree of symmetrization produced.

The skew of the electron velocity distributions toward electron velocities $v_e > v_i$ can be qualitatively and quantitatively explained by considering the velocity dependence of the electron scattering cross section in solids [26]. We correct the observed cusp shape by a velocity-dependent factor ($\sim v^{-1.6}$) reflecting the exponential attenuation resulting from electron scattering within the bulk. The result of this procedure is shown in Fig. 11 for 16 MeV O^{3+} traversing carbon. The resultant symmetric, peak-normalized curve is similar to experimental electron loss cusps from C^{q+} and O^{q+} traversing Ne and Ar as shown in Figs. 5 and 6 (except for the additional low-energy Auger lines which appear in the cusp wings). Since no fitting procedure beyond overall peak normalization is used, the ability to quantitatively symmetrize the skewed peak is viewed as support for the bulk production of convoy electrons, a small fraction of which escape through the surface.

Despite the many appealing features of the model we have described, there remains an enigma. Since binary electron capture and loss processes are both Z- and q-dependent, it seems reasonable to assume a similar dependence for convoy production. Since 10^{-17} cm^2 charge changing cross sections appropriate to the ion velocities used [13] imply $\sim$ 200 Å mean free paths for capture and loss at solid target densities, it is then difficult to account for the complete destruction of correlation between convoy production and emergent ion charge, if the convoys observed indeed originate in the final $\sim$ 10 Å of target passage. What destroys the correlation so effectively? Is it possible that the mean free path for free electron scattering is very small compared to that for convoy electron scattering owing to the strong drag provided by the projectile ion? Is it possible that there is rapid capture and loss of excited electrons with mean free paths $\lesssim$ 10 Å which re-equilibrate the charge in the last layers? Is it possible that shielding of projectile K-shell electrons by the solid medium is so complete that convoy production is q-independent? What other models do as well at accounting for the shape and channeled ion yield data as the model proposed while avoiding this enigma? It seems clear, substantial clues to a complete understanding of convoy production are provided by the present data. There remain, however, perplexing problems in understanding the unexpected lack of correlation of shape and yield with q_e.

ACKNOWLEDGEMENTS

A number of colleagues have contributed greatly to the carrying out of the experiments described, ably aided by generous help from staff at Oak Ridge National Laboratory, Lawrence Berkeley Laboratory, and Brookhaven National Laboratory. The most substantial contributions - also those sustained over the longest period of time - have been provided by Charles R. Vane (with whose able colleagues Professor Richard Marrus and Dr. Harvey Gould a joint collaboration at Lawrence Berkeley Laboratory has been possible) and by my faculty colleague Stuart B. Elston. Marianne Breinig has played a leading role in recent analysis and interpretation of the data. Substantial contributions have also been made by our other University of Tennessee-Oak Ridge National Laboratory colleagues Gerald Alton, Leif Liljeby, Sven Huldt, and Robert Thoe. Thoe's design and fabrication of our most recent electrostatic analyzer (replacing one kindly lent us by M. O. Krause), together with Scott Berry's design and fabrication of our magnetic analyzer, have been essential to the work. Finally, the continued collaboration of one of us (R. Laubert) has been aided by the East Carolina University Research Foundation and by Oak Ridge Associated Universities.

REFERENCES

1. M. E. Rudd, J. H. Macek: Case Studies in Atomic Physics 3, 125 (1972)
2. R. Shakeshaft: Phys. Rev. A 18, 1930 (1978)
3. G. B. Crooks, M. E. Rudd: Phys. Rev. Lett. 25, 1599 (1970)
4. K. Dettmann, K. G. Harrison, M. W. Lucas: J. Phys. B 7, 269 (1974)
   K. G. Harrison, M. W. Lucas: Phys. Lett. 33A, 142 (1970), and 35A, 402 (1971)
5. R. W. Cranage, M. W. Lucas: J. Phys. B 9, 445 (1976)
6. R. Strong, M. W. Lucas: Phys. Rev. Lett. 39, 1350 (1977)
7. W. Steckelmacher, R. Strong, M. N. Khan, M. W. Lucas: J. Phys. B 11, 2711 (1978)
8. M. G. Menendez, M. M. Duncan: In *Beam-Foil Spectroscopy*, ed. by I. A. Sellin and D. J. Pegg (Plenum, New York 1976), Vol. 2, p. 623
   M. G. Menendez, M. M. Duncan: Phys. Lett. 54A, 409 (1975); Phys. Rev. A 13, 566 (1976); Phys. Rev. Lett. 56A, 177 (1976); and Phys. Rev. Lett 40, 1642 (1978)
   M. G. Menendez, M. M. Duncan, F. L. Eisele, B. R. Junker: Phys. Rev. A 15, 80 (1977)
   M. M. Duncan, M. G. Menendez, F. L. Eisele, J. Macek: Phys. Rev. A 15, 1785 (1977)
   M. M. Duncan, M. G. Menendez: Phys. Rev. A 16, 1799 (1977)
9. W. Meckbach, K. C. R. Chiu, H. H. Brongersma, J. W. McGowan: J. Phys. B 10, 3255 (1977), and references therein
   W. Meckbach, N. Arista, W. Brandt: Phys. Rev. Lett. 65A, 113 (1978)
   W. Meckbach, V. Ponce, to be published in Comments in Atomic and Molecular Physics
10. K. C. R. Chiu, J. W. McGowan, J. B. A. Mitchell: J. Phys. B 11, L117 (1978)
11. W. Brandt, R. H. Ritchie: Phys. Rev. Lett. 62A, 374 (1977)
12. R. Shakeshaft, L. Spruch: Rev. Mod. Phys. 51, 369 (1979), and many references therein
13. J. R. Macdonald, F. W. Martin: Phys. Rev. A 4, 1965 (1971)
14. For reviews see H. D. Betz: Rev. Mod. Phys. 44, 665 (1972), and V. P. Shevelko, Z. Phys. A 287, 19 (1978)

15. I. A. Sellin: Journal de Physique, Colloque C1, 40, C1-225 (1979)
    C. R. Vane, I. A. Sellin, M. Suter, G. D. Alton, S. B. Elston, P. M. Griffin, R. S. Thoe: Phys. Rev. Lett. 40, 1020 (1978)
16. N. Stolterfoht, D. Schneider, P. Ziem: Phys. Rev. A 10, 81 (1974)
17. M. Rødbro, F. D. Andersen: In *Abstracts of the Tenth International Conference on the Physics of Electronic and Atomic Collisions*, ed. by G. Watel (Comissariat à l'Energie Atomique, Paris 1977), p. 1012; and J. Phys. B 12, 2883 (1979)
18. F. Drepper, J. S. Briggs: J. Phys. B 9, 2063 (1976)
    J. S. Briggs, F. Drepper: J. Phys. B 11, 4033 (1978)
19. J. Macek: Phys. Rev. A 1, 235 (1970)
20. A. Salin: J. Phys. B 2, 631 and 1255 (1969), and 5, 979 (1972)
21. M. Suter, C. R. Vane, S. B. Elston, G. D. Alton, P. M. Griffin, R. S. Thoe, L. Williams, I. A. Sellin: Z. Phys. A 289, 433 (1979)
22. S. B. Elston: In *Proceedings of the X-80 Conference*, Stirling, Scotland, August 25-29, 1980, to be published
23. M. H. Day: Phys. Rev. Lett. 44, 752 (1980)
24. R. Laubert, S. Huldt, M. Breinig, L. Liljeby, S. Elston, R. S. Thoe, I. A. Sellin: Submitted for publication to J. Phys. B
    R. Laubert, I. A. Sellin, C. R. Vane, M. Suter, S. B. Elston, G. D. Alton, R. S. Thoe: Nucl. Inst. Meth. 70, 557 (1980)
    R. Laubert, I. A. Sellin, C. R. Vane, M. Suter, S. B. Elston, G. D. Alton, R. S. Thoe: Phys. Rev. Lett. 41, 712 (1978)
25. M. D. Brown, L. D. Ellsworth, J. A. Guffey, T. Chiao, E. W. Pettus, L. M. Winters, J. R. Macdonald: Phys. Rev. A 10, 1255 (1974)
26. J. C. Ashley, C. J. Tung, R. H. Ritchie: Surface Science 81, 409 (1979), and references therein

New Experiments with Electron Capture Spectroscopy at Ni, Gd and Cr Surfaces

Carl Rau and Simon Eicher
Sektion Physik der Universität München
D-8000 München 40, Fed. Rep. of Germany

1. Introduction

The electron spin polarization ESP at ferromagnetic metal surfaces can be investigated by various experimental methods. Here we report on experiments performed with electron capture spectroscopy ECS at surfaces of oligatomic epitaxial Ni(100)-films, of polycrystalline thin films of Gd, of bulk Gd and of bulk Cr(100)crystals. The results are that the ESP at surfaces of oligatomic (2 - 64 layers) epitaxial Ni(100)-layers depends on the Ni-layer thickness. Already two atomic layers Ni(100) on Cu(100) substrate crystals show ferromagnetic behaviour. This finding clearly excludes the existence of so-called "intrinsic magnetic dead layers" at Ni(100) surfaces. Measuring the ESP at atomically clean surfaces of bulk and thin film Gd material it is found that the ESP depends drastically on the temperature T. The T-dependence of the ESP is significantly different from bulk behaviour. Nonzero ESP values above the bulk Curie temperature T_{Cb} = 292 K demonstrate that ferromagnetic order at the surface exists beyond T_{Cb}. ECS measurements at surfaces of antiferromagnetic bulk Cr(100) crystals clearly show that the topmost atomic layer is ferromagnetic.

2. Experimental

The experimental set-up and the technique of the ECS method are described in [1-3]: Spin-polarized electrons are captured by 150 keV deuterons during grazing angle reflection at magnetized surfaces of ferro- and antiferromagnetic materials. The distance of closest approach of the ions is at about 2 Å for a reflection angle of 0.2°. The maximum interaction length of the ions with the surface is a few hundred atomic distances. After an adiabatic transition the nuclei of the neutralized ions become also polarized in the field of the captured polarized electrons. The nuclear polarization then serves as a measure of the ESP. The measurements are performed in UHV at $5 \cdot 10^{-11}$ mbar. The Ni(100)-films (2-64 layers) were prepared on Cu(100) and NaCl(100) substrate crystals using electron beam evaporation at $1 \cdot 10^{-9}$ mbar. The absence of holes in these oligatomic films and the homogeneity of these layers was measured by monitoring the Ni- and Cu-Auger electron peak heights as function of the Ni-layer thickness as described in [4,5]. The monocrystalline state of the surface was checked by LEED investigations. The Gd samples were prepared by electron beam evaporation of Gd from cooled Gd material on Cu and Gd bulk substrates. Atomically clean and flat surfaces of Cr(100) crystals were prepared by standard preparation techniques as described in [1-3]. The deviation of the geometric surface orientation from the respective crystallographic surface orientation was less than $0.01°$. The temperature of the samples was kept constant within $0.03°$ using an automatic control device. The absolute temperature of the samples was measured within $o.5°$. The cleanness of the investigated surfaces was checked by AES-measurements. C and O coverages were less than 1/100 monolayer.

3. Results and Discussion

Fig.1 gives the experimental results for the ESP at surfaces of thin epitaxial Ni(100)-layers [6].

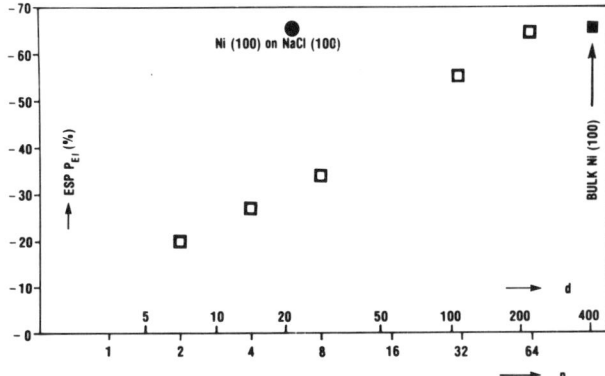

Fig.1 ESP at surfaces of n monolayers Ni(100) on a Cu(100) substrate (□) (d = layer thickness in Å)

The ESP is measured at room temperature . The ESP at the surface of two atomic layers Ni(100) on Cu(100) amounts to -19%. This finding clearly excludes the existence of so-called "intrinsic magnetic dead layers" at Ni(100)-surfaces. Increasing the thickness of the Ni(100)-layers on Cu(100) up to 64 layers the ESP increases up to -65%. This ESP value is close to the surface-ESP of bulk Ni(100) which amounts to -64% [3]. This thickness effect is not present using Ni(100)-layers on NaCl(100) substrate crystals. 7 atomic layers Ni(100) on NaCl(100) yields and ESP of -64% (see Fig.1). Therefore the observed reduction of the ESP at the free Ni(100)-surface using Cu(100) substrate crystals might be caused by the influence of non-polarized Cu-electrons.

Unfortunately, at present there are no theoretical calculations available on the ESP of the Ni(100)/Cu(100) system. We note, however, that first theoretical calculations on the ESP at the surface of a 9 layer thick unsupported Ni(100)-film are available [7]. From these self-consistent calculations the surface-ESP amounts to -57%. This value is close to the ESP at the surface of 7 atomic layers Ni(100) on NaCl(100) which amounts to -64%. We remark that in these calculations [7] the influence of surface states on the ESP is taken into consideration.

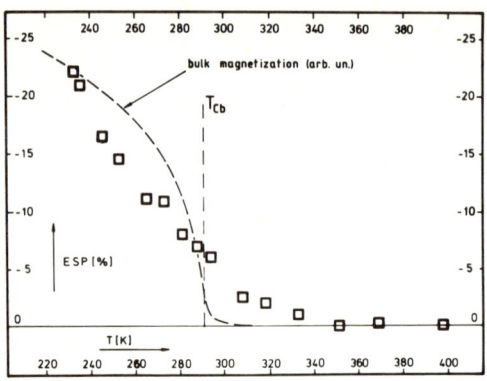

Fig.2 ESP at Gd-surfaces as function of temperature T
(magnetizing field : 47.76 kA/m)

Fig.2 gives experimental results for the ESP at surfaces of
5000 Å thick polycrystalline Gd-films which were evaporated on
polycrystalline Cu substrates and magnetized in a magnetic field
of H= 47.76 kA/m. The dashed line in Fig.2 shows the temperature
dependence of the bulk magnetization of Gd [8]. We note three
interesting aspects of these ESP measurements at Gd-surfaces :
1. The existence of a nonzero ESP below the bulk Curie temperature T_{Cb} = 292 K is in accordance with the model of indirect
exchange (RKKY theory) and with ferromagnetic band structure
calculations of Harmon and Freeman [9] which also predict a nonzero ESP of the conduction electrons. 2. The temperature dependence of the surface-ESP is significantly different from
bulk behaviour. 3. The surface-ESP of Gd is unequal to zero at
and above T_{Cb}.
Fig.3 shows the H- and T-dependence of the ESP at polycrystalline
surfaces of bulk Gd (5000 Å Gd evaporated on bulk Gd material).
From the H-dependence of the ESP at T_{Cb} = 292 K we deduce using
a linear extrapolation for the ESP at H = 0 a polarization of
-(4.6±0.4) % (see triangle in Fig.3a). This indicates that the
surface Curie temperature T_{Cs} is beyond T_{Cb}. Fig.3b gives the
T-dependence of the ESP near T_{Cb} for a magnetizing field of
H = 47.76 kA/m. From a linear extrapolation we get zero ESP at
T_{Cs}= 311 K.

Perfect antiferromagnetic Cr consists of two compensating ferromagnetic sublattices producing zero bulk magnetization. Since
the spin structure in the (100)-direction is sinusoidal one
might expect ferromagnetic behaviour of the topmost surface
layer of Cr(100). We remark that it is also theoretically shown
that the topmost (100)-surface plane of antiferromagnetic Cr has
a ferromagnetic behaviour [10]. We have investigated the ESP at
surfaces of Cr(100) at T = 293 K which is below the Néel temperature T_N = 310 K . Our preliminary results yield an ESP of
- (18±2) % which clearly shows that the Cr(100)-surface is ferromagnetic. At T = 365 K zero ESP is detected.

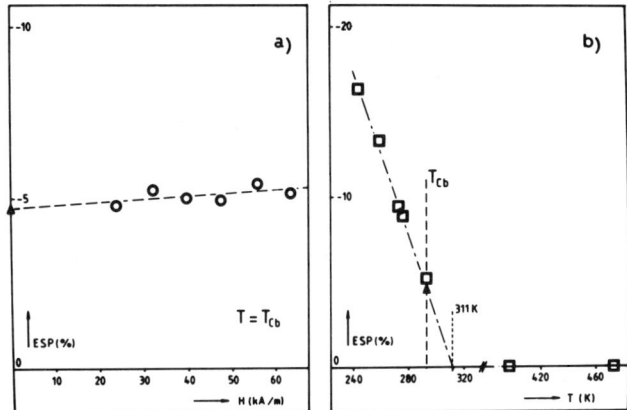

Fig.3 (a): H-dependence of the ESP at Gd-surfaces at T= T_{Cb} = 292 K; (b): T-dependence of the ESP at Gd-surfaces magnetized in a magnetic field of H = 47.76 kA/m (triangle: see Fig.3a)

This research is supported by the "Bundesministerium für Forschung und Technologie".

References

[1] C. Rau and R. Sizmann, Phys. Lett. 43A,317(1973)
[2] C. Rau and R. Sizmann, in *Atomic Collisions in Solids*, Vol.1 ed. S. Datz et. al. (Plenum Press, N. Y.) p. 295(1975)
[3] C. Rau, Comments on Solid State Physics 9,177(1980)
[4] M. P. Seah, Surf. Sci. 32,703(1972)
[5] C. C. Chang, Surf. Sci. 48,9(1975)
[6] G. Eckl, Diplom Thesis, University of Munich (1979)
[7] C. S. Wang and A. J. Freeman, to be published
[8] H. E. Nigh, S. Legvold and F. H. Spedding, Phys. Rev. 132 1092(1963)
[9] B. N. Harmon and A. J. Freeman, Phys. Rev. B10,1979(1974)
[10] G. Allan, Surf. Sci. 74,79(1978)

Electronic Excitation in Ion-Molecule Collisions

Sheldon Datz
Chemistry Division, Oak Ridge National Laboratory
Oak Ridge, TN 37830, USA

1. Introduction

Attempts to connect phenomena occurring in gas phase single collision processes with those taking place on the surface of a solid have led in the past to many fruitful insights. Almost all of the connections made have been with information obtained from two body ion-atom collisions via an intuitive leap to a solid which in some cases may be treated as a molecule with an almost infinite number of atoms. Here we will attempt to introduce some information obtained from single collision processes involving simple molecules and see if this can add useful concepts in dealing with the solid surface. In some cases the connection will be obvious while in some others experiments with solid surfaces analogous to gas phase experiments have not yet been carried out. The discussion is subdivided into three parts: first, high energy collisions where the primary event is predominantly a binary encounter; second, low energy collisions where interactions with the whole molecule must be taken into account; and finally, a comparison of the behavior of positive ion collisions with recently developing studies in the collisions of negative ions with molecules.

2. High Energy Collisions

For violent collisions at high energies the range of interaction is almost always very short compared to interatomic distances in molecules or lattice spacings in solids. For these collisions there is good justification in assuming binary atomic collisions, but the presence of adjacent atoms in the target can cause major effects because the time between sequential binary collisions is so short that, e.g., electronically excited states persist. Thus, one of the molecular (or solid state) effects is the preparation of excited states in the projectile or target which can then be affected by a collision in close time proximity.

Another proximity effect is that of position correlation. In a solid for every collision that occurs with impact parameter greater than half a lattice spacing there must also be a collision with an atom with an impact parameter of less than half a lattice spacing. On the other hand, deflection of a projectile ion by one atom in a target molecule (or solid) can lead to a shadowing effect for a subsequent collision. Both of these effects are well known in Ion Scattering Spectroscopy (ISS) from surfaces, e.g., "double scattering" events which are not fundamentally different from scattering from a diatomic molecule are known to give greater neutralization of the backscattered rare gas ion than "single scattering" events.

In one of the first experiments to demonstrate that scattering by surface atoms could indeed be described by a two particle mechanism, DATZ and SNOEK [1] bombarded a Cu single crystal surface with 50-100 keV Ar$^+$ ions. They indeed found that the energies of the scattered particles followed binary collision mechanics (see Fig.1). Actually, the main point of the experiments was to investigate electronic excitation in such collisions and how it compared with binary collisions in gases. The total electronic inelasticity [2] and the degree of final ionization of the scattered particles was studied as a function of impact parameter. It was found that, although the total inelasticity was the same, the degree of ionization (i.e., the charge fraction of highly charged particles) was much lower for a surface target atom than for a gas target atom for collisions of equivalent violence. This discrepancy was shown by VAN DER WEG and BIERMAN [3] to be due to charge capture from adjacent atoms in the surface and to be sensitive to the distance to the adjacent surface atom.

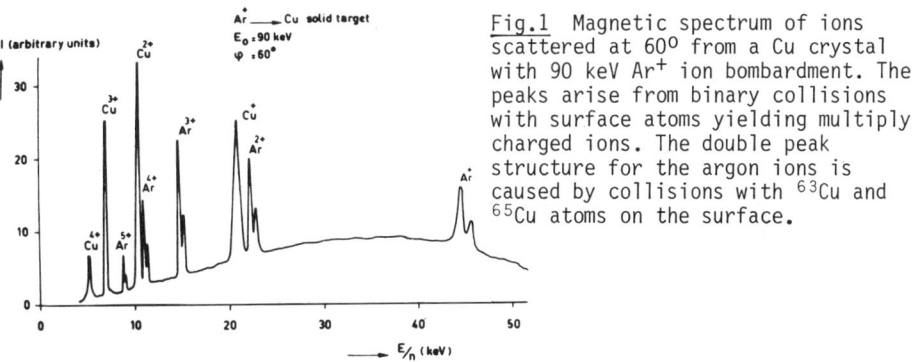

Fig.1 Magnetic spectrum of ions scattered at 60° from a Cu crystal with 90 keV Ar$^+$ ion bombardment. The peaks arise from binary collisions with surface atoms yielding multiply charged ions. The double peak structure for the argon ions is caused by collisions with ^{63}Cu and ^{65}Cu atoms on the surface.

Such charge capture effects should also occur in ion molecule collisions and can be seen in the results of the work by KAUFFMAN, et al., [4] on inner-shell ionization of Si in solid silicon and gaseous SiH$_4$ by ∼50 MeV Cl ions and by HOPKINS, et al. [5] who extended the work to include SiF$_4$ gaseous targets. Examination of the high resolution Si K$_\alpha$ x-ray spectrum from 45 MeV Cl ion collisions with solid Si showed a typical case of multiple L shell ionization accompanying K shell ionization which is found in heavy ion bombardment. The spectrum shown for solid Si in Fig.2 shows lines arising from 2p-1s radiative transitions. When a single L electron is also missing, the energy of the transition is somewhat increased; this state is designated as (KL^1). When 2 L electrons are missing (KL^2) the energy is even higher, etc. This spectrum therefore indicates (neglecting differences in fluorescence yield) that the most probable number of L vacancies created in the same collision which created the K vacancy is ∼ 2.5. For the SiH$_4$ target, however, the spectrum showed that in addition to L shell vacancies all of the M (valence) shell electrons had been removed in the K vacancy forming collision. The degree of inner shell ionization created in both cases should be the same but in the case of solid Si, electrons from the adjacent Si atoms had filled the M shell. HOPKINS, et al. [5] extended this work to include SiF$_4$; comparison of the Si K$_\alpha$ x-ray spectra obtained by collision of 53.4 MeV Cl ions with SiH$_4$ and SiF$_4$ is shown in Fig.3. From the shifts in the satellite spectrum it can be seen that substantial M shell relaxation has occurred by shifting electrons from the fluorine atoms remaining in the vicinity.

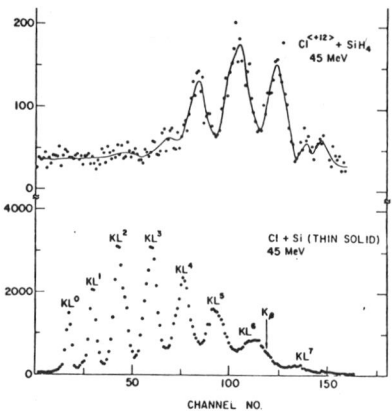
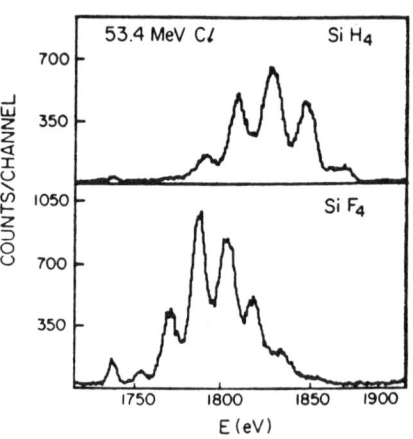

Fig.2 Silicon K-α x-ray satellite spectra obtained from the bombardment of solid Si and gaseous SiH$_4$ by 45 MeV Cl ions.

Fig.3 Silicon K-α x-ray satellite spectra obtained from the bombardment of gaseous SiH$_4$ and gaseous SiF$_4$ with 53.4 MeV Cl ions.

Another molecular effect of innershell ionization was found by SARIS, et al. [6] who investigated differences in K x-ray production and x-ray spectra in Ar$^+$(0.25 to 2.2 MeV) collisions with Ar, HCl, and Cl$_2$ targets. In particular, they found a large enhancement of "molecular orbital" (MO) x-rays when Cl$_2$ was used as a target rather than HCl. An MO x-ray is emitted when an ion containing an innershell vacancy collides with a target atom. During the collision the vacancy follows a molecular orbital whose energy depends upon the internuclear spacing. If the vacancy is relaxed during the collision, an x-ray with energy corresponding to the MO energy level spacing is emitted. In the particular case investigated, Ar 2p vacancies created in the collision with the first Cl atom led, in collision with the second Cl atom, to: promotion of Cl-K electrons via 2pπ-2pσ rotational coupling; the promotion of Ar-K electrons via 2pπ-1sσ rotational coupling; and to the production of MO x rays. None of these processes were possible in the Ar and HCl targets. (The existence of steady state inner shell vacancies in ions penetrating solids is well documented (see, e.g., [7]). In this case the proximity effect is predominant and leads to higher excitation and ionization. Such processes could also occur in double surface scattering events but this possibility has not yet been investigated.

In recent experiments reported by ALTON, et al. [8], single and multiple electron loss cross sections for 20 MeV Fe^{4+} in atomic and molecular gaseous targets were measured. Effects due to both shadowing and multiple collision effects were observed. A comparison of the total cross sections for single, double, triple,... electron loss in various rare gas targets is shown in Fig.4. Each electron loss cross section shows the anticipated monotonically increasing dependence on target Z. However, when molecular targets made up of approximately neon-like atoms were used, quite a different effect was obtained (see Fig.5). As the number of atoms in the molecule increases there is a significant decrease in the cross section per atom for one, two, and three electron loss and an increase for five and six electron loss.

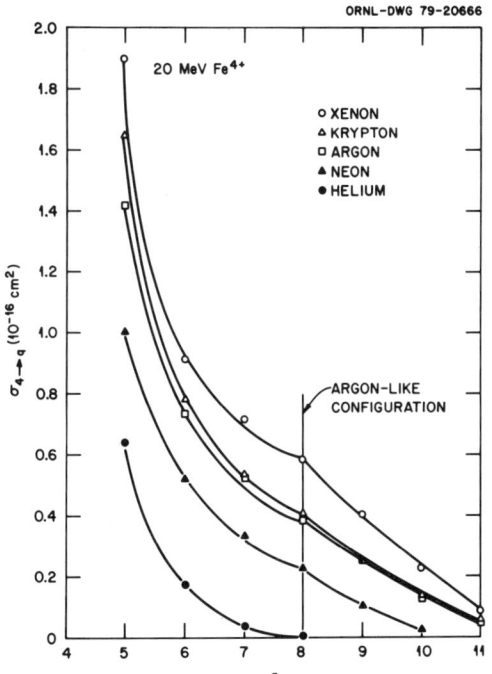

Fig.4 Cross sections for multiple electron loss of 20 MeV Fe^{4+} in various inert gases vs charge state following collision (e.g., q = 7 corresponds to 3 electron loss).

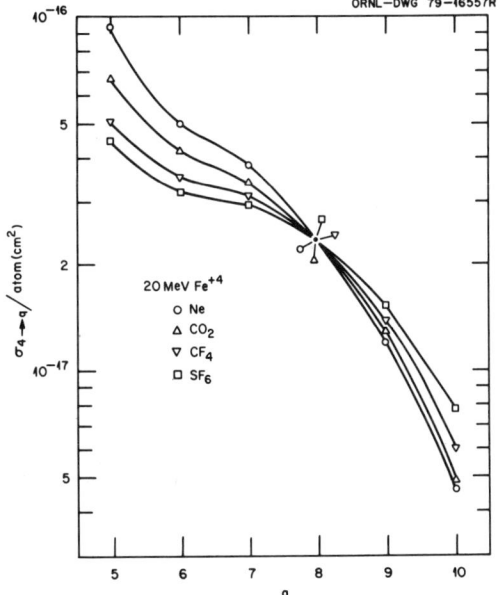

Fig.5 Cross sections for multiple electron loss of 20 MeV Fe^{4+} in Ne and various molecular gases.

The physical picture invoked to explain the data is as follows. Single electron loss occurs at predominantly large impact parameters; for many orientations of the molecule, large impact parameters with respect to one of the constituent atoms requires a small impact parameter with another atom in the molecule, hence a reduction in the single loss cross section by shadowing and correlated interaction effects. The picture changes for the smaller impact parameter events which lead to a high degree of ionization. Here multiple electron loss may be enhanced by collisional excitation followed by a second collision within the molecule before radiative decay takes place. In this aspect the result is similar to that of SARIS, et al. [6] mentioned above.

3. Low Energy Collisions

For collisions at low energies (<<10 keV) energy transfer to the entire molecule as an entity must be taken into account. Unlike ion-atom collisions where the only inelastic process is electronic excitation, ion molecule collisions can also involve rotational and vibrational excitation, collisional dissociation, or chemical reaction. Often these processes occur simultaneously and a proper description requires the use of multidimensional hypersurfaces rather than the simple one-dimensional potential curves used in ion-atom scattering. If, for example, a collision occurs in which a diatomic molecule is electronically excited, there is a manifold of vibrational states of the excited molecule which may be populated (vibronic excitation). If the collision is rapid compared to motion in the vibrational coordinate, the vibrational levels of the excited state formed will be determined by Frank-Condon factors in much the same manner as excitation by electron-molecule collisions. If the collision is slower, re-adjustment of molecular internuclear distances will cause excitation in other parts of the vibronic manifold.

It is clear that even a cursory survey of all of the processes which can occur in ion-molecule collisions is too extensive to be covered in this paper. Instead we will present some selected cases which serve to illustrate some of the simpler parts of the interactions.

The experiments which will be discussed all fall into the classification of "translational spectroscopy," i.e., the measurement of the energy loss of a beam of projectiles scattered from an almost stationary target as a function of initial kinetic energy E and total scattering angle θ. In this sense they are similar to a large class of experiments involving low energy ion beam scattering from solid surfaces.

3.1 Energy Loss Scaling in Low Energy Ion-Diatomic Molecule Scattering

The treatment of even the simplest collision dynamics in ion-diatomic molecule systems requires a knowledge of whether one is dealing with collisions with the molecule as a whole or whether the interaction can be treated as a binary collision with a single atom.

Consider a collision between a projectile C of mass M_p and a homonuclear diatomic molecule AB of mass $M = M_A = M_B$. The momentum transfer is

$$\vec{P} = \vec{P}_A + \vec{P}_B.$$

The amount of energy transferred to internal excitation of the molecule is the sum of the kinetic energies of A and B in the fixed molecular frame

$$Q = [2(P_A^2 + P_B^2) - P^2]/4M.$$

In the case of a pure binary encounter between A and C (high velocity) the momentum transfers are $P_A = P$ and $P_B = 0$, hence $Q = P^2/4M$. The transfer from the projectile to internal degrees of freedom then takes place by elastic transfer to one atom of the molecule which then is transferred into vibro-rotational energy of the molecule and translation of the center-of-mass. In this case for small scattering angle θ

$$\Delta E_{bin} \simeq (M_p/M)E_0\theta^2 \tag{1}$$

A second limiting case is a purely elastic collision with the entire molecule (low velocity). Here

$$\Delta E_{el} \simeq (M_p/2M)E_0\theta^2. \tag{2}$$

The dependencies of (1) and (2) suggest that $E\theta^2$ would be an appropriate scaling function and indeed its use is widespread. However, for cases intermediate between these two the scaling does not work.

In general, a molecular target will give an energy loss distribution corresponding to a distribution of orientation and impact parameters, and SIGMUND[9] has recently developed an expression for the peak position for this distribution.

$$\Delta E = (M_p/M)E\theta^2 f(E\theta) = (M_p/M)(\tau^2 E)f(\tau)$$

where $\tau \equiv E\theta$ is the reduced scattering angle and $f(\tau)$ is a function of the scattering potential. For small τ values $f(\tau) = 1/2$ (i.e., molecular elastic) and for high τ values $f(\tau) = 1$ (i.e., binary elastic). In this context M/f can be thought of as an effective scattering mass which is a function of the reduced mass only.

3.2 Positive Ion-Molecule Collisions

The efficacy of Sigmund's procedure has been demonstrated recently by ANDERSEN, et al. [10] and is shown in Figs. 6 and 7. Figure 6 shows a traditional plot of ΔE vs $E\theta^2$ for $Ne^+ + D_2$ for three Ne^+ energies (1.5, 2.5, and 3.5 keV). The deviation from the lower line represents the inelasticity of the process which obviously does not scale with $E\theta^2$. The dashed curve in Fig.7 shows the composite curve for Ne^+ (3.0, 2.5, 2.0, 1.5 keV) on D_2 scaled according to Sigmund's procedure. The solid curve with points is that obtained with Ne^0 at 0.5, 1.0, and 1.5 keV on D_2. The success of the scaling procedure is obvious. The difference in the τ positions of the knees in the curves for Ne^+ and Ne^0 is interesting and implies significant differences in the forces for the ion and neutral cases.

Very little electronic excitation is seen with either Ne^+ or Ne^0 on H_2 (Fig.8). The same result was found in the earlier experiments of DITTNER and DATZ [11] who studied inelasticity in backscattering (~180° center-of-mass) of K and Na neutrals and ions from H_2 and D_2 (since the projectile mass is greater than that of the target, backscattering is observed in the forward direction in the laboratory system). The results for Na 180° backscattering are shown in Fig.9. Here the abscissa is the relative energy in the $Na-D_2(H_2)$ center-of-mass system and the ordinate is the inelasticity in relation to the center-of-mass molecular system. The dashed lines indicate

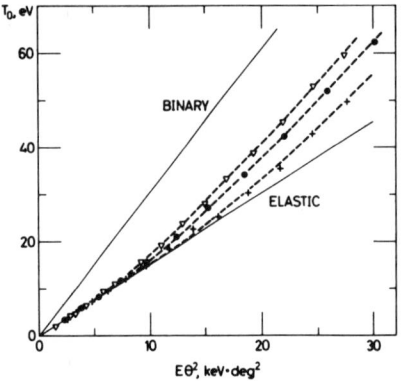

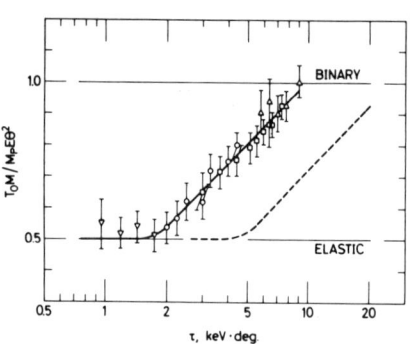

Fig.6 Most probable energy loss plotted vs $E\theta^2$ for $E = 1.5$ (×), 2.5 (●), and 3.5 (∇) keV Ne^+ scattered from D_2.

Fig. 7 f values for (∇) 0.5 keV, (○) 1.0 keV, (□) 1.5 keV, and (△) 2.0 keV Ne + D collisions. The dashed line shows the f function for the Ne^+ + D_2, H_2 systems.

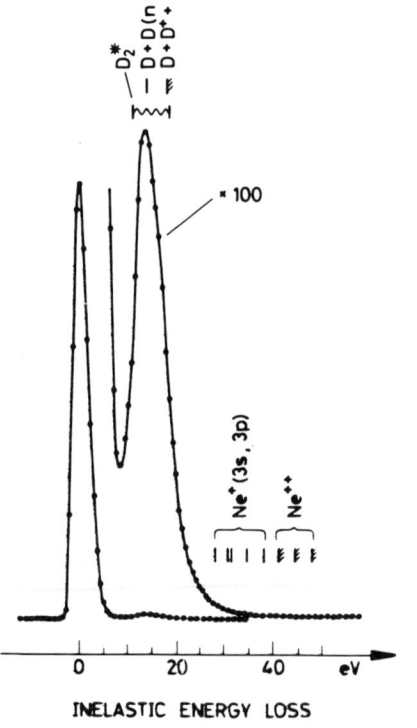

Fig.8 Energy loss spectrum of Ne^+ ($E=3.5$ keV) scattered from D_2 at $\theta = 0.53°$. Only small amounts of electronic excitation (of the D_2) are observed.

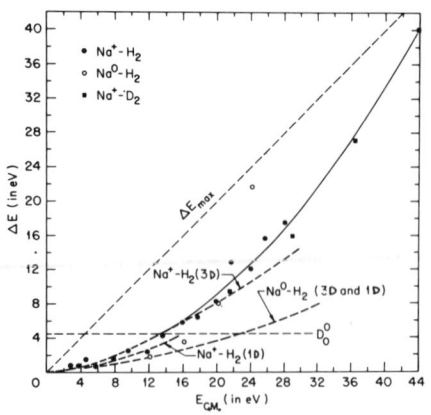

Fig.9 Inelastic energy loss (i.e., departure from "elastic limit") for collisions of Na^+ and Na^0 with H_2 versus collision energy in the center-of-mass system. D_0^0 is the dissociation energy of H_2. Dashed lines represent three dimensional (3D) and one dimensional (1D) calculations.

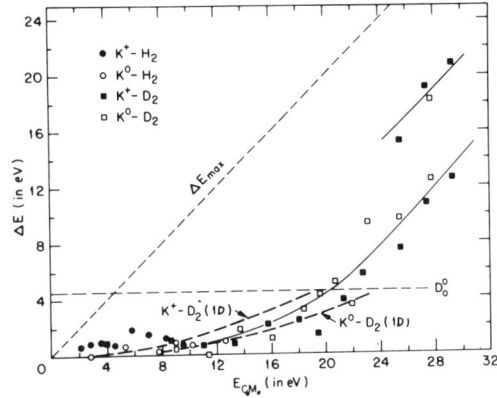

Fig.10 Inelastic energy loss (i.e., departure from "elastic limit") for collisions of K^+ and K^0 with H_2 and D_2 versus collision energy in the center-of-mass system. D_0^0 is the dissociation energy of H_2. Dashed lines represent three dimensional 3D) and one dimensional (1D) calculations.

calculated vibrational inelasticities based on one dimensional and three dimensional models for pure mechanical coupling between a repulsive ion-molecule potential and a Morse molecular oscillator. The results for the $K + H_2$ system shown in Fig.10 show another feature; namely a second higher energy loss peak appearing at about $E_{rel} = 24$ eV and becoming dominant at higher energies. This peak is not due to electronic excitation; instead it represents the onset of binary collision contributions to the scattering (i.e., the proper center-of-mass system here is that for K-H rather than K-H_2).

The small difference between inelasticity in the ion-molecule and atom-molecule cases here compared with that observed in the Ne-H_2 system may be due to the difference in scattering angle range. The repulsive potential is only slightly different for ion and atom cases [11]. For 180° backscattering no rotational excitation is possible and only the purely repulsive part of the potential is important. The lack of electronic excitation in scattering of Ne, Na, and K ions and atoms from H is not reproduced in the $Cl^- + H$ system (see below).

As was demonstrated in the experiments of FERNANDEZ, et al. [12], when N_2 is used as a target molecule for Ar^+ bombardment, electronic excitation can play a dominant role. Energy loss spectra for 1 and 2 keV Ar^+ scattered from N_2 are shown as a function of scattering angle in Fig.11. Peak A is is associated with vibro-rotational excitation of N_2 in its electronic ground state, Peak B with electronic excitation of N_2 to the B $^3\pi_g$ state, and Peak C with ionization of N_2 into the ground X $^2\Sigma_g^+$ state of N_2^+. The relative cross sections appear to scale with $E\theta$; e.g., compare 2.5° scattering at 1 keV with 1.25° scattering at 2 keV. Noteworthy is the fact that at $E\theta > 2$ keV deg the electronic excitation of the N_2 is dominant. A plot of the energy losses associated with these peaks is shown in Fig.12 as a function of $E\theta^2$ for Ar^+ incident energies of 1, 2, and 3 keV.

Two features are notable here. First is the lack of scaling of the 3 keV points on Peak A with the 1 and 2 keV points (scaling using SIGMUND's procedure [9] may bring these into line). Second is that the energy losses of Peak B and Peak C are parallel to that of Peak A implying that the degree of vibro-rotational excitation is the same for collisions which involve electronic excitation as for those which do not (Peak A). The authors [12] state that this implies the validity of the Frank-Condon assumption in the electronic excitation channel. Similar, but not identical, results were

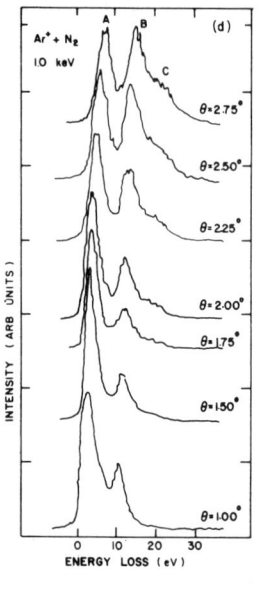
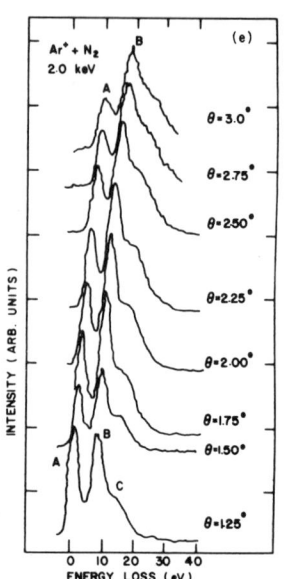

Fig.11 Energy loss spectra for Ar^+ (1 keV and 2 keV) incident on N_2 as a function of scattering angle θ in the laboratory system.

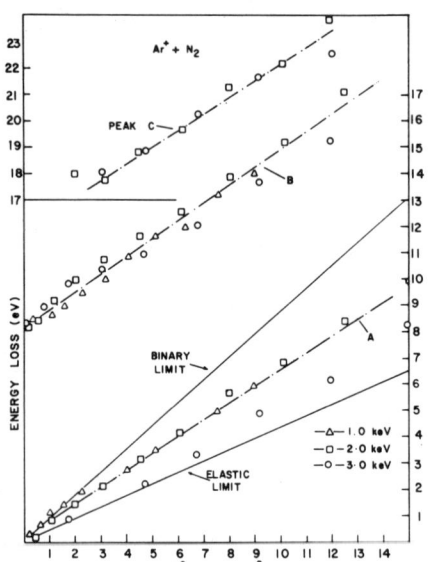

Fig.12 Inelastic energy losses for Ar^+ on N_2 vs $E\theta^2$. Peaks A, B, and C refer to data of Fig. 11.

obtained for $K^+ + N_2$ by INOUYE et al. [13], whose measurements extended to higher scattering angles, Fig.13.

If parallels are to be made with ISS, several perplexing questions arise. First, from the data of these papers [12,13], the energy losses are far from the binary limit even at $\tau = 10$ keV deg (Fig.12), e.g., 20 degree scattering at 500 eV. Yet "single scattering" is an accepted concept for surface scattering

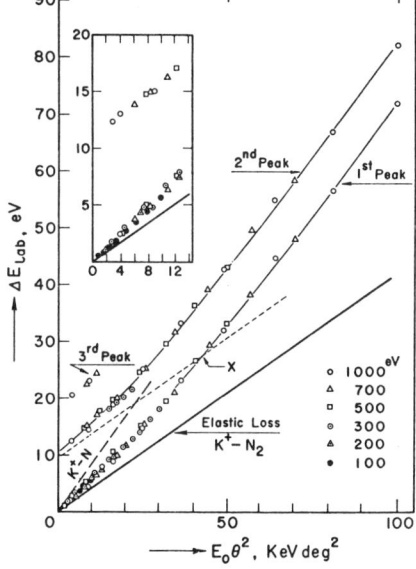

Fig.13 Inelastic energy losses for K^+ on N_2 vs $E\theta^2$.

in this $E\theta$ region. Second, the probability of electronic excitation of the molecule is greater than that for electronically elastic scattering at $E\theta \gtrsim 3$ keV deg and in the case of the $K^+ + N_2$ system the energy loss resembles binary scattering. It is intriguing to speculate on the possible observation of such effects in scattering from adsorbed diatomic molecules, molecular solids, and large band gap insulators. Will similar effects be observable in inelastic energy loss or will some of the photon emission effects observed be attributable to similar mechanisms?

3.3 Negative Ion-Molecule Interactions

The subject of electronic excitation in negative ion-molecule collisions is one which has just recently come under study. In keeping with the previous sections we will discuss only collisions of Cl^- with H_2 and N_2. The Cl^- ion is almost the same mass as the K^+ and Ar^+ ions discussed earlier; it is isoelectronic with K^+ with a $3p^6$ outer electron configuration and it has an ionization potential of 3.6 eV (i.e., Cl^0 has an electron affinity of 3.6 eV). This latter fact means that electron detachment to form neutral Cl atoms enters as a factor in low energy collision processes. At energies below 100 eV, collisional detachment of Cl^- with rare gases has been satisfactorily analyzed in terms of a complex potential model [14] which treats the process as a curve crossing between a bound state ($Cl^- + M$) and a continuum state ($Cl^0 + M + e^-$) somewhere along the repulsive wall of the incoming state. The energy loss of the Cl^0 formed by this process should be ~4 eV. At higher energies, electronic excitation of either the rare gas atom or excitation of Cl^- ion to autodetaching states is observed and FAYETON et al. [15] have proposed that quasi-molecular negative ion resonances may be involved.

For molecular targets, however, the situation is much more complex. In Figs. 14 and 15 we compare energy loss spectra obtained by CHEUNG and DATZ [16] for the processes

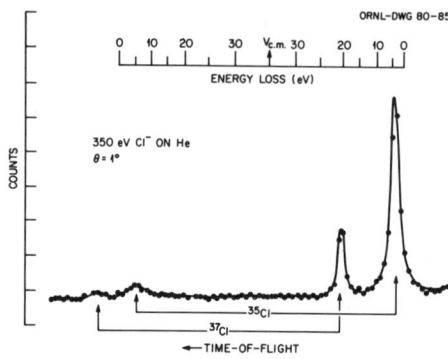

Fig.14 Energy loss spectrum of Cl^0 formed by electron detachment of Cl^- in collision with He; E_{Cl} = 350 eV, $\theta = 1°$. The peaks at the left of the figure are caused by particles which are backscattered in the center-of-mass system. Since no isotope separation is used, peaks are for ^{35}Cl and ^{37}Cl isotopes.

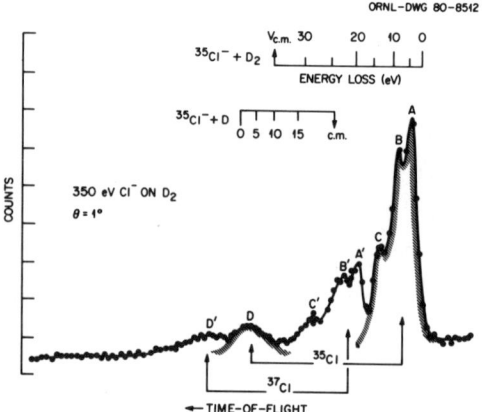

Fig.15 Energy loss spectrum of Cl^0 formed by electron detachment of Cl^- with D_2; E = 350 eV, $\theta = 1°$.

$$Cl^- + He \rightarrow Cl^0 + He + e$$

and

$$Cl^- + D_2 \rightarrow Cl^0 + (D_2) + e$$

respectively. Both are taken for the same Cl^- ion beam energy and Cl^0 scattering angle. In the He target case (Fig.14) only "simple detachment" is observed in both forward and backward scattering. With the D_2 target, two additional processes involving excited states give rise to higher energy loss peaks B and C. Peak C is caused by an electron transfer reaction forming an H_2^- molecular ion resonance in the repulsive $^2\Sigma_g^+$ state and has a similarity to chemi-ionizing collisions of the type $K + O_2 \rightarrow K^+ + O_2^{-*}$ where O_2^- is formed in an autodetaching state [17]. Peak B has an energy loss corresponding to dissociative attachment to H_2, i.e.,

$$Cl + H_2 \rightarrow Cl^0 + H + H^-$$
or $\rightarrow Cl^0 + H + H + e^-$.

This mechanism bears a strong resemblance to one which was observed in the inelastic but non-detaching collisions of Cl^- with H_2 [18]. In that experiment, energy losses associated with the $Cl^- + H_2$ ($v \neq 0$) process showed a sudden onset of a 4 eV inelastic loss at a relative collision energy of 7 eV (a feature which is absent in the $K^+ + H_2$ case shown in Fig.10). Both experimental results could be explained on the basis of the same intermediate excited state

$$Cl^- + H_2 \rightarrow [HCl^- + H] \rightarrow H + H + Cl^-$$
$$\rightarrow H + H + Cl + e^-.$$

Peak D results from small impact parameter backward scattering and appears to follow binary collision dynamics accompanied by simple detachment.

With low energy $Cl^- + N_2$ collisions, electronically excited states are even more predominant as was shown in a recent study by ANNIS et al. [19] on the inelasticity of

$$Cl^- + N_2 \rightarrow Cl^0 + N_2 + e^- \quad (a)$$
$$\rightarrow Cl^- + N_2 \quad (b)$$

Typical energy loss spectra for the detachment process (a) are shown in Fig.16.

The ordinates indicate the inelasticity, Q, for the reaction (the minimum inelasticity is 3.6 eV, the electron affinity of Cl). In order to illustrate the apparatus energy resolution, the spectra for the elastic scattering of Cl^- by argon are also indicated in the figure as the dashed lines. The solid curve in the lowest figure shows a spectrum for $Cl^- + Ar \rightarrow Cl + Ar + e$; this spectrum has a peak at $Q \simeq 4$ eV which is typical of direct detachment and it exhibits a high energy-loss tail as found by FAYETON et al. [15].

In the data shown for detachment on N_2, two energy loss peaks are clearly resolved corresponding to energy losses of approximately 6 and 12 eV. For both peaks the most probable energy loss determined at 150, 200, and 300 eV to angles of $\theta \simeq 8°$ is almost independent of τ which would be expected if the inelasticity resides in electronically, rather than vibro-rotational, excited states of the target.

A remarkable feature in the data is the dominance of the high energy loss channel at low τ. The fact that the high Q channel is dominant at low scattering angle may arise from an attractive region in the scattering potential in the exit channel. It is possible that the $Cl^- + N_2$ repulsive state might detach an electron at short distances (as a quasi-triatomic molecule) and then exit along strongly attractive ion pair states ($Cl^- + N_2^+$) which are strongly coupled to excited states ($Cl + N_2$ $^1\Pi_g$) so that little direct ion-pair formation occurs. The observed energy loss is commensurate with such a process. The asymptotic $Cl^- + N_2^+$ curve lies $\sim$4 eV above the $Cl + N_2(^1\Pi_g)$ but a curve crossing should occur at separations of $\sim$4 Å. Thus the integrated deflection function for such a process could lead to relatively low scattering angles for small impact parameter collisions.

The low energy loss (6 eV) process is the dominant one in the total detachment cross section (see Fig.17). The inelastic energy loss for this

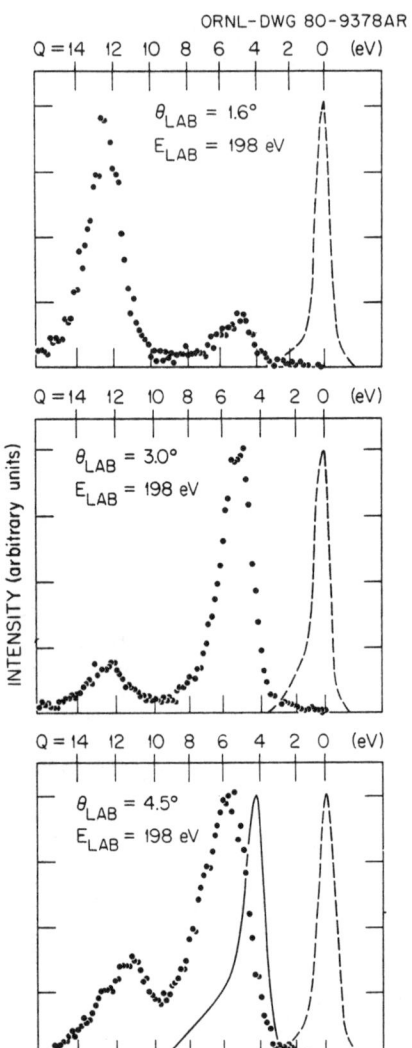

Fig.16 Energy loss spectra of Cl⁰ formed in Cl⁻ (E_{lab}=198 eV) + N_2 collisions at θ=1.6, 3.0, and 4.5°. Also shown are spectra for elastic scattering from Ar(---) and collisional detachment from Ar(——).

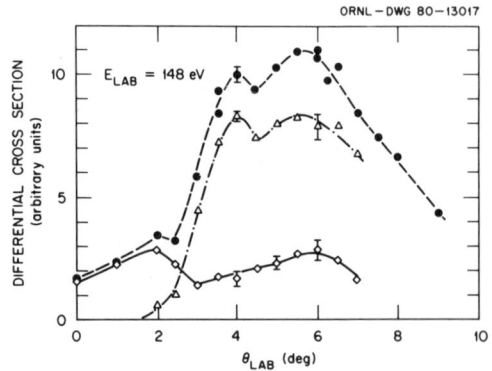

Fig.17 Relative differential cross sections for Cl⁻ (E_{lab}=148 eV) electron detachment in collision with N_2. ● = total cross section, △ = partial cross section for low loss (ΔE=5-7 eV) peak, ◇ = partial cross section for high loss (ΔE=11.5 to 12.5 eV).

process also scales with τ, is significantly greater than that for simple detachment (compare, e.g., with Ar target peak in Fig.1), and it is also significantly greater than the vibro-rotational excitation obtained with K^+ or Ar^+ on N_2 [12,13] in the same region of τ. The observed Q is, however, entirely consistent with charge transfer to the $^2\Pi_g$ resonance [20] state of N_2^- whose lowest vibrational state lies 1.8 eV above the ground state of N_2. The rise in the inelasticity for τ ≳ 1000 eV-deg and the asymmetric broadening in the energy spectra toward higher Q seen at the higher τ-values indicate that a band of vibrational states with the $^2\Pi_g$ electronic state of N_2^- is involved in the detachment mechanism. The reason for the dominance of this process may involve an attractive well for the (Cl - N_2 $^2\Pi_g$)⁻ system which drops the crossing point at the repulsive wall below that for simple detachment to the continuum.

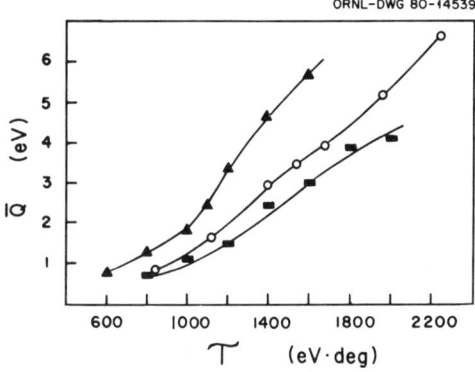

Fig.18 The most probable energy loss for Cl⁻ inelastically scattered by N_2 versus τ ($\tau = E\theta$) for E_{Cl^-} = 100 eV (▲), 140 eV (o), and 200 eV (■).

The doubly differential cross sections for the inelastic non-detaching channel, (b), were also measured [19] and a summary of the results for several collision energies are seen in Fig.18. Unlike the energy losses associated with detachment these do not scale with τ. Instead they scale reasonably well with $E\theta^2$ and are quite similar in magnitude with those observed for vibro-rotational excitation in the isoelectronic system ($K^+ + N_2$) [13] and for ($Ar^+ + N_2$) [12]. However, for the $Ar^+ + N_2$ and the $K^+ + N_2$ systems additional energy loss peaks corresponding to electronic excitation of N_2 were found. For example, in the $Ar^+ + N_2$ system the energy loss peak corresponding to electronic excitation of N_2 was more than 0.5 the height of the simple vibro-rotational excitation peak at τ = 2200 eV-deg (see Fig.11). The absence of such peaks in the non-detaching channel with $Cl^- + N_2$ indicates that when electronically excited states of the target molecule are involved in collision, electron detachment also occurs.

Of particular interest to possible application to surface scattering is the fact that neutralization of a high electron affinity negative ion by collisional detachment requires rather violent collisions. For example, in the case of $Cl^- + N_2$, the non-detaching (b) process is dominant up to $\tau \simeq$ 1000 eV-deg [21], and for Cl⁻ scattering from Xe at 750 eV, the ratio of the non-detaching elastic cross section to the detachment cross section is still 10% at 3 keV deg [15]. Thus for low energy surface scattering there may be a greater propensity for maintaining the projectile charge than is generally true for positive ions. For high work function metallic surfaces, the use of alkali ions [22,23] rather than rare gas ions has been shown to be useful because of the increased probability of being reflected from the surface collision as a positive ion rather than a neutral atom. For low work function surfaces (e.g., Cs covered Ni), ECKSTEIN, et al., [24] have shown very high yields of H⁻, a relatively weakly bound (0.8 eV) negative ion, from surfaces bombarded with protons. Negative ions of the halogens may therefore find a future use as surface probes for ISS for such systems.

Acknowledgment

This research was sponsored by the Division of Chemical Sciences, Office of Basic Energy Sciences, U. S. Department of Energy, under contract W-7405-eng-26 with Union Carbide Corporation.

References

[1] S. Datz and C. Snoek, Phys. Rev. 134, 347 (1964).
[2] C. Snoek, W. F. van der Weg, R. Geballe, and P. K. Roe, Physica 34, 1 (1967).
[3] W. F. van der Weg and D. J. Bierman, Physica 44, 177 (1969).
[4] R. L. Kaufman, K. A. Jamison, T. J. Gray, and P. Richard, Phys. Rev. Lett. 36, 1074 (1976).
[5] F. Hopkins, A. Little, N. Cue, and V. Dutkiewitz, Phys. Rev. Lett. 37, 1100 (1976).
[6] F. W. Saris, C. Foster, A. Langenberg, and J. van Eck, J. Phys. B 7, 1494 (1974).
[7] S. Datz, Atomic Collisions in Solids, in *Structure and Collisions of Ions and Atoms*, ed. by I.A. Sellin, Topics in Current Physics, Vol. 5 (Springer, Berlin, Heidelberg, New York 1978) pp. 309-340
[8] G. D. Alton, L. B. Bridwell, M. Lucas, C. D. Moak, P. D. Miller, C. M. Jones, Q. C. Kessel, A. A. Antar, and M. D. Brown, Phys. Rev. A (in press).
[9] P. Sigmund, J. Phys. B 11, L145 (1978).
[10] N. Andersen, M. Vedder, A. Russek, and E. Pollack, Phys. Rev. A 21, 782 (1980).
[11] P. F. Dittner and S. Datz, J. Chem. Phys. 54, 4228 (1971).
[12] S. M. Fernandez, F. J. Erikson, A. V. Bray, and E. Pollack, Phys. Rev. A 12, 1252 (1975).
[12] H. Inouye, K. Niurao, and Y. Sato, J. Chem. Phys. 64, 1250 (1976).
[14] R. L. Champion and L. D. Doverspike, Phys. Rev. A 13, 609 (1976).
[15] J. Fayeton, D. Dhuicq, and M. Barat, J. Phys. B11, 1267 (1978).
[16] J. T. Cheung and S. Datz, J. Chem. Phys. 73, 3159 (1980).
[17] N. Kashihira, F. Schmidt-Bleek, and S. Datz, J. Chem. Phys. 61, 160 (1974).
[18] J. T. Cheung and S. Datz, J. Chem. Phys. 71, 1814 (1979).
[19] B. K. Annis, S. Datz, R. L. Champion and L. D. Doverspike, Phys. Rev. Lett. 45, 1554 (1980).
[20] G. J. Shultz, Rev. Mod. Phys. 45, 423 (1973)
[21] R. L. Champion and L. D. Doverspike (private communication).
[22] E. Taglauer, W. Englert, W. Heiland, and D. Jackson, Phys. Rev. Lett. 45, 740 (1980).
[23] S. H. Overbury, W. Heiland, D. M. Zehner, and S. Datz, Bull. Am. Phys. Soc. 25, 427 (1980).
[24] W. Eckstein, H. Verbeek, and R. S. Bhattacharya, Surf. Sci. (in press).

Charge Fractions of Reflected Particles

W. Eckstein

Max-Planck-Institut für Plasmaphysik, EURATOM Association
D-8046 Garching, Fed. Rep. of Germany

A. Introduction

The charge state fractions of reflected particles have attracted increased attention in the last ten years. The major interest is to understand how the charged state fractions are formed and which processes are most important. But apart from these more fundamental aspects there are several practical reasons for having data on the charge state fractions of backscattered particles. For example, low energy ion scattering could be made more quantitative, if the positive fractions were known. In fusion research the problem of recycling requires knowledge of the fraction of neutrals backscattered into the discharge. In the same connection, an understanding of the mechanisms important in negative hydrogen ion sources is of relevance in obtaining intense neutral beams for heating discharges to fusion temperatures.

The first workshop in this series has discussed many of the charge changing processes of backscattered particles especially Auger neutralization, resonance neutralization and quasi-resonant processes for example. Therefore the emphasis of this article is to review recent progress in the field of charged fractions at low energies and also to discuss such topics, which had not been covered in the preceding workshops as the difference in the charged fractions of hydrogen and helium and the negative charge fractions.

B. Definitions

If a particle in a given charge state approaches a solid surface it may be reflected by surface atoms or it may penetrate the solid and be reflected after moving some distance in the solid. The emerging particle may be in a neutral, positive or negative charge state. The positive (and the negative) charge state may be a multiple one and for most charge states it is possible that the reflected particle is in an excited state. In this paper we will concentrate on the charge fractions neglecting excited states, although it has been shown that such excitation occurs [1]. Another restriction is that the discussion is limited mainly to particles with velocities $v \leq 2.19 \times 10^8$ cm/s, which corresponds to an energy of 25 keV for hydrogen.

The charged fraction is defined as

$$n^{m+(n-)} = N^{m+(n-)}/N_{tot} ,$$

where $N^{m+(n-)}$ is the number of reflected particles with a multiple positive (negative) charge m(n).

$$N_{tot} = N^0 + \sum_m N^{m+} + \sum_n N^{n-} .$$

It is necessary to discuss only the charged fractions, $\eta^{m+(n-)}$, because the neutral fraction is

$$\eta^0 = 1 - \sum_m \eta^{m+} - \sum_n \eta^{n-} .$$

The charged fraction η is dependent on a number of variables which are partly given in Fig.1

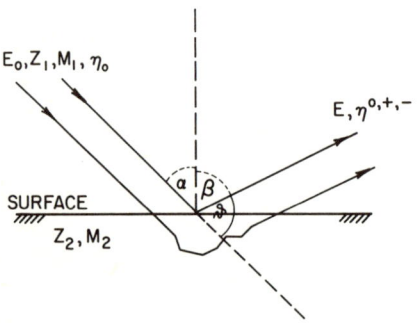

Fig.1   Schematic of the scattering geometry

$$\eta = \eta(Z_1, M_1, Z_2, M_2; E_0, E; \theta, \alpha, \beta, \phi; \eta_0; d) ,$$

where
- Z_1 - atomic number of the incident particle
- M_1 - mass of the incident particle
- Z_2 - atomic number of the target atom
- M_2 - mass of the target atom
- E - energy of the incident particle
- E_0 - energy of the reflected particle
- θ - scattering angle
- α - angle of incidence (relative to the surface normal)
- β - exit angle (relative to the surface normal)
- ϕ - azimuthal angle (single crystal target)
- η_0 - charge state of the incident particle
- d - distance traveled in the solid

At low energies m and n are essentially unity and for the rare gases the negative fraction can be considered to be zero.

C. Experimental Methods

Before a discussion of the charged fractions a brief description of the experimental methods will be given in order to indicate possible errors and problems in the determination of the charged fractions.

Usually a beam of ions (or neutral particles) is directed on to a target and then the reflected particles in a specific charge state and in a specified solid angle are energy analyzed and detected. Two methods have been used for the energy analysis 1: a time of flight technique applicable for all charge states [2-5] and 2: ionizing the neutrals in a gas stripping cell with subsequent energy analysis in an electrostatic condensor [4, 6-8].

Both methods have advantages and disadvantages, which one should keep in mind, when considering experimental data. The main differences between methods may be summarized as follows. Time of flight: This method can not usually distinguish between singly and multiply charged ions; when the charged fraction is small the uncertainty is large since the value is determined from a difference between two large numbers; the energy resolution depends on the velocity of the particles and the detection efficiency depends on the particle energy and may be different for particles in different charge states, at least for low energies. Stripping and electrostatic analysis: the ionization efficiency in the stripping cell has be to calibrated absolutely; this efficiency depends strongly on energy and the kind of particles and the stripping gas. For both detection systems the absolute error in the charged fractions is usually not smaller than 20 % for energies below about 20 keV.

At higher energies surface barrier detectors can be used, which give both energy and intensity information (see, for example, [9]), but with poorer energy resolution.

An example of a measurement, the positive and neutral spectra after scattering of He from Ni, is shown in Fig.2. The neutral spectra are broad extending from a maximum energy down to zero energy, whereas the positive spec-

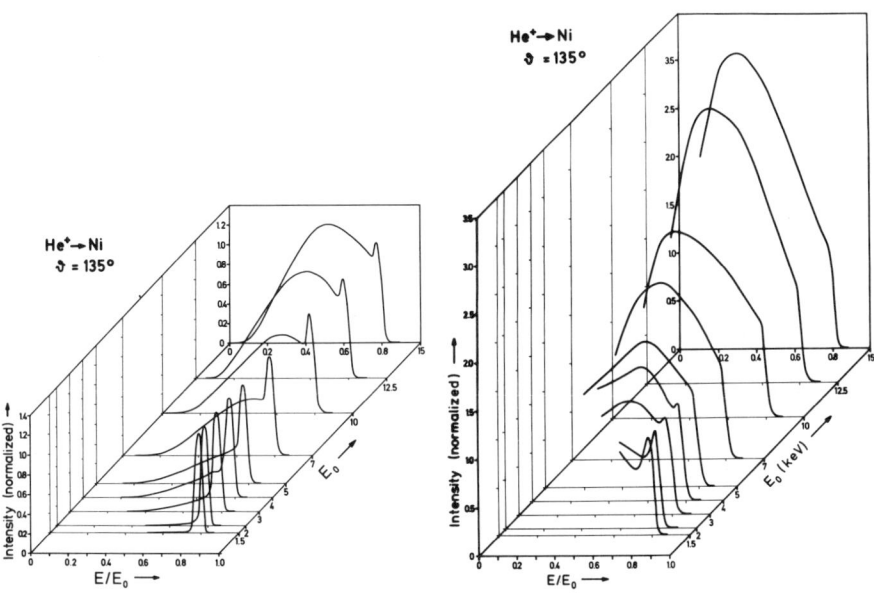

Fig.2 Energy spectra of He backscattered from clean polycrystalline Ni for normal incidence ($\alpha = 0$) and 9 different incident energies E_0: a) ions, b) neutrals [43].

tra show a so-called surface peak presumably originating from particles which have been scattered by surface atoms. At lower incident energies this surface peak is the only feature remaining in the positive spectrum. This

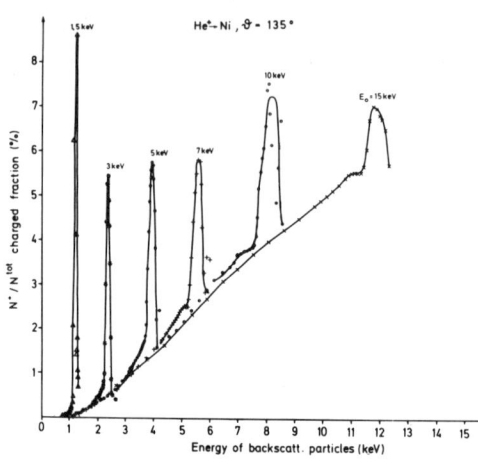

Fig.3 Positive fraction η^+ for He backscattered from Ni, determined from the data shown in Fig.2 [43]

example shows clearly that there is a large difference in the probability for a particle escaping as an ion, depending on whether a particle is reflected from near surface atoms or only after significant penetration into the solid. This is very apparent in the charged fraction (Fig.3), which has been determined by dividing the positive spectrum by the sum of the positive and neutral spectra in Fig.2. The charge fraction in the surface peak

Table 1   References for ion target combinations for which charge fractions have been determined in reflection measurements

Ions \ Targets	H	D	He	Li	B	C	N	F	Ne	Ar
Be	9									
C	18, 87	18	18							
Si	18, 87, 96		96, 98							
V	9									
Ti	18, 87	18								
Ni	18, 49, 77, 87	18, 77	18, 43, 77						43, 44, 77	
Cu	9, 97		97						4, 45, 46	4, 46
Nb	9									
Mo	9, 87		18							
Cs	35		35							
Ta	9									
W	18, 87		18							
Pt										47
Au	17, 36, 49, 52, 53, 87, 97, 98, 99	36	17, 38, 87, 97, 98, 100	39	39	39	39, 42	39	38	6, 37

is very sensitive to the correct determination of the high energy edge of the positive and neutral spectra. It should be mentioned here, that the charge fraction may be given as a function of the energy E of the reflected particles for a single incident energy E_0 or in those cases, where the positive spectra exhibit only a surface peak, as a function of the incident energy E_0.

Therefore, the discussion (I) of the positive fraction is subdivided in three sections: (a) particles reflected after penetration of the solid, (b) particles reflected in surface collisions and (c) particles experiencing additional ionizations. The negative fractions are discussed in a separate chapter (II).

Ion-target combinations, for which charged fractions have been determined in reflection measurements are given in Table 1.

D. Comparison between Theoretical and Experimental Results and Discussion

I. Positive and Neutral Fractions

a) Charge Fractions of Particles which traveled in the Solid

1) Capture and Loss

The idea is that a particle experiences electron capture and loss processes during its travel through the solid and after some collisions charge state equilibrium is obtained. This picture has been used extensively for gas phase collisions and for transmission of high energy particles through thin foils [10]. Assuming that only one positive and a neutral state of the particle is possible in the target, one can write down the change in the positive fraction n^+ with pathlength as

$$\frac{dn^+}{dx} = N(n^0 \sigma_{01} - n^+ \sigma_{10}),$$

where N is the number density, σ_{01} is the electron capture cross section and σ_{10} is the electron loss cross section.

With the assumptions $n^0 = 1 - n^+$ and $n^+(x = 0) = n_0$ and by integration

$$n^+ = n^+_{eq} + (n_0 - n^+_{eq}) \exp(-x/\lambda) = n^+_{eq} + \Delta n^+ \quad (1)$$

with

$$\lambda = [N(\sigma_{01} + \sigma_{10})]^{-1} \quad \text{and} \quad n^+_{eq} = [1 + \sigma_{10}/\sigma_{01}]^{-1} \quad (2)$$

If the second term is small compared to the first term in (1), the charged fraction n^+ becomes independent on the pathlength x and reaches an equilibrium value n^+_{eq}, which is only dependent on the ratio of the capture and loss cross sections. The depth needed to reach equilibrium depends mainly on the value λ, that means on the target density N and the sum of the two cross sections. Typical measured cross sections per atom for H → C, N_2 and O_2 as a function of energy are given in Fig.4 [1 - 13]. Calculations of σ_{10} for H → Cr, Fe, Ni, Mo, W and Au show a similar energy dependence [14]. Cross section curves for He are similar in shape but shifted to higher energies due to an approximate velocity scaling. Using the cross sections for H → N from Fig.4, λ (assuming N = 10^{23} atoms/cm^3) and the ratios of the cross sections are shown in Fig.5. A 10 % contribution of the non-

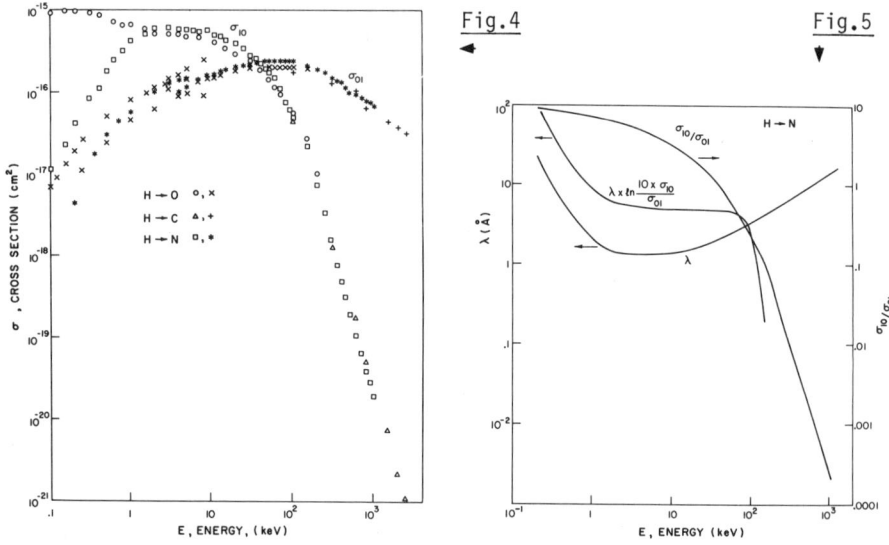

Fig.4 Capture (σ_{01}) and loss (σ_{10}) cross sections (per atom) versus energy for H on C, N and O [11 - 13]

Fig.5 The cross section ratio σ_{01}/σ_{10}, the value λ and $\Delta n^+ = 10$ % versus energy are given for H $\rightarrow$ N, using the data from Fig.4 and N = 10^{23} atoms/cm^3

equilibrium part Δn^+ is given by the curve $\lambda \cdot \ln(10 \cdot \sigma_{10}/\sigma_{01})$ for an incident proton ($n_0 = 1$). λ is of the order of a few monolayers in the energy range from 1 to 30 keV and is increasing with increasing energy.

Many measurements have been done with oxygen and/or carbon covered surfaces so that it should be possible to compare these experimental data with the equilibrium charged fraction determined by the measured gas cross sections given in Fig.4. The experimental neutral fraction data measured in reflection [9] and in transmission [15] show good agreement with $n_{eq}^0 = 1 - n_{eq}^\pm$ in a wide range, see Fig.6. Even at energies below 10 keV the equilibrium fractions $n_{eq}^\pm$ for H and He $\rightarrow$ O_2 (cross sections from [16]) show a similar velocity dependence as the experimental data see Fig.7. Here the comparison is very uncertain due to the large scatter in the experimental cross section data.

This simple model does not give any dependence of the charged fraction on the exit angle β of the reflected particles. Such a dependence has been observed for H $\rightarrow$ Au but these results were not well reproducible due to possible surface structure changes [17]. For He $\rightarrow$ Au no such exit dependence has been seen [17], but such structure changes will appear at lower dose than for H. The agreement of neutral fractions measured in reflection [18] and transmission [19] suggests that a possible exit angle dependence may be weak, since the transmission measurements are usually done for an exit angle β ≈ 0°, whereas all reflection measurements are done for β ≥ 25°. Table 2 presents ion target combinations, for which charged fractions have been determined in transmission experiments.

Table 2 References for ion target combinations for which charge fractions have been determined in transmission measurements

Ions / Targets	H	D	He	B-Fe
Be				
C	15 29 101	19 29	29	86 104 105
Mg		19		
Al	15 29 102 103	29	29	
Si	103			
Ca	103			
Cr	15			
Ni	15			
Nr		19		
Ag	102 103			
Cs	29		29	
Au	15 29 102 103	29	29	

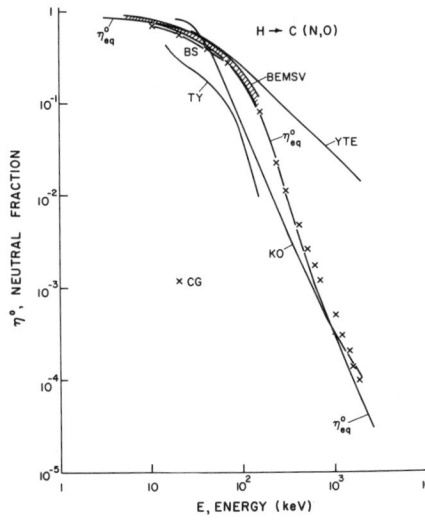

Fig.6 A comparison of calculated (BS, TY, YTE, KO)[20, 30-32] and experimental (BEMSV, CG) [9, 15] data for the neutral fraction η^0 versus energy E for H on C(N,O). η^0_{eq} is the equilibrium fraction determined from the data shown in Fig.4

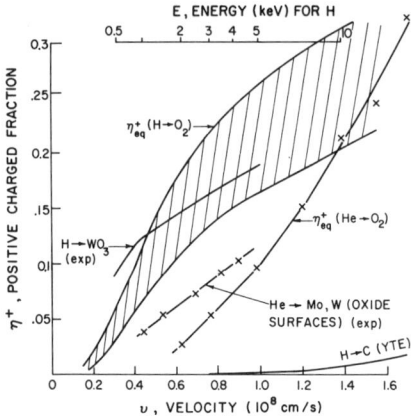

Fig.7 A comparison of calculated and experimental data [18] for the positive fraction η^+ versus velocity for H and He on O (oxides). η^+_{eq} is the equilibrium fraction

It should be noted that the concept of capture and loss inside the solid may be meaningless for velocities below the Fermi-velocity v_F of the electrons in a solid. In this velocity range the conduction electrons will screen an ion completely and it may be impossible for an ion to capture an electron into a bound state. In this sense the capture and loss picture breaks down below v_F. For heavier ions the electron clouds of both the incident particle and the target atom will overlap so that a molecular orbital model (or promotion model) would be possibly more appropriate.

2) Surface Effects

A completely different approach is the assumption that the charge state of the reflected particles is determined at the surface. It was pointed out by YAVLINSKII et al. [20] and by BRANDT [21] that there exists no bound state on a proton in a solid due to collective screening of the ion potential by the target electrons and/or due to collision broadening of the bound state. Cross [22, 23] had disputed these statements and at present the question of the existence of bound hydrogen states in a solid is not yet finally resolved.

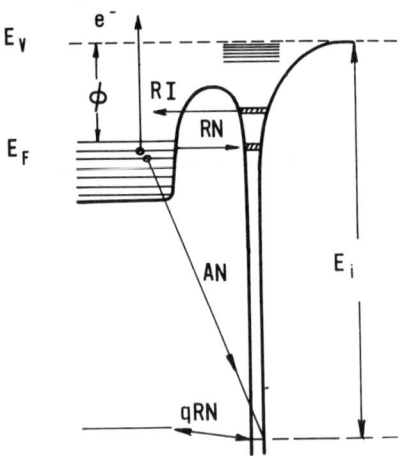

Fig.8 Schematic diagram for surface processes: AN = Auger-neutralization, RN = resonance-neutralization, RI = resonance-ionization, qRN = quasi-resonant neutralization; E_i = ionization energy of the ground state of the atom, ϕ = work function, E_F = Fermi-energy and E_V = vacuum level

But if one assumes that hydrogen particles have to move in a solid as screened protons then the charge states of the emergent hydrogen particles are completely governed by the surface.

At the surface different processes such as Auger-neutralization (AN), resonance-neutralization (tunneling) and resonance-ionization (RI) [24], as indicated in Fig.8, can take place. Triple-collision recombination may also be present [20].

Auger-neutralization has been discussed extensively by HAGSTRUM [25]. For a particle leaving the surface the positive fraction η^+ is given by

$$\eta^+ = \exp(-A/av_\perp) \tag{3}$$

where $v_\perp = v \cos \beta$ is the velocity component normal to the surface. A and a are constants, which depend on the ion-target combination. In Fig.9 the positive fraction versus $v_\perp^{-1}$ is shown for 7 ion-target combinations [18]. Here the curves have been determined for different exit energies at a constant exit angle. Only two examples show a linear dependence of $\ln \eta^+$ versus $v_\perp^{-1}$. If the exit energy is kept constant and the exit angle is varied [17] the parameter A/a is not constant but depends on the exit angle β and/or on energy, see Fig.10. These results show that the simple formula (3) does not describe the experimental results in this energy range.

It has been pointed out by HAGSTRUM [25] that the exponential transition rate function $R = Ae^{-as}$ from which (3) is derived is not as good a description for outgoing particles as for incident ones, because the transition rate is highest close to the surface, where a plane potential distribution is not as good an assumption as it might be for larger distances from the surface. HORIGUCHI et al. [26] have shown, that the neutralization probability per unit time for slow protons has the following form:

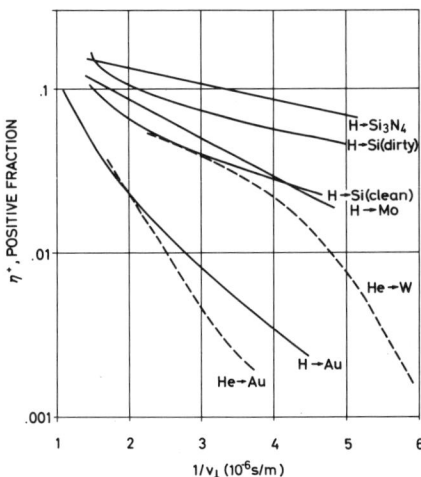 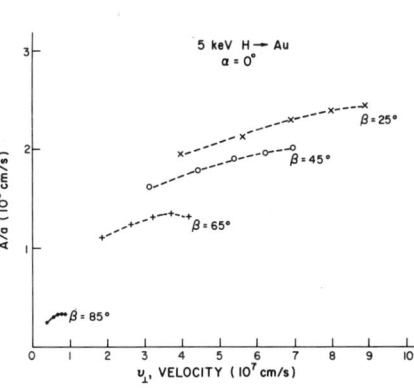

Fig.9 Positive fractions versus the reciprocal velocity normal to the surface for 7 ion target combinations [18]. The exit angle β is kept constant at 45°

Fig.10 The value A/a versus the velocity normal to the surface for 4 different exit angles [17]. The broken lines are drawn to guide the eye

$$P(s) = A\{\exp(-2\kappa s) + (Bs + C)\exp(-(\kappa + a_0^{-1})s) +$$
$$(Ds^2 + Es + F)\exp(-2a_0^{-1}s)\} \tag{4}$$

Only the first term is similar to Hagstrum's approach and the other terms can be regarded as corrections. The result is that the neutralization per unit time is reduced at small distances but approaches the simple exponential dependence at larger distances. A comparison of formula (4) with experimental data for H on Au is shown in Fig.11. The calculated neutralization is much too low, which might indicate that Auger neutralization is only one part of the total neutralization process.

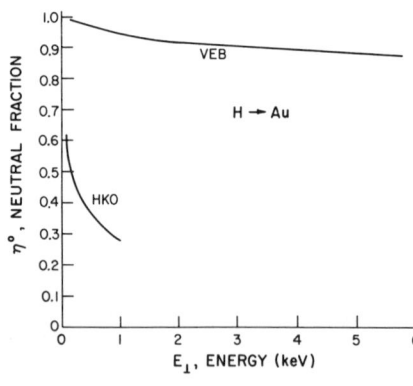

Fig.11 A Comparison of the calculated [26] and measured [18, 87] neutral fraction η^0 versus the energy normal to the surface $E_\perp$ for H on Au

It has been proposed [27] to combine the capture and loss picture and the Auger neutralization model by multiplying both probabilities. The positive fraction n^+ for atoms with an energy level below the Fermi level E_F of the metal is given by

$$n^+(E) = \frac{E}{E+E_1} + \frac{E_1}{E+E_1} \exp[-N\ell(\frac{K_1}{\sqrt{E}} + K_2\sqrt{E})] \exp[-A/(av_\perp)] \tag{5}$$

where ℓ is the pathlength after the violent collision inside the solid in a double scattering model and $E_1 = K_1/K_2$ is a theory parameter for which no estimate has been given. The same criticism as mentioned before for the applicability of the capture and loss picture to low energies and the simple Auger formula (3) is valid here. In the same framework PARILIS and coworkers [28] determined charge fractions for hydrogen backscattered from single crystals.

A similar approach has been used by KREUSSLER [29]. He also combines capture and loss processes with Auger neutralization. In his model a distinction is made between capture and loss processes at target atom core and valence electrons. Using theoretical estimates for the corresponding cross sections he gets agreement with his measured neutral fractions for H → C, Al, Cs and Au in the energy range where his model is applicable ($v \geq v_0 = 2.19 \times 10^8$ cm/s).

Resonance or tunneling-neutralization has been discussed by TRUBNIKOV and YAVLINSKII [30]. Their calculated neutral fraction is independent of the exit angle and disagrees markedly with experimental results, see Fig.6. This kind of process is very unlikely because the first excited level will be shifted above the Fermi-level for most materials.

KITAGAWA and OHTSUKI [31] used a Bates-type quantum-mechanical recombination theory, to describe the electron pickup at the surface. Their results show an exit angular dependence and their neutral fraction gives agreement at high energies ($\sim$ 1 MeV) but deviates at lower energies from measured values, see Fig.6.

Still another approach has been tried by YAVLINSKII et al. [20]. Their triple-collision recombination theory gives results which are independent of exit angle. The calculated energy dependence shows agreement between 30 and 80 keV but deviates strongly at higher energies.

A formula given by BRANDT and SIZMANN [32] for the neutral fraction

$$\eta^0 = (1 + 40E)^{-1} \qquad (6)$$

(E, energy in MeV), which should be valid between 0.1 and 0.01 MeV gives values which are slightly below measured ones, see Fig.6. These authors argue that the neutral fraction in this energy range should be nearly the same for different surfaces. Also their formula shows no exit angle dependence.

CROSS has discussed most of the papers mentioned thus far from a theoretical point of view and the reader is referred to Cross review at the first workshop [22].

3) Zwiegel and Kleber's Approach

A different approach was taken by ZWIEGEL and KLEBER (these proceedings and [33]).This theory will be discussed in the following contribution, so that here only the essential points are given. A moving proton or helium ion, moving inside a metal with low velocity ($v \ll v_0$) is assumed to be screened completely by an electron cloud. He is supposed to have one electron bound. The moving particle sees an electron current in the opposite direction. This negative current is interrupted nonadiabatically when the ion reaches the surface. Only those electrons from the cloud screening the ion with suitable velocity components can follow the ion are bound, when the ion leaves the surface. A detailed quantum-mechanical calculation leads to the following formula

$$\eta^+ = 0.2834 (1 - |R|^2) \qquad (7)$$

where

$$|R| = |(1 + k_{FT}^2 \, g(q)/(2q^2))^{-1} - 1|$$

$$g(q) = 1 + \alpha^{-1} (1 - \alpha^2/4) \ln|(1 + \alpha/2)/(1 - \alpha/2)|$$

$$\alpha = q/k_F \; ; \; q = 0.2 \sqrt{E \text{ (in keV)}}; \; k_f = 1.919/r_s$$

$$k_{FT}^2 = 1.695 \, (\hbar\omega_p)^{2/3} = 2.44 \, r_s^{-1} \; .$$

At low energies formula (7) reduces to

$$\eta^+ = 0.00928 \, r_s \, E/M \qquad (8)$$

or $\quad \eta^+ = 0.0134 \, (\hbar\omega_p)^{-2/3} \, E/M$

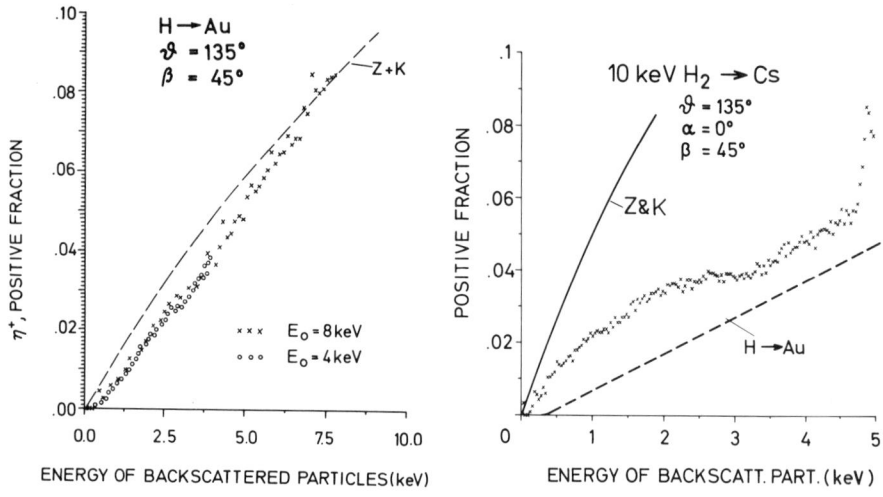

Fig.12 A comparison of the calculated [33] and measured [18, 35] positive fraction versus the energy E for normal incidence: a) H → Au, b) H → Cs

with the target constants k_{FT}, r_s and $\hbar\omega_p$ in atomic units, the mass M of the leaving particle (M = 1 for H) and its energy E in keV. The positive fraction depends on the particle velocity q and on target constants such as the plasma frequency ω_p or the Wigner-Seitz radius r_s. Using experimental data for $\hbar\omega_p$ [34] formula (7) gives good agreement with experimental positive fractions for H → Au [18], see Fig.12a. For H on Cs, see Fig. 12b, the agreement is not good but the tendency for a more curved dependence is seen in the experimental as well as in the calculated curve [35]. The disagreement may be due to the possibility that hydrogen can have a bound neutral state in Cs; this will be discussed in the next chapter. Formula (7, 8) also predicts the same charge fraction for D at the same velocity which has been established experimentally for clean surfaces [18, 36]. The same formula applies if helium is moving in a metal as a screened He^+. Again this has been found for reflection of H and He from a clean Au surface [18], see Fig.13. In formula (8) E means the energy normal to the surface. Therefore this theory gives an exit angle dependence of the charged fractions. This has not been taken into account in Fig.12; it would make the agreement better for H → Cs but worse for H → Au for higher energies. The surface microstructure which is usually not known for polycrystalline material may have a large influence on the charged fraction.

So far all theories describe some experimental aspects but no theory is able to give an understanding of all experimental data. It is even not solved if bulk or surface effects are more important but it is generally accepted that the surface processes dominate at low velocities. For particles heavier than H and He the situation is even more complicated due to the higher possibilities of excitations. Neutral fractions for incident particles such as Li to Ar have been determined [37-39]; the data of POEHLMAN et al. [39] are shown in Fig.14.

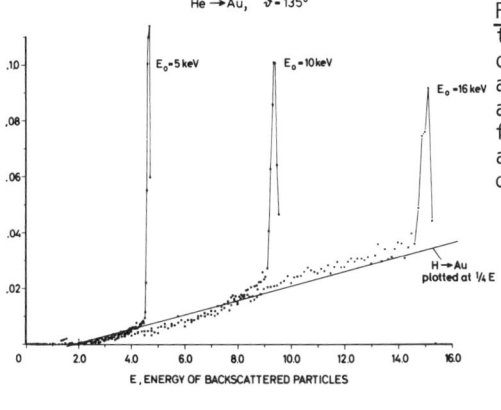

Fig.13 Energy dependence of the positive fraction of He backscattered from clean Au for three incident energies E_0 and normal incidence ($\alpha = 0$) and an exit angle $\beta = 45°$. The positive fraction for H backscattered from the same target and for the same exit velocity is indicated by the solid line [18]

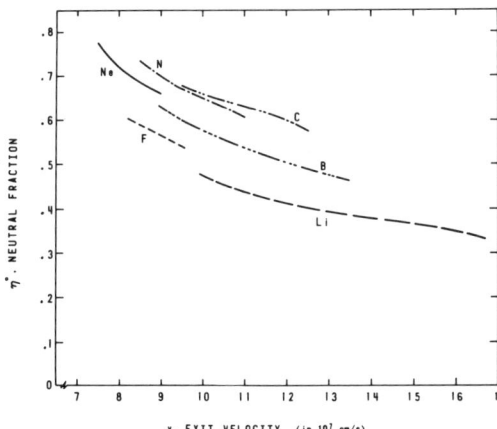

Fig.14 Neutral fraction η^0 versus the exit velocity v for Li, B, C, N, F and Ne reflected from Au [39]. $E_0 = 125 - 130$ keV, $\theta = 150°$, $\alpha = 0°$, $\beta = 30°$

b) <u>Charged Fractions of Particles Reflected by Surface Collisions</u>

It is obvious from Fig.2 that He particles scattered from surface atoms have a higher probability to be reflected as ions than particles backscattered from the bulk. This effect is so pronounced that at low energies (for example E < 2 keV for He) only the surface peak is visible in the positive energy spectrum, which forms the basis for the ISS technique [40]. For higher energies the surface peak becomes less important until it disappears. This higher probability to escape as an ion for particles scattered at surface atoms occurs regardless of whether there is a surface peak in the neutral spectrum or not, see Figs. 2 and 3. The same situation has been observed for other incident particles such as Li [41], N [42], Ne [4, 38, 43-46], Ar [4, 6, 37, 46, 47], but it has not been seen for hydrogen [18, 36, 48-50], see Fig.15. This remarkable difference will be discussed first.

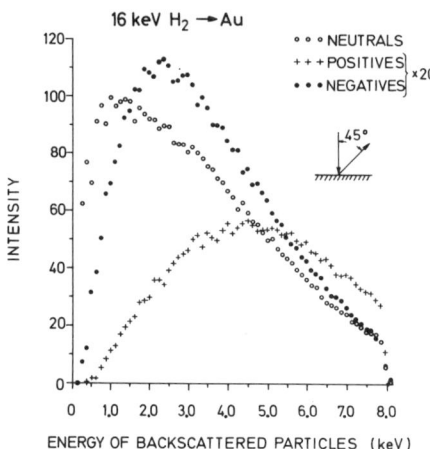

Fig.15 Measured energy spectra of the three charge states of H backscattered from clean Au, when bombarded with 16 keV H_2^+ [18]

1) Differences in Charge Changing Surface Processes for Hydrogen and Rare Gases

The experimental data show that the surface peak in η^+ for Ne and Ar are still high [37, 38, 43] at higher energies whereas it has nearly vanished for He [38, 43]. From simple kinematics it is clear, that at a given energy surface collisions are more important for heavier incoming projectiles because the cross sections are higher. Furthermore the penetration depth of reflected particles is increasing with decreasing projectile mass. This does not explain an enhancement in ion yield for surface scattering, but if there is such an enhancement one would expect to see it more clearly for heavier projectiles because the ratio of particles scattered from surface atoms to those scattered from the bulk is much smaller for H than for heavier projectiles.

These arguments might suggest that the surface peak is possibly too small or too narrow to be seen in the experimental conditions used so far. This cannot be ruled out completely but it is unlikely since in the backscattering of hydrogen from Cs a surface peak in the positive spectrum and in the charged fraction has been observed [35]. But then the question arises why is there a surface peak for H backscattered from Cs but not from other targets studied so far? There are some experimental results which also show surface peaks for H, but these are shadowing effects due to a single crystal structure [51-54].

In addition to the kinematic differences between H and He, there are also differences in the electronic structure. It has been argued [21] that because of screening (or collisional broadening) hydrogen in a solid cannot have a bound state. While this is not generally accepted [22] it would be a major difference for He. The ground state is so much deeper for He that a bound state should be possible, see Fig.16.

Another reason is that the difference of closest approach is so small for H down to 100 eV that even in the electron selvage at the surface the screening is complete and there should be no difference for H scattered from the surface or the bulk. But this argument would be true also for He.

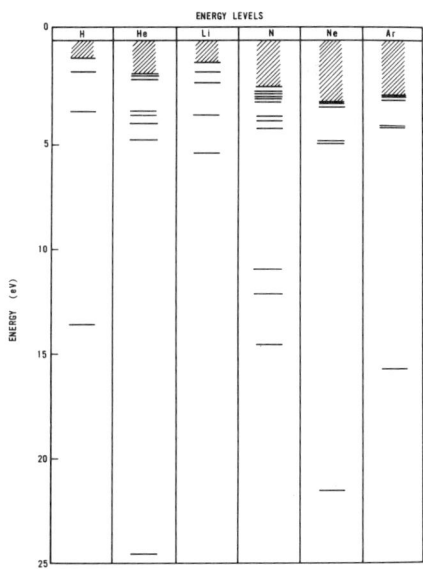

Fig.16 Energy level diagram for the most commonly used incident particles

An explanation for the occurrence of a surface peak for H on Cs could be that in Cs a bound state may be possible. The number k of bound states of H in a solid has been estimated [29, 55]:

$$k = 0.583 + 0.499 \cdot h \quad \text{or} \tag{9}$$

$$k' = [2h \exp(-1)]^{1/2} \quad \text{with} \tag{10}$$

$$h = r_s^{1/2}/1.56$$

where r_s is the Wigner-Seitz radius, which depends on the target electron gas density. In the first estimate (9) a bound state is possible for $r_s \geq 2$, in the second estimate (10) for $r_s \geq 4.5$. In both estimates a bound state is possible in Cs ($r_s = 5.88$), but not in most other metals. It may also be possible that hydrogen has a bound state in some crystal channels [52, 53].

Another possibility is a difference in the interaction of H and He with the electronic structure of the surface as indicated in Fig.8. In this respect there is not much difference between the ground states but the excited levels may lie above or below the Fermi-level, allowing for the possibility of resonance-neutralization or -ionization. But for clean Ni, Cs, Au surfaces studied with both H and He there is no principal difference between H and He. A strong dependence of the surface peaks on the work function has not been observed although this effect has been seen [35]. It should be noted that a discussion of resonance neutralization and ionization in this simple way is rather speculative, since the energy levels are shifted and broadened during the approach to the surface.

Examining the energy levels of different projectiles, Fig.16, it seems reasonable that He, Ne, Ar behave similarly according to their electronic configurations. Li is expected not to be strongly neutralized due to its low ionization energy. Nitrogen might be more strongly neutralized because it has more low lying states.

After all the arguments mentioned above, it seems very probable, that in the energy range discussed here the charge changing process in a binary collision of the projectile with a surface atom plays an important role. This charge changing collision has also been proposed as a reason for the difference between H and He [18]. After the so-called adiabatic maximum rule [56] the maximum for a neutralization collision is in the keV range for H whereas it is at higher energies for He. This argument has the disadvantage that it can only be applied for total cross sections and that these cross sections might be changed due to the presence of the surface. Differential cross sections would be needed. So far it is only known that neutralization occurs predominantly at small scattering angles [57] and ionization at larger scattering angles [13]. It should be mentioned that many of these neutralization collisions can result in excited states: for example, the cross section for $H^+ + Cs \rightarrow H^0 + Cs^+$ is dominated by the creation of the H(2s, 2p) states [58] at low energies. For heavier ions multiple excitation may occur, which finally results in auto-ionization [59].

A dip in the negative fraction at the energy, where the surface peak in the positive fraction occurs for H on Cs [35], indicates that the number of neutrals in surface collisions is less if one assumes that the negative ions are created not directly from an ion but from a neutral particle at larger distances from the surface. This experimental result points also to the importance of a charge changing collision at the surface.

One of the best demonstrations of the importance of a charge changing binary collision at the surface is the comparison of the ion yield for the scattering of He from a Pb atomic beam and from a Pb covered surface. Both scattering events give similar oscillations in the ion yield [60].

2) Sequence of Charge Changing Processes

A model, which describes the charge fraction by a sequence of processes, was given by VERHEIJ [61, 62]. In this approach it is assumed that an incoming ion is at first subjected to Auger neutralization, then to an ionizing or neutralizing close collision and finally again to Auger neutralization. This model also includes proposals by VAN DER WEG et al. [63] and BRONGERSMA et al. [64].

The charged fraction is then given by

$$n^+ = f_2 \{ f_1 (1 - P_n - P_i) + P_i \} \tag{11}$$

for an incident ion, and by

$$n^+ = f_2 P_i \tag{12}$$

for an incident neutral,
where P_i is the ionization probability of a neutral particle in a close collision
P_n is the neutralization probability of an ion in a close collision
f_1 is the probability for an ion to escape Auger neutralization on the incoming path
f_2 is the probability for an ion to escape Auger neutralization on the exit path.

With the assumption that the simple exponential function holds for the Auger neutralization it is possible to determine P_i, if the ion yield or n^+ for an incoming neutral is measured. At first the constant $v_c = A/a$ is calculated from experimental results for different exit (β) and entrance angles (α) at a fixed scattering angle θ

$$\frac{n^+(\alpha_1)}{n^+(\alpha_2)} = \frac{\exp[-v_c/(v_f\cos\beta_1)]}{\exp[-v_c/(v_f\cos\beta_2)]}$$

with v_f the exit velocity normal to the surface.

Choosing α_1 and α_2 so that $\alpha_1 + \alpha_2 = 180° - \theta$ it then follows

$$\ln \frac{n^+(\alpha_1)}{n^+(180-\theta-\alpha_1)} = \frac{v_c}{v_f}((\cos\alpha_1)^{-1} + (\cos(180-\theta-\alpha_1))^{-1} ,$$

because $\alpha + \beta = 180 - \theta$.

A plot of this relation yields v_c/v_f as the slope and thus v_c can be determined. For this result it is only necessary to measure the ion yield because only the ratio of yields is needed. With the knowledge of f_2 also the probability P_i can be determined.

The last technique has been used by Verheij et al. for He on Cu [61, 62] and by OVERBURY et al. [50] for H on C to determine v_c and P_i. For H the "kinematical edge" instead of a surface peak is used. Both authors observe an increasing ionization probability P_i with increasing energy. Whereas P_i is reaching 30 % for He at 10 keV, for H it is below 10 % for energies between 1 and 2 keV. In the He case the authors [62] assumed a neutralization probability $P_n \approx 0$, whereas in the H case [50] P_n is very near to unity. Also the velocities v_c differ by a factor of 10, where the v_c values in the H case seem to be rather low. Contrary to what one might expect, the hydrogen data do not show a clear correlation between P_i and the scattering angle. These results are further evidence that charge changing atomic collisions are important in determining the charge fraction of the exciting particles. Again it should be mentioned, that in many cases the simple exponential dependence for the Auger escape neutralization probability is not found [40, 50]. This could impose relatively large errors to the estimated P_i values.

Charge changing collisions are important at collision distances of the order of 1 Å or less as estimated from typical cross sections. The Auger neutralization probability for an outgoing ion decreases with increasing distance from the surface. This means that both processes appear to happen in the same distance range. It is not clear, if it is possible or indeed reasonable to separate these processes.

The hydrogen data [50] could be fitted surprisingly well with a relation

$$\frac{n^+}{v_{i_\perp}} = c_1 \frac{v_{f_\perp}}{v_{i_\perp}} + c_2$$

where $v_{i_\perp}$, $v_{f_\perp}$ are the initial and final velocities normal to the surface respectively. However this empirical dependence is not a result from any theoretical approach thus far proposed.

3) Atomic Collisions

TULLY [65, 66] tried to combine all possible processes in a quantummechanical treatment which includes adiabatic (resonance and Auger neutralization) as well as non-adiabatic processes. Another quantummechanical approach was made by BLOSS and HONE [67]. Both theories were developed to explain the oscillatory behaviour of the ion yield as a function of energy, which has been observed by a number of authors [40, 60, 68-76]. TULLY in particular, got reasonable agreement with experimental results. The oscillatory ion yield is attributed to a near resonant charge transfer from a 3d, 4d or 5d level of the target atom to the ground state of He. Both theories treat the Auger neutralization in the simple exponential form. This topic has been extensively discussed in the first workshop [65, 73] and no charge fractions but only ion yields have been measured. A fraction of a monolayer of Pb on Ni also gives rise to oscillatory ion yields for Ni [60] that are not seen for clean Ni surfaces. This is explained by multiple scattering effects.

As a conclusion from Section b) it appears that the positive fraction of particles reflected from surface atoms can only be understood, if a charge changing collision with a surface atom is taken into account in addition to ion solid interactions. If that is the case, then the charged fraction becomes dependent on the ion path. This is indicated by experimental data, which show a lower positive fraction for multiply scattered particles (which have higher energies than those scattered in a binary collision) than for particles scattered in a binary collision [4, 43-45], see Fig.17. At this point another problem arises, namely which particles really contribute to the surface peak and to the higher energy tail or peaks (the so-called quasi-double and quasi-triple in single crystal results). In single crystal measurements the surface may have a superstructure exhibiting semichannels. In these semichannels the incoming projectiles experience a larger neutralization although they have not penetrated the target and are scattered from surface atoms, but which do not belong to the first layer [44]. For single crystal surfaces the charge fraction becomes strongly dependent on azimuth [4, 77], see Fig.18.

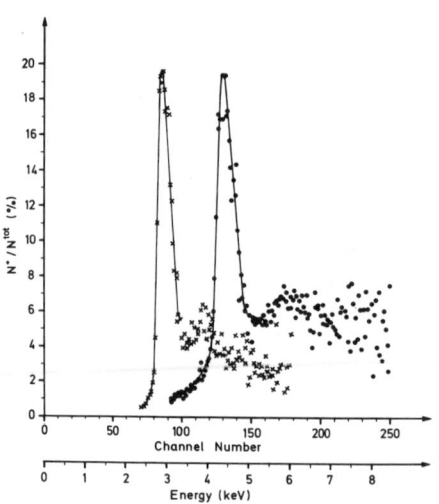

Fig.17 Positive fraction η^+ versus the energy E of the backscattered particles for Ne on clean Ni and for two incident energies E_0 = 10 and 15 keV. α = 0°, β = 45° [43]

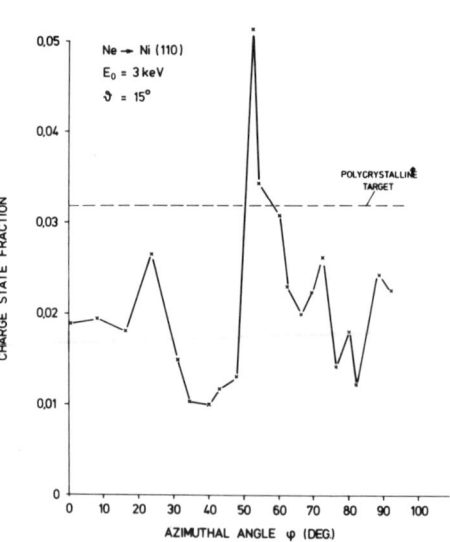

Fig.18 Positive fraction η^+ versus the azimuthal angle ϕ for 3 keV Ne on Ni(110). α = β = 82.5° [77]

A way, where one has not to rely on these unknown charged fractions in ISS, is to use alkali-metal ions as incident ions. For these alkali ions the charged fraction for surface collisions is assumed to be near unity [41].

c) Multiply Ionization

Ions which have a positive charge larger than one have not yet been seen for energies below 10 keV, but they appear at higher energies [63, 78]. Even at incident energies as low as 15 keV, significant n^{2+} values arise from binary collisions at the surface: eg $n^{2+} \approx 1/3\ n^+$ for Ne on Ni, see Fig.19 [43]; due to the use of electrostatic energy analysis the Ne^{++} peak appears at half the energy of the Ne$^+$ peak. From the calculated inelastic energy losses in a binary collision by the FIRSOV and OEN-ROBINSON model one would expect a threshold for the creation of the Ne^{++} ion at about 2 keV, because the ionization energy for Ne$^+$ is 41.07 eV. From this point of view one would expect also a scattering angle dependence of the multiply charged ion fraction. Multiply charged ions should contribute to the total positive fraction at lower energies for Ar and N than for Ne and only at higher energies for He and Li.

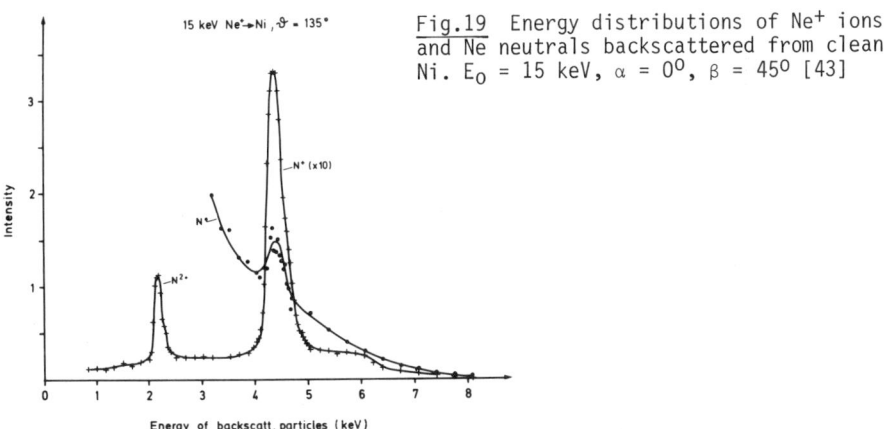

Fig.19 Energy distributions of Ne$^+$ ions and Ne neutrals backscattered from clean Ni. $E_0 = 15$ keV, $\alpha = 0°$, $\beta = 45°$ [43]

KISHINEVSKI et al. [79] gave a theoretical description of the occurrence of higher charge states. This model is based on the promotion of M and L shell electrons to excited states in a violent collision and subsequent relaxation by Auger transitions. The calculated results depend on the trajectory of the particle and they are in reasonable agreement with experimental data [37, 47, 63].

II) Negative Charged Fractions

Negatively charged ions in reflection have been reported so far only for H and D [80-85]. In transmission through thin foils there have been seen also negative carbon, oxygen and fluorine ions [86]. For the rare gases the negative fraction of reflected particles is well below 0.1 % in the keV-energy range, whereas the negative fraction for hydrogen can be larger than the positive fraction for exit energies below 5 keV [87], see Fig.20. The major difference compared to the positive fraction is a broad maximum in

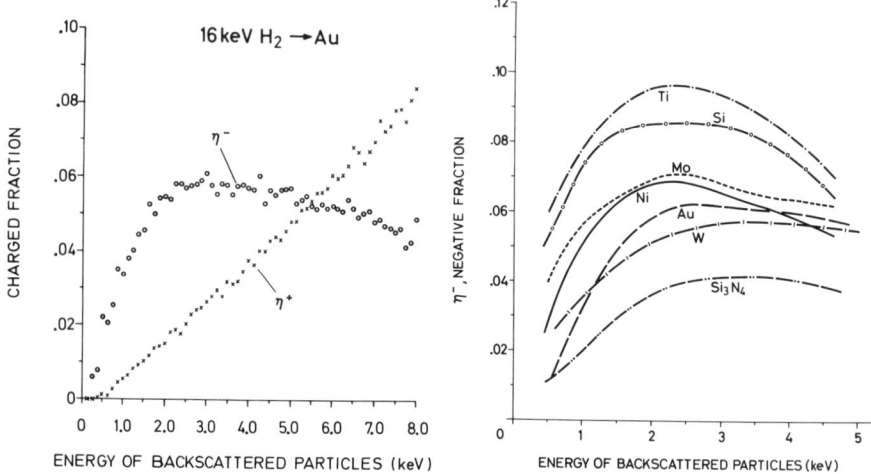

Fig.20 Charged fractions of hydrogen backscattered from clean Au versus the energy E. E_0 = 8 keV, $\alpha = 0°$, $\beta = 45°$ [87]

Fig.21 Negative fractions η^- of hydrogen versus the energy E for seven ion-target combinations. E_0 = 5 keV, $\alpha = 0$, $\beta = 45°$ [87]

the negative fraction in the dependence on the energy of the reflected particles [87, 88]. This maximum appears for many elements and compounds (oxides, nitrides) at an energy of between 2 and 3 keV, see Fig.21. The existence of a maximum implies that two competing processes are at work with a formation probability, which is decreasing with increasing energy and a survival probability of the formed negative ions, which is increasing with increasing energy. The height of the maxima of the negative fractions seems to be related to the work function of the target material. The maximum height is increasing with decreasing work function and the position of the maximum is shifted to lower energies with decreasing work function, which is clearly demonstrated by the negative fraction for hydrogen backscattered from Cs [35], where a possible maximum must be below 300 eV, see Fig.22. The measured data also indicate that the negative fraction depends only on the exit velocity and on the exit angle β [35, 87].

The first theoretical approach to treat the problem of negative ions dates back to 1938 [89]. The theory available so far has been developed for hydrogen backscattered from alkali surfaces [90-92]. The idea is that a proton backscattered from surface atoms or coming out from the bulk is first neutralized, then at even larger distances from the surface there might be a probability for the formation of a negative ion, which again can lose one electron by resonant neutralization (tunneling of the electron to the target). Following HISKES [91, 93], that the formation and destruction of negative ions occur in two distinct regions and neglecting positive ions ($\eta^+ \ll 1$), the formation of the negative fraction η_f^- with the distance from the surface is

$$\frac{d\eta_f^-}{dz} = \frac{\dot{F}(z)}{v_\perp} \eta^0 = \frac{\dot{F}(z)}{v_\perp} (1 - \eta_f^-) \quad ,$$

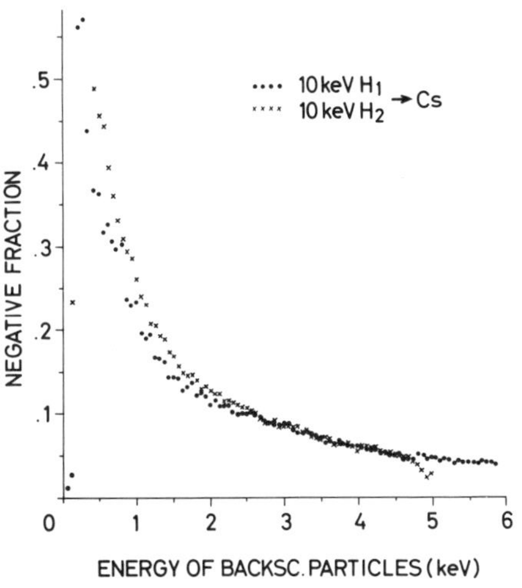

Fig.22 Negative fraction n^- versus the energy E for 10 keV H^+ and H_2^+ on Cs. $\alpha = 0°$, $\beta = 45°$ [35]

where $\dot{F}(z)$ is the transition rate of electrons transferred from the solid to the neutral atoms moving with a constant velocity $v_\perp$ normal to the surface. Hence by integration

$$n_f^- = 1 - \exp(-f/v_\perp) \text{ with } f = \int_{z_1}^{z_2} \dot{F}(z)dz$$

Correspondingly the survival fraction n_s^- is derived from

$$\frac{dn_s^-}{dz} = - \frac{\dot{L}(t)}{v_\perp} n_s^-$$

where $\dot{L}(z)$ is the electron loss rate. Integration gives

$$n_s^- = \exp(-s/v_\perp) \text{ with } s = \int_{z_2}^{z_3} \dot{L}(z)dz \quad .$$

Finally the measurable negative fraction is given by the product of the formation and survival probabilities

$$n^- \approx (1 - \exp(-f/v_\perp))\exp(-s/v_\perp) \tag{13}$$

The theory described does not take into account a broadening of the energy levels and the small probability of filled energy levels of the metal above the Fermi-level for temperatures above zero. From the adiabaticity of resonant processes it follows that the theory will not work at higher energies (E > 10 keV).

In Fig.23 the shift of the H^- electron level as a function of the distance from the surface is shown. The level position is determined as the sum of the image potential and the negative affinity, where the affinity is the differ-

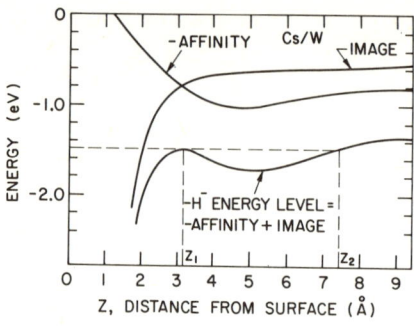

Fig.23 The image potential, the affinity and the resulting active electron energy level as a function of the distance from the surface (after [91])

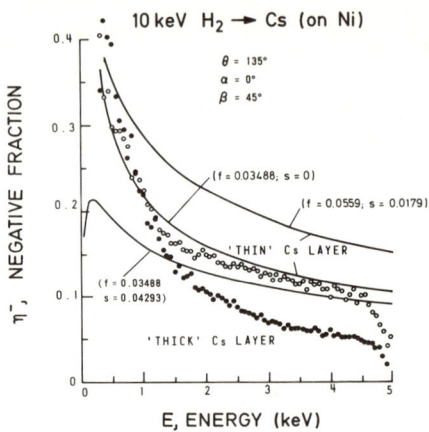

Fig.24 Comparison of measured negative fractions η^- for a 'thin' and a 'thick' layer of Cs on Ni [35] with calculated curves using formula (13). E_0 = 5 keV, $\alpha = 0°$, $\beta = 45°$

ence between the two potential curves of CsH and CsH⁻. Figure 23 shows that a resonant electron capture to form a H⁻ is possible between 3 and 8 Å from the surface, if the work function is low enough. For larger distances the electron may tunnel back.

An estimate for f using the WKB formalism for the determination of the tunneling probability has been given [94]

$$f = 2 \cdot 10^{14} \cdot 10 a_0 \simeq 10^7 \text{ cm/s}.$$

The value s is determined by a quantum mechanical treatment using a truncated image potential for the electronic potential between the negative ion and the surface [91]. Calculated results show a strong dependence of the survival probability on the work function. The survival probability for a partial monolayer of Cs on a metallic substrate is much higher due to the formation of a dipole layer at the surface; this greatly reduces the tunneling probability, resulting in a survival probability near unity down to low energies (10 eV).

In fact a higher negative fraction has been seen for a 'thin' Cs layer (in the order of a monolayer) than for a thick Cs film [35], see Fig.24. From measurements of the negative ion yield in the reflection of hydrogen from alkali surfaces semiempirical values for f and s have been determined [93]:

$$f = 3.49 \cdot 10^{-2}$$
$$s = 4.29 \cdot 10^{-2}$$

in atomic units (v_0 = 1 corresponds to 25 keV for H).

Figure 25 shows the dependence of the survival and formation probabilities and η^- on the energy of the reflected particles normal to the surface. A comparison with experimental values is given in Fig.24. For the 'thick' Cs film the semiempirical data [93] put into formula (13) give a reasonable trend

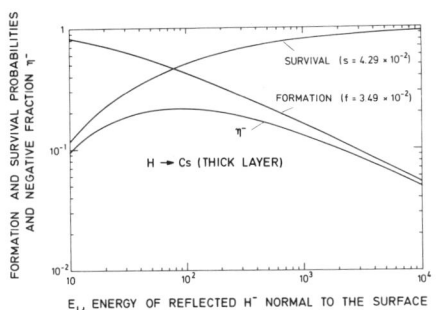
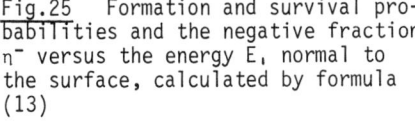
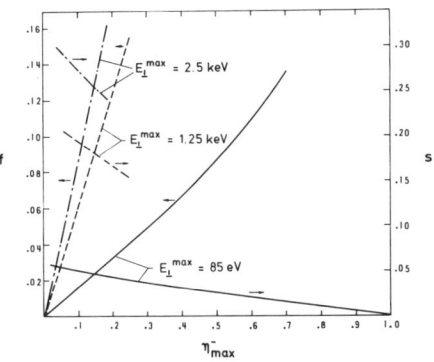

Fig.25 Formation and survival probabilities and the negative fraction η^- versus the energy $E_\perp$ normal to the surface, calculated by formula (13)

Fig.26 The parameters f and s versus the maximum value η^-_{max} of the negative fraction η^-. The value $E_\perp^{max}$ gives the energy position for the maximum in the negative fraction η^- ($\eta^-(E_\perp^{max}) = \eta^-_{max}$)

but the decrease of η^- with increasing energy is appreciably stronger in the experimental data. The agreement between theory is better for the 'thin' Cs layer, if the same value f as for the 'thick' Cs film and a survival probability of unity is used. Semiempirical values for f and s for the 'thin' Cs layer [95] shift the theoretical curve to higher energies. The decrease of the negative fraction at energies which are due to surface collisions seem to indicate a lower neutralization probability for surface collisions.

Resonant processes are applicable to the alkali metal surfaces. For other metal surfaces also a formation and a reneutralization process must be responsible for the energy dependence of η^-. But the processes will not be resonant and therefore η^- will be smaller in general. One could assume for that reason the same formula for η^- would be applicable with other values of f and s. In Fig.26 f and s are plotted versus the maximum values η^-_{max} of the negative fraction for three positions of the maximum in respect to the energy normal to the surface. This plot shows that f depends more strongly on the maximum value η^-_{max} of η^-, whereas s is more strongly dependent on the position of η^-_{max}. The application for the experimental curves in Fig.21 shows that the decrease of η^- on the high energy and low energy side of the maximum is too slow. Also the experimental exit angle dependence [87] does not fit to formula (13), where f and s should be independent of β.

E. Conclusion and Outlook

This survey shows clearly that the charge fractions in the energy range covered are not completely understood. The physical processes, which determine the charge states of the reflected particles are generally known, however the relative contributions of each to the final charged fractions are not known. The comparisons between the experimental data and calculated results make it obvious that in most cases no single process dominates the observed charged fractions. Even in the simplest case of hydrogen, many experimental results are not described quantitatively by any theory. For heavier ions where the number of participating electrons is larger, the quantitative theoretical description of measured data is very difficult in the intermediate velocity range.

Nevertheless more theoretical effort toward a quantitative understanding of the charge fractions would be highly desirable. Also from an experimental point of view there are many topics which have just been touched on and these require a more thorough investigation. In this regard measurements with incident neutrals would be interesting for many cases. A systematic study of the charged fraction dependence on the different angles (α,β,ϕ) for the reflection from a well defined single crystal surface should be undertaken. The investigation of the excitation of reflected particles, especially angle resolved, would give some insight into the processes involved. Systematic studies of the influence of adsorbed atoms on the charged fractions of reflected particles would be important for ISS.

Acknowledgements

The author wishes to thank D. Jackson, H. Verbeek and M. Kleber for many helpful discussions, D. Jackson also for the calculations of some binary collision data, J.A. Davies, T. Jackman and D. Jackson for critical reading of the manuscript. The hospitality for a three month stay at Chalk River Laboratories (AECL) is gratefully acknowledged.

References

1. C.W. White, E.W. Thomas, W.F. van der Weg and N.H. Tolk, in Inelastic Ion-Surface Collisions, Eds. N.H. Tolk, J.C. Tully, W. Heiland and C.W. White; Academic Press, New York, 1977, p. 201
2. Y.-S. Chen, G.L. Miller, D.A.H. Robinson, G.H. Wheatley and T.M. Buck, Surf.Sci. 62 (1977) 133
3. J.E. Robinson and S.A. Agamy, Nucl.Instr.Meth. 149 (1978) 595
4. S.B. Luitjens, Rijksuniversiteit te Groningen, Thesis, Groningen 1980
5. S.B. Luitjens, A.J. Algra, E.P.Th.M. Suurmeijer and A.L. Boers, Appl. Phys. 21 (1980) 205
6. V.M. Chicherov, JETP Lett. 16 (1972) 328
7. H. Verbeek, W. Eckstein and F.E.P. Matschke, J. Phys. E10 (1977) 944
8. S.B. Luitjens, A.J. Algra, E.P.Th.M. Suurmeijer and A.L. Boers, J. Phys. E13 (1980) 665
9. R. Behrisch, W. Eckstein, P. Meischner, B.M.U. Scherzer and H. Verbeek, in Atomic Collisions in Solids, Eds. S. Datz, B.R. Appleton and C.D. Moak, Plenum Press, New York, 1975, Vol. 1, p. 315
10. H.D. Betz, Rev. Mod. Phys. 44 (1972) 465
11. S.K. Allison, Rev. Mod. Phys. 30 (1958) 1137
12. L.H. Toburen, M.Y. Nakai and R.A. Langley, Phys. Rev. 171 (1968) 114
13. H. Tawara, Nagoya University, Nagoya, Japan, Report IPPF-AM-1, 1977
14. S. Hiraide, Y. Kigoshi and M. Matsuzawa, Nagoya University, Nagoya, Japan, Report IPPJ-AM-5, 1978

15 A. Chateau-Thierry and A. Gladieux, in Atomic Collisions in Solids, Eds. S. Datz, B.R. Appleton and C.D. Moak, Plenum Press, New York, 1975, Vol. 1, p. 307
16 K. Okuno, Nagoya University, Nagoya, Japan, Report IPPF-AM-9, Report IPPF-AM-11, 1978
17 H. Verbeek, W. Eckstein and R.S. Bhattacharya, Nucl.Instr.Meth. 170 (1980) 539
18 R.S. Bhattacharya, W. Eckstein and H. Verbeek, Surf. Sci. 93 (1980) 563
19 K.H. Berkner, I. Bornstein, R.V. Pyle and J.W. Stearns, Phys. Rev. A6 (1972) 278
20 Y.N. Yavlinski, B.A. Trubnikov and V.F. Elesin, Bull. Acad. Sci. USSR, Phys. Ser. 30 (1966) 1996
21 W. Brandt, in Atomic Collisions in Solids, Eds. S. Datz, B.R. Appleton and C.D. Moak, Plenum Press, New York, 1975, Vol. 1, p. 261
22 M.C. Cross, in Inelastic Ion-Surface Collisions, Eds. N.H. Tolk, J.C. Tully, W. Heiland and C.W. White, Academic Press, New York, 1977, p. 253
23 M.C. Cross, Phys. Rev. B15 (1977) 602
24 H.D. Hagstrum, Phys. Rev. 96 (1954) 336
25 H.D. Hagstrum, in Inelastic Ion-Surface Collisions, Eds. N.H. Tolk, J.C. Tully, W.Heiland and C.W.White, Academic Press, New York, 1977, p.1
26 S. Horiguchi, K. Koyama and Y.H. Ohtsuki, phys. stat. solidi (b) 87 (1978) 757
27 E.S. Parilis and V.K. Verleger, J.Nucl.Mat. 94&95 (1980), to be published
28 S.L. Nizhnaya, E.S. Parilis and V.K. Verleger, Rad. Eff. 40 (1979) 23
29 S. Kreussler, University of Munich, Thesis, Munich 1980
30 B.A. Trubnikov and Y.N. Yavlinski, Soviet Phys. - JETP 25 (1967) 1089
31 M. Kitagawa and Y.H. Ohtsuki, Phys. Rev. B13 (1976) 4682
32 W. Brandt and R. Sizmann, Phys. Lett. 37A (1971) 115
33 J. Zwiegel, Technical University of Munich, Thesis, Munich 1979
34 D. Isaacson, "Compilation of r_s Values", New York University, Document No. 02698, National Auxiliary Publication Service, New York, 1975
35 W. Eckstein, H. Verbeek and R.S. Bhattacharya, Surf. Sci. 99 (1980) 356
36 W. Eckstein and F.E.P. Matschke, Phys. Rev. B14 (1976) 3231
37 T.M. Buck, Y.-S. Chen, G.H. Wheatley and W.F. van der Weg, Surf. Sci. 47 (1975) 244
38 T.M. Buck, in Inelastic Ion-Surface Collisions, Eds. N.H. Tolk, J.C. Tully, W. Heiland and C.W. White, Academic Press, New York, 1977, p. 47
39 W.F.S. Poehlman, R.S. Bhattacharya and D.A. Thompson, Nucl.Instr.Meth. 170 (1980) 549
40 W. Heiland and E. Taglauer, Nucl.Instr.Meth. 132 (1976) 535
41 E. Taglauer, W. Englert, W. Heiland and D. Jackson, Phys. Rev. Lett. 45 (1980) 740

42 W. Eckstein and H. Verbeek, 1980, not published
43 W. Eckstein, V.A. Molchanov and H. Verbeek, Nucl.Instr.Meth. 149 (1978) 599
44 T.M. Buck, G.H. Wheatley and L.K. Verheij, Surf. Sci. 90 (1979) 635
45 S.B. Luitjens, A.J. Algra and A.L. Boers, Surf. Sci. 80 (1979) 566
46 S.B. Luitjens, A.J. Algra, E.P.Th.M. Suurmeijer and A.L. Boers, Surf. Sci. 99 (1980) 631 and 652
47 S.Yu. Lukyanov and V.M. Chicherov, JETP Lett. 17 (1973) 360
48 D.P. Smith, Surf. Sci. 25 (1971) 171
49 H. Verbeek and W. Eckstein, Proc. 7th Intern.Vac.Congr. and 3rd Internat.Conf. Solid Surfaces, Vienna, 1977, Vol. 2, p. 1309
50 S.H. Overbury, P.F. Dittner and S. Datz, Nucl.Instr.Meth. 170 (1980) 543
51 W. Eckstein, H.G. Schäffler and H. Verbeek, Rad. Eff. 18 (1973) 263
52 F.E.P. Matschke, Technical University of Munich, Thesis, Munich, 1977
53 F.E.P. Matschke, W. Eckstein and H. Verbeek, Proc. 7th Int. Conf. on Atomic Collisions in Solids, Moscow, 1977
54 E.S. Mashkova, V.A. Molchanov, A.D. Pavlova and V.A. Snisar, Rad. Eff. 41 (1979) 187
55 F.J. Rogers, H.C. Graboske and D.J. Harwood, Phys.Rev. A1 (1970) 1577
56 J.B. Hasted, Physics of Atomic Collisions, Butterworths, London, 1972, p. 621ff
57 P. Pradel, G. Spiess, V. Sidis and C. Kubach, J. Phys. B12 (1979) 1485
58 V. Sidis and C. Kubach, J. Phys. B11 (1978) 2687
59 W. Heiland and E. Taglauer, in Inelastic Ion-Surface Collisions, Eds. N.H. Tolk, J.C. Tully, W. Heiland and C.W. White, Academic Press, New York, 1977, p. 27
60 A. Zartner, Max-Planck-Institut für Plasmaphysik, Report IPP 9/31 (1979)
61 L.K. Verheij, Rijksuniversiteit te Groningen, Thesis, Groningen 1976
62 L.K. Verheij, B. Poelsema and A.L. Boers, Nucl.Instr.Meth. 132 (1976) 565
63 W.F. van der Weg and D.J. Biermann, Physica 44 (1969) 177
64 H.H. Brongersma, N. Hazewindus, J.M. van Nieuwland, A.H.M. Otten and A.J. Smets, J. Vac. Sci. Technol. 13 (1976) 670
65 J.C. Tully and N.H. Tolk, in Inelastic Ion-Surface Collisions, Eds. N.H. Tolk, J.C. Tully, W. Heiland and C.W. White, Academic Press, New York, 1977, p. 105
66 J.C. Tully, Phys. Rev. B16 (1977) 4324
67 W. Bloss and D. Hone, Surf. Sci. 72 (1978) 277
68 R.L. Erickson and D.P. Smith, Phys. Rev. Lett. 34 (1975) 297
69 H.H. Brongersma and T.M. Buck, Nucl.Instr.Meth. 132 (1976) 559
70 T.W. Rusch and R.L. Erickson, J. Vac. Sci. Technol. 13 (1976) 574
71 D. Christensen, V. Mosotti, T. Rusch and R. Erickson, Chem. Phys. Lett. 44 (1976) 8

72 N.H. Tolk, J.C. Tully, J. Kraus, C.W. White and S.N. Neff, Phys. Rev. Lett. 36 (1976) 747
73 T.W. Rusch and R.L. Erickson, in Inelastic Ion-Surface Collisions, Eds. N.H. Tolk, J.C. Tully, W. Heiland and C.W. White; Academic Press, New York, 1977, p. 73
74 H.F. Helbig and P. Adelman, J. Vac. Sci. Technol. 14 (1977) 488
75 A. Zartner, E. Taglauer and W.Heiland, Phys. Rev. Lett. 40 (1978) 1259
76 H.F. Helbig and K.J. Orvek, Nucl. Instr. Meth. 170 (1980) 505
77 M. Hou, W. Eckstein and H. Verbeek, Rad. Eff. 39 (1978) 107
78 W. Eckstein, H.G. Schäffler and H. Verbeek, Max-Planck-Institut für Plasmaphysik, Report IPP 9/16 (1974)
79 L.M. Kishinevsky, E.S. Parilis and V.K. Verleger, Rad.Eff.29 (1976) 215
80 L.P. Levine and H.W. Berry, Phys. Rev. 118 (1960) 158
81 D.V. McCaughan, R.H. Sloane and J. Geddes, Rev.Sci.Instr.44 (1973) 605
82 E.R. Cawthron, Proc. Roy. Soc. London, A341 (1975) 213
83 H. Verbeek, W. Eckstein and S. Datz, J. Appl. Phys. 47 (1976) 1785
84 W. Eckstein, F.E.P. Matschke and H. Verbeek, J.Nucl.Mat. 63 (1976) 199
85 G.I. Zhabrev and V.A. Kurnaev, Sov. Techn. Phys. Lett. 4 (1978) 429
86 W.N. Lennard, D. Phillips and D.A.S. Walker, J.Phys. B, to be published
87 H. Verbeek, W. Eckstein and R.S. Bhattacharya, Surf. Sci. 95 (1980) 380
88 H. Verbeek and W. Eckstein, Proc. SASP, Maria Alm, 1980, p. 14
89 R.A. Smith, Proc. Roy. Soc. (London) A168 (1938) 19
90 M.E. Kishinevskii, Sov. J. Techn. Phys. 48 (1978) 773
91 J.R. Hiskes, J. Physique 40 (1979), C7-179
92 R.K. Janev, Surf. Sci. 45 (1974) 609; J. Phys. B7 (1974) 1506 and L359
93 J.R. Hiskes and P.J. Schneider, Phys. Rev. B, to be published
94 J.R. Hiskes and Karo, Lawrence Livermore Laboratory, Report UCRL-79512, 1977
95 J.R. Hiskes and P.J. Schneider, UCRL-94262 and J.Nucl.Mat.

96 S.A. Agamy and J.E. Robinson, Surf.Sci. 90 (1979) 648
97 T.M. Buck, G.H. Wheatley and L.C. Feldman, Surf. Sci. 35 (1973) 345
98 T.M. Buck, L.C. Feldman and G.H. Wheatley, in Atomic Collisions in Solids, Eds. S.Datz, B.R. Appleton and C.D. Moak, Plenum Press, New York, 1975, Vol. 1, p. 331
99 H. Verbeek and W. Eckstein, Proc. SASP, Tirol, 1978, p. 43
100 D.A. Thompson and W.F.S. Poehlman, Nucl. Instr. Meth. 168 (1980) 63
101 R.L. Wax and W. Bernstein, Rev. Sci. Instr. 38 (1967) 1612
102 T. Hall, Phys. Rev. 79 (1950) 504
103 J.A. Phillips, Phys. Rev. 97 (1955) 404
104 P.L. Smith and W. Whaling, Phys. Rev. 36 (1969) 188
105 P. Hvelplund, E. Laegsgård, J.O. Olson and E.H. Pedersen, Nucl. Instr. Meth. 90 (1970) 315

Plasma Effects in the Theory of Charge Fractions

M. Kleber and J. Zwiegel
Physik-Department, Technische Universität München
D-8046 Garching, Fed. Rep. of Germany

1. Introduction

Recently, the positive charge fractions of slow protons and deuterons backscattered from clean metallic surfaces have been measured [1-3]. The positive charge fraction of protons

$$\eta^+ = n_+/(n_+ + n_0 + n_-) \tag{1.1}$$

is defined in terms of the number of backscattered protons, n_+, of hydrogen atoms, n_0, and of negative hydrogen atoms, n_-. It was found that for low velocities η^+ is proportional to the kinetic energy of the backscattered particles. This important result holds true for deep-inelastically scattered $Z = 1$ particles which penetrate into the solid, loose energy and, finally, emerge from the surface with low velocity $v<v_F$, where v_F is the Fermi velocity of the metal. Such processes play an important role in the context of plasma-wall interaction.

In this contribution we want to point out a mechanism which may cause the positive charge fractions of slow protons, deuterons or He^+ to depend linearly on their kinetic energy. It is not easy to understand this behaviour. In case of a stationary proton at a distance R outside a metallic surface (Fig.1) there is a potential barrier between metal and proton: An empty proton state with binding energy I can get filled via electron tunnelling. Energy matching requires

$$\phi \leq I \leq \phi + E_F, \tag{1.2}$$

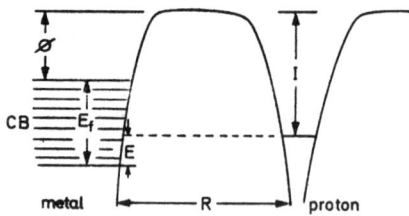

Fig.1 Stationary proton at a distance R from a metallic surface. CB denotes the conduction band; all other symbols are explained in the text.

where ϕ is the work function and where $E_F = p_F^2/2m$ denotes the Fermi energy of the metal with Fermi momentum p_F. For an estimate of the tunnelling probability T we have [4]

$$T = \gamma \frac{E}{\hbar} \exp[-\frac{2R}{\hbar}\sqrt{2mI}]. \qquad (1.3)$$

Here $E/\hbar = (\phi+E_F-I)/\hbar$ is the "bouncing" frequency of that electron which is going to penetrate the barrier and γ stands for an undetermined dimensionless quantity of order unity. For a slowly moving proton, $v \ll v_F = p_F/m$, one finds from (1.3) that tunnel recombination is very effective. The positive charge fraction should be exponentially small;

$$n_+ \cong \text{const} \cdot \exp(-A/v_\perp) \qquad (1.4)$$

with A some positive constant of dimension velocity and $v_\perp$ the proton velocity normal to the surface. As was discussed in the preceding contribution, (1.4) is clearly at variance with the experimental findings.

For a proton emerging from a surface there are, however, two serious problems with the concept of tunnel recombination:
(i) When the proton is leaving the surface, there is no tunnelling because there is no barrier. When the slowly moving proton is still very close to the surface the barrier will be very small. Then, a first-order tunnelling calculation is not adequate because the electron can tunnel back and forth.
(ii) The proton will reach and leave the surface as a screened proton. Right at the surface protons cannot be treated as isolated particles. Energy levels and wave function are quite different from their asymptotic values.
The screening of an ion represents an object which requires a very complicated wave function.

It is evident that screening will have some influence on the charge state of an ion both inside and outside the metal. This is a plasma effect which is beyond the independent particle picture. In what follows we shall expend some effort on the many-body aspect of the charge state problem. To isolate the screening effects, let us focus upon slow Z = 1 particles because they are found [5,6] to have no bound states in metals. Therefore, only dynamical screening takes place which we briefly recall in section 3. Then, in section 4 plasma effects upon charge states are studied. The next section deals with the kinematics of the moving proton.

2. Kinematics

Let us consider a proton of mass m_p and given velocity $\underline{v}$ in a metal (Fig.2). Then, the position of the proton will be smeared out over the de Broglie wavelength $\lambda = h/(m_p v)$. For the range of velocities considered here, λ is both small compared with the atomic scale (10^{-8}cm) and with the classical minimum distance between proton and atoms of the metal. Therefore, the motion of the proton will be treated classically.

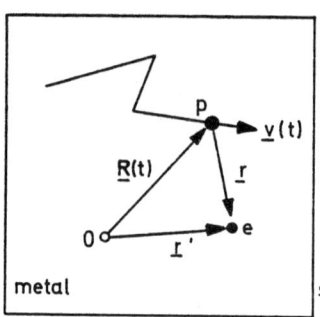

Fig.2 Coordinates of proton (p) and electron (e)

We are interested in electron states which asymptotically correlate to the proton. Hence, it will be useful to define the electron states with respect to the travelling proton. The corresponding system of reference will be a non-inertial one because the proton does not necessarily move on a straight line at constant speed. But what do wave function and Hamiltonian look like in an arbitrarily accelerated system? The answer is quite simple. If $\phi(\underline{r},t)$ stands for any wave function which travels with the proton, then by writing

$$\phi(\underline{r},t) = \exp[\frac{im}{\hbar} (\underline{v}\cdot\underline{r} + \frac{1}{2} \int^t v^2(t')dt')] \chi(\underline{r},t) \qquad (2.1)$$

the wave function $\chi(\underline{r},t)$ may be shown (Appendix A) to satisfy the time-dependent Schrödinger equation

$$i\hbar\partial_t \chi(\underline{r},t) = (H + m \dot{\underline{v}}\cdot\underline{r}) \chi(\underline{r},t). \qquad (2.2)$$

H is the Hamiltonian of an electron in the stationary (v = 0) system of the proton. The additional term $m\dot{\underline{v}}\cdot\underline{r}$ in (2.2) reflects the inertial force on an electron at position $\underline{r}$ (Fig.2). Finally, $\underline{v}(t)$ is the velocity of the proton relative to an inertial system.

The inertial force

$$\underline{F}_i = - \underline{\nabla}(m\dot{\underline{v}}\cdot\underline{r}) = - m\dot{\underline{v}} \qquad (2.3)$$

is due to the acceleration of the proton and it may shake off an electron which travels with the proton. Due to the heavy proton mass, $\dot{\underline{v}}$ will be small except for head-on collisions between proton and atoms of the metal. This process has, however, a very small cross section. So we shall neglect the inertial term $m\dot{\underline{v}}\cdot\underline{r}$, assuming that the proton moves with constant velocity.

The problem we are going to attack is that of a proton moving in a "soup" of conduction electrons. Such a plasma may be a complicated object, but we assume that we are dealing with jellium. In this model electrons move like a fluid against a uniform positively charged medium (Fig.3). Most plasma electrons do not see

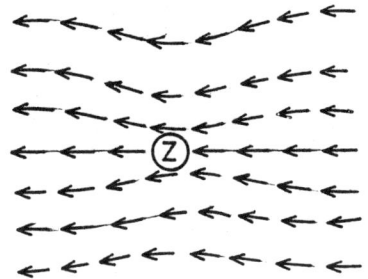

Fig.3 Flow of electrons against a positively charged impurity Z

the moving proton directly. Instead, they feel the presence of the ion indirectly through a change in the position of their neighbouring electrons. Thus in jellium the proton behaves like a recoiling particle which can shake off its electronic screening. If $\phi(\underline{r},t)$ is the screening wave function which travels with the proton, then from (2.1)

$$\chi(\underline{r},t) = \exp(-i\,\underline{q}\cdot\underline{r})\,\phi(\underline{r},t) \qquad (2.3)$$

describes the same screening wave function which now, however, is stationary. In (2.3) we introduced the recoil momentum $\underline{q} = m\underline{v}/\hbar$ and skipped an unimportant time-dependent phase. The amplitude that the free electron cloud will stick to the moving proton is just the overlap between travelling and stationary wave function;

$$\begin{aligned}T_o &= \int d^3 r\,\chi^*(\underline{r},t)\,\phi(\underline{r},t) \\ &= \int d^3 r\,|\phi(\underline{r},t)|^2\,\exp(i\,\underline{q}\cdot\underline{r}).\end{aligned} \qquad (2.4)$$

Note that in a fluid any steady motion is "sudden". This behaviour is reflected in (2.4) and it implies that electron shake-off occurs at distances $r \simeq 1/q$ from the proton. In passing we note that uniform motion of an isolated atom will, of course, never lead to electron shake-off.

3. Dynamical Screening

It is well known [7,8] that inside a metal a moving proton with charge density

$$e\,\rho_p(\underline{r}') = e\,\delta(\underline{r}'-\underline{v}t) \qquad (3.1)$$

will induce an electronic charge density

$$-e\,\rho(\underline{r}') = \frac{-e}{(2\pi)^3}\int d^3k\,[1-1/\varepsilon(\underline{k},\omega=\underline{k}\cdot\underline{v})]\,e^{i\underline{k}\cdot(\underline{r}'-\underline{v}t)} \qquad (3.2)$$

with $\varepsilon(\underline{k},\omega)$ the dielectric "constant" of the medium. In metals there is complete screening of ions:

$$\int d^3r' \, \rho(r') = \lim_{k \to 0} [1 - 1/\varepsilon(\underline{k}, \omega = \underline{k} \cdot \underline{v})] = 1 \qquad (3.3)$$

because for $\underline{k} \to \underline{0}$ the dielectric constant $\varepsilon(\underline{k}, \omega = \underline{k} \cdot \underline{v})$ goes to infinity.

If we wish to study screening for very slow protons, then we may use the static limit of the dielectric constant. In Thomas Fermi approximation

$$\varepsilon_{TF}(\underline{k}, 0) = 1 + \left(\frac{k_{TF}}{k}\right)^2 . \qquad (3.4)$$

In this limit (3.2) gives the well-known Debye or Yukawa screening

$$\rho_{TF}(\underline{r}') = \frac{k_{TF}^2}{4\pi} \frac{1}{r} \exp(-k_{TF} r),$$

$$\underline{r} = \underline{r}' - \underline{v} t . \qquad (3.5)$$

In contrast to the adiabatic limit (3.5) the induced charge (3.2) will lack behind the proton for $v \neq 0$. A small fraction of the screening may scatter off the proton, but will be replaced as long as the proton moves in the metal. It has been mentioned before that within the fluid-dynamical approach (2.4) the electron loss will mainly take place at distances $r \geq 1/q$ from the proton. A more detailed analysis will follow in the next section.

The free-electron plasma is characterized by a single quantity, the plasma frequency ω_p. It may be expressed in terms of the Fermi wave vector k_F

$$\omega_p^2 = \frac{4}{3\pi} \frac{e^2}{m} k_F^3 . \qquad (3.6)$$

The Thomas-Fermi wave vector k_{TF} is related to k_F by

$$k_{TF}^2 = \frac{4}{\pi} \frac{k_F}{a_o} , \qquad (3.7)$$

where a_o is the Bohr radius.

4. Fluid-dynamical Approach to Sticking Amplitude

Let us study the effect of a moving heavy particle where $\omega = vq$ is small compared to the plasma frequency ω_p. We denote the particle's nuclear charge by Z and assume that it is screened by $N \leq Z$ electrons with wave function $\phi(\underline{r}_1, \underline{r}_2, \ldots \underline{r}_N, t)$. We do not know ϕ, but we do know two important things:

(i) The one-body screening density $\rho(\underline{r}, t)$ is given by the matrix element

$$\rho(\underline{r}, t) = \langle \phi(t) | \psi^+(\underline{r}) \psi(\underline{r}) | \phi(t) \rangle \qquad (4.1)$$

a quantity much easier accessible than ϕ itself. In Eq.(4.1) $\psi^+(\underline{r})$ and $\psi(\underline{r})$ create and annihilate an electron at point $\underline{r}$. We have used the Dirac notation for bra and ket vectors, suppressing the electron spin.

(ii) Outside the metal, for $v \to 0$, ϕ correlates to an atomic state of the adiabatically moving particle. In this limit the ion will leave the metal as a neutral atom in the ground state.

In the free-electron gas approximation electrons are streaming against the moving ion. Note that this simple classical picture is able to account for the phenomenon of wake riding [9,10] of an ion which moves fast through a fluid. Here we apply the fluid-dynamical aspect to low velocities.

In the system of the moving ion the electrons of the metal flow back. Since the screening electrons are free, they will somehow flow away, but will be replaced by other electrons of the metal. However, as soon as the ion reaches the surface, replacement stops. We now calculate the probability that all screening electrons stick to the moving ion. From the results in section 2 the sticking amplitude T_o is seen to be

$$T_o = <\phi(t)| \exp(i\sum_{j=1}^{N} \underline{q}\cdot\underline{r}_j)|\phi(t)> . \tag{4.2}$$

In general there is no simple expression for T_o. However, within the Hartree approximation T_o can be calculated from $\rho(\underline{r},t)$ in the low velocity limit $v = \hbar q/m \ll v_F$. In Hartree approximation the screening wave function is a product of single-particle wave functions

$$\phi(\underline{r}_1, \ldots \underline{r}_N, t) = \varphi_1(\underline{r}_1, t) \cdot \ldots \varphi_N(\underline{r}_N, t). \tag{4.3}$$

The corresponding charge density (4.1) then becomes

$$\rho(\underline{r},t) = \sum_{j=1}^{N} |\varphi_j(\underline{r},t)|^2 . \tag{4.4}$$

For small q the exponent in (4.2) may be expanded to second power of q with the result

$$T_o \cong 1 - \frac{1}{2}\sum_{j=1}^{N} <\phi(t)|(\underline{q}\cdot\underline{r}_j)^2|\phi(t)> . \tag{4.5}$$

In deriving (4.5) we made use of

$$<\phi|\underline{r}_j|\phi> = \underline{0} \text{ and } <\phi|(\underline{q}\cdot\underline{r}_j)(\underline{q}\cdot\underline{r}_\ell)|\phi> = 0 \text{ for } j \ne \ell.$$

Now

$$<\phi(t)|(\underline{q}\cdot\underline{r}_j)^2|\phi(t)> = \int d^3r (\underline{q}\cdot\underline{r})^2 |\varphi_j(\underline{r},t)|^2 \tag{4.6}$$

so that

$$T_o \cong 1 - \frac{1}{2} \int d^3r (\underline{q} \cdot \underline{r})^2 \rho(\underline{r},t). \tag{4.7}$$

This is the desired relation between sticking amplitude and density. Note that N, the number of electrons which participate in ion screening, does not appear explicitly in (4.7). In this equation N is no longer restricted to integers.

Let us return to Z = 1 screening. Here N = 1 and (4.7) may be written in the form ($q \ll k_F$)

$$T_o = \int d^3r \, \rho(\underline{r},t) \exp(i \underline{q} \cdot \underline{r}). \tag{4.8}$$

If we put (3.2) into (4.8), we find

$$T_o = [1 - 1/\varepsilon(-\underline{q},\omega=-\underline{q}\cdot\underline{v})] \exp(i\underline{q}\cdot\underline{v}\, t). \tag{4.9}$$

Equation (4.9) is only valid at small velocities. It is well known [11] that in the random phase approximation (RPA) we have

$$\varepsilon(-\underline{q},\omega=-\underline{q}\cdot\underline{v}) \xrightarrow[q \ll k_F]{} \varepsilon_{TF}(q,0) = 1 + \left(\frac{k_{TF}}{q}\right)^2. \tag{4.10}$$

Hence the long-wavelength screening is identical to that calculated in Thomas-Fermi approximation.

The screening electron sticks with probability $P_o = |T_o|^2$ to the proton (or deuteron). We assume a sharp surface and take into account that for adiabatic motion the screening configuration correlates to the atomic ground state. Then, when the proton leaves the metal, P_o is the probability of finding hydrogen in its ground state. From (4.9) and (4.10) we get

$$P_o = 1 - \frac{\varepsilon + \varepsilon^* - 1}{\varepsilon \varepsilon^*} \xrightarrow[q \ll k_F]{} 1 - 2\left(\frac{q}{k_{TF}}\right)^2 \tag{4.11}$$

with $\varepsilon = \varepsilon_{TF}(q,0)$.

One must, of course, also consider formation of hydrogen in excited states. This again is a complicated problem of computation. The quantity we are really after is the positive charge state fraction, i.e. the ionization probability P_{ion}

$$\eta^+ = P_{ion} = 1 - P_o - P_{ex}, \tag{4.12}$$

where P_{ex} denotes the formation probability in an excited state. In appendix B we calculate P_{ion} in hydrogen approximation. The result is

$$\eta^+ = f(v)(1 - P_o) \tag{4.13}$$

with $f(v)$ a velocity-dependent function and $f(v) \cong 0.28$ for $v \ll v_F$. In this limit an expansion of (4.13) in powers of q/k_{TF} yields

$$\eta^+ = 2 f(0) \left(\frac{q}{k_{TF}}\right)^2 \qquad (4.14)$$

$$= 2 f(0) \frac{\pi}{4} \left(\frac{4}{3\pi}\right)^{1/3} \frac{E_{kin}[keV]/25}{(\hbar\omega_p[a.u.])^{2/3}} .$$

This is our final result for the positive charge state fractions of slowly moving protons. It is seen that the ionization probability depends linearly on the kinetic energy E_{kin} of the proton; the properties of the metal enter into η^+ only through the plasma frequency ω_p.

In passing we note that ECKSTEIN [3] uses the dielectric constant of the electron gas in the static limit [11]

$$\varepsilon(\underline{q},\omega = 0) = 1 + \left(\frac{k_{TF}}{q}\right)^2 g(x)$$

$$g(x) = \frac{1}{2} \left[1 + \frac{1}{2x} (1-x^2) \ln\left|\frac{1+x}{1-x}\right|\right] \qquad (4.15)$$

with $x = q/(2k_F)$. For $q \ll k_F$ we get the same results as derived above.

5. Miscellaneous Comments

In this contribution we worked out a fluid-dynamical approach to the problem of positive charge states of slow $Z = 1$ particles. The theory ought to be considered as a first step towards an understanding of the experimental findings. A comparison with experimental results is found in ECKSTEIN's contribution. Here we discuss some open problems.

(a) Role of work function: Let Δx be a typical length which determines the thickness of the surface region. Then the image force $F \cong \phi/\Delta x$ tends to pull a single electron into the metal. For the combined system, electron plus proton, the image force will be considerably smaller. We neglected this force, but a more detailed analysis would be desirable.

(b) Realistic screening: Close to a real surface the proton-induced screening should be treated in a more realistic way [12].

(c) Surface plasmons: It is difficult to decide whether ω_p is the plasma frequency of the bulk or of the surface.

(d) Angular dependence: In the approximation (4.14) η^+ is independent of the proton's exit angle. For normal exit (4.14) should work best. In case of near parallel exit, tunnelling will become important and (4.14) is not applicable.

(e) Bound states: In some metals electron screening does not prevent the proton from having a bound state. In Cs, for example, $\hbar\omega_p$ = 3.4 eV, which means [5] that a proton can bind an electron with 2.4 eV. In this case the observed positive charge fraction will be smaller than predicted by (4.14).

(f) Microscopic theory: It would be very useful to calculate correlation diagrams for energies and wave functions of a proton as a function of its distance from the metal. One then could calculate transition amplitudes by utilizing the time-dependent version of Schwinger's variational method. Up to now such a program has only been carried out for single-collision phenomena [13,14].

Acknowledgments: Part of this work was supported by the Bundesministerium für Forschung und Technologie. We would like to thank W. Brenig, W. Eckstein and H. Verbeek for stimulating discussions.

Appendix A

The transition from an arbitrary inertial system 0 (Fig.2) to the system of the moving proton is usually achieved [15] by a canonical transformation of the Lagrangian. There is, however, an easier line of argument. We know that the numerical value of the electron wave function at a fixed point (e) cannot depend on the coordinate system. Therefore, if primed variables refer to the inertial system 0 and unprimed variables to the proton, we have from Fig.2

$$\phi(\underline{r},t) = \phi'(\underline{r}',t'). \tag{A1}$$

Now, for rigid nonrelativistic transformations

$$\underline{r}' = \underline{r} + \underline{R}(t) ; \quad t' = t. \tag{A2}$$

In the inertial system 0 the time-dependent Schrödinger equation is given by

$$i\hbar\, \partial_{t'}\, \phi'(\underline{r}',t') = H'\, \phi'(\underline{r}',t'). \tag{A3}$$

By use of (A1) and (A2) we find

$$\partial_{t'}\, \phi'(\underline{r}',t') = \partial_{t'}\, \phi(\underline{r},t)$$
$$= [\partial_t - \underline{\dot{R}}(t)\cdot\underline{\nabla}]\, \phi(\underline{r},t). \tag{A4}$$

The Hamiltonian is invariant against translation. From the last two equations we obtain

$$i\hbar\, \partial_t\, \phi(\underline{r},t) = (H - \underline{\dot{R}}(t)\underline{p})\, \phi(\underline{r},t) \tag{A5}$$

with $\underline{p} = -i\hbar\underline{\nabla}$. Now, all quantities are defined relative to the proton. If H does not contain velocity-dependent forces, then one can easily verify that the unitary transformation (2.1) will cast (A5) into the more convenient form (2.2).

Appendix B

In Thomas-Fermi approximation the screening density does not have nodes. Hence the wave function which corresponds to $\rho(\underline{r},t)$ is given by

$$\chi(\underline{r},t) = [\rho(\underline{r},t)]^{1/2}. \tag{B1}$$

Note that $\chi(\underline{r},t)$ looks very much like a hydrogen wave function. Nevertheless, we do not claim that the ionization probability

$$P_{ion} = \int d^3k |<\underline{k}| \exp(i\underline{q}\cdot\underline{r}) | \chi >|^2 \tag{B2}$$

may be calculated from hydrogenic functions. Instead, we only assume the ratio between ionization and total excitation

$$f = \frac{P_{ion}}{P_{ion} + P_{ex}} = \frac{P_{ion}}{1 - P_o} \tag{B3}$$

to be the same as for atomic hydrogen. For the ground state of hydrogen one has

$$P_o = |<1s| \exp(i\underline{q}\cdot\underline{r}) | 1s>|^2$$
$$= 1 - (q a_o)^2 \tag{B4}$$

if $q a_o \ll 1$ holds. In this limit one also finds [16]

$$P_{ion} = \frac{256}{3} (q a_o)^2 \int_o^\infty dx \frac{x^7 \exp[-4x \text{ arc cotx}]}{(1-\exp(-2\pi x))(1+x^2)^5}$$
$$= 0.2834 (q a_o)^2. \tag{B5}$$

Hence, for proton velocities which are considerably less than the Fermi velocity and the Bohr velocity we obtain

$$f = 0.2834; \qquad v \ll v_F, v_o. \tag{B6}$$

In this limit direct ionization is only 28 percent of the total excitation. For higher v the ratio f will increase. One therefore has a velocity-dependent ratio as indicated in (4.13).

References

[1] R.S. Bhattacharya, W. Eckstein and H. Verbeek, Surf. Sci. 93 (1980) 563
[2] W. Eckstein, H. Verbeek and R.S. Bhattacharya, Surf. Sci. 99 (1980) 356
[3] W. Eckstein, contributed paper to this conference.
[4] B.A. Trubnikov and Yu.N. Yavlinskii, Sov. Phys.-JETP 25 (1967) 1089
[5] F.J. Rogers, H.C. Graboske and D.J. Harwood, Phys. Rev. A1 (1970) 1577
[6] W. Brandt in "Atomic Collisions in Solids", ed. S. Datz, B.R. Appleton and C.D. Moak; Plenum Press, N.Y. (1975), p. 261
[7] J. Neufeld and R.H. Ritchie, Phys. Rev. 98 (1955) 1632
[8] R. Brout and P. Carruthers, "Lectures on the Many-Electron Problem"; Wiley, N.Y. (1963), ch. 5
[9] N. Bohr, Mat. Fys. Medd. Dan. Vid. Selsk 18 (1948) 71; fig.6
[10] P.M. Echenique, R.H. Ritchie and W. Brandt, Phys. Rev. B20 (1979) 2567
[11] J. Lindhard, Mat. Fys. Med. Dan. Vid. Selsk. 28 (1954) No.8
[12] N.D. Lang and A.R. Williams, Phys. Rev. B16 (1977) 2408
[13] M. Kleber and K. Unterseer, Z. Physik A292 (1979) 311
[14] M. Kleber and J. Zwiegel, Phys. Rev. A19 (1979) 579
[15] D.M. Greenberger and A.W. Overhauser, Rev. Mod. Phys. 51 (1979) 43
[16] M.R.C. McDowell and J.P. Coleman, "Introduction to the Theory of Ion-Atom Collisions", North Holland, Amsterdam (1969), p.322

Polarized Light Emission

Polarization of Balmer Radiation from Grazing Incidence Collisions of Protons on Surfaces

J.C. Tully, N.H. Tolk, J.S. Kraus, C. Rau* and R.J. Morris[†]
Bell Laboratories
Murray Hill, NJ 07974, USA

Introduction

We are carrying out experimental and theoretical studies of optical radiation emitted during grazing incidence collisions of ions with surfaces. This effort is directed at elucidating the mechanisms of formation and destruction of anisotropically populated excited states at surfaces. This has relevance not only to grazing incidence experiments, but also to ion backscattering, sputtering, transmission through foils, and in general the interactions of excited states with surfaces. The experiment is illustrated schematically in Fig. 1. Ions are incident at grazing angles to the surface. After interacting with the surface, neutralized particles proceed through an applied electric field region and then into the observation region where optical radiation emitted perpendicular to the collision plane is detected. The polarization of this radiation can provide information about the nature of the electronic state produced during the encounter with the surface.

To illustrate how polarization might contain clues about the underlying electronic processes, consider the shapes of the $3p_x$ and $3p_y$ states of the hydrogen atom, Fig. 2. We might imagine the probability of electron pickup into the $3p_x$ orbital to be different from that into the $3p_y$ orbital, due perhaps to different overlaps with surface electronic wave functions. Then Balmer α radiation emitted in the z direction (in the absence of complications from s and d states) would exhibit linear polarization perpendicular (parallel) to the surface if $3p_x$ ($3p_y$) were preferentially populated.

Strong optical polarization has, indeed, been observed in grazing incidence collisions of ions with surfaces. However, it is not of the simple linear nature described above. Andrä and coworkers [1,2] and Berry and coworkers [3] have reported circular polarization in 40 to 2500 keV collisions. At somewhat lower energies, 1-9 keV, Tolk and coworkers [4,5] have observed very high polarization fractions exceeding 60%. Furthermore, Tolk et al. have observed that the polarization of Balmer α radiation from H atoms is not pure circular but elliptical, and that the major axis is directed, surprisingly, at about 45° with respect to the surface normal, not at 0° or 90° as suggested by the simple picture above.

---

* Present address: Sektion Physik der Universität München, D8 Munich 40, FRG.
† Present address: Department of Physics, Columbia University, New York, N.Y. 10027, USA

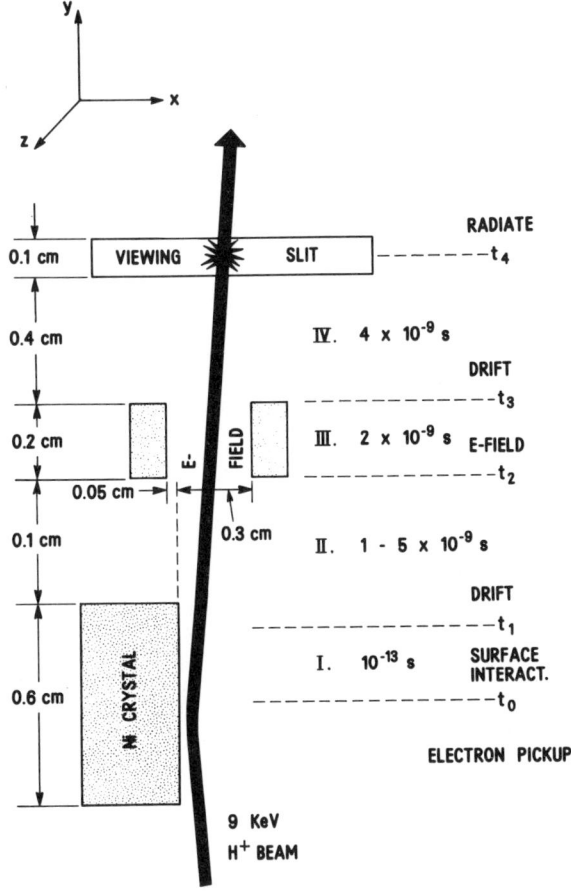

Fig. 1. Schematic illustration of grazing incidence experiment.

These observations have spurred considerable speculation about the mechanism of production of excited states at surfaces. The following is a list, probably not exhaustive, of processes which have been proposed to play a role in the formation of oriented and aligned excited states at surfaces:

a. Anisotropic electron pickup [4-8].

b. Evolution in surface fields [9-11].

c. Impact parameter selection [12].

d. Scattering from surface plasmons [13].

e. Anisotropic de-excitation or ionization [5].

f. Spin-aligned pickup [14].

Any of these mechanisms, or combinations of them, might be involved. Existing data cannot distinguish among them, mainly because measurement of the three polarization parameters is usually not sufficient to determine the excited state wave function. The objective of the present study is to provide further data which will be sufficient to uniquely define the excited state produced during interaction with the surface. We are attempting this, as described below, by forcing the coherent state to re-orient in applied electric fields (Fig. 1.) and observing the subsequent change in optical polarization. Although we have not yet achieved the final objective, our new results have yielded interesting qualitative information that bears directly on several of the possible mechanisms listed above.

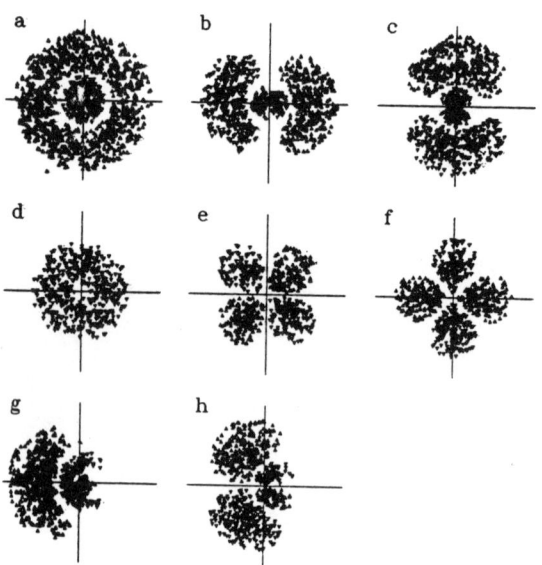

Fig. 2. Hydrogenic $n=3$ wavefunctions. Density of dots represents absolute value of amplitude of wave function in the xy plane ($z=0$). a. s; b. p_x; c. p_y; d. d_z^2; e. d_{xy}; f. $d_{x^2-y^2}$; g. $\sqrt{8}s - \sqrt{12}p_x + d_{z^2} + \sqrt{3}d_{x^2-y^2}$; h. $p_y - d_{xy}$.

Experiment

Experiments were carried out as described previously [4,5]. A 1 to 9 keV beam of H^+ or H_2^+ was directed at the (100) face of a single Ni crystal at grazing incidence angle $\sim 4°$. The scattering chamber was maintained at a pressure of 2×10^{-10} torr, and the target was cleaned by ion bombardment and heating. As demonstrated in earlier work [4,5], a fraction of a monolayer of oxygen on the surface results in a drastic decrease in observed optical polarization.

The experimental geometry is illustrated in Fig. 1. Balmer α radiation arising from the $n=3$ to $n=2$ transition of the H atom emitted in the z direction (out of the page in Fig. 1) was detected. Four quantities were determined, the relative intensity I and the three Stokes parameters [15] S/I, M/I and C/I defined in the usual way:

$$S/I = (I_{RHC} - I_{LHC})/(I_{RHC} + I_{LHC})$$
$$M/I = (I_0 - I_{90})/(I_0 + I_{90}) \qquad (1)$$
$$C/I = (I_{45} - I_{135})/(I_{45} + I_{135}).$$

I_{RHC} and I_{LHC} are the intensities of right-hand and left-hand circular polarization, respectively. I_0, I_{45}, I_{90}, and I_{135} are the intensities of radiation with direction of polarization 0°, 45°, 90° and 135°, respectively, where 0° is along the x direction and 90° along the y direction. The three Stokes parameters are sufficient to define the total polarization fraction

$$P = [(S/I)^2 + (M/I)^2 + (C/I)^2]^{1/2}, \qquad (2)$$

as well as the magnitudes and directions of the major and minor axes of the polarization ellipse.

In addition to observing radiation emitted along the z direction, we also performed experiments with the optical detector rotated to observe radiation emitted in the x direction of Fig. 1. We denote the intensity of radiation in the x direction by I', and the linear polarization by M'/I', defined as

$$M'/I' = (I'_\| - I'_\perp)/(I'_\| + I'_\perp) \qquad (3)$$

where $I'_\|$ and $I'_\perp$ are the intensities of radiation with direction of polarization parallel and perpendicular to the projection of the beam direction on the surface plane. By symmetry, the other two Stokes parameters vanish for radiation emitted in the x direction.

The experiments as described above yield four pieces of information, S/I, M/I, C/I and M'/I'. The relative intensities I and I' are of only qualitative significance. The n=3 hydrogenic level comprises 9 states (one s, three p and five d states). Accounting for the two spin orientations, the density matrix that defines the n=3 level is of order 18×18, with both diagonal and off-diagonal elements to be determined. The four polarization parameters, even including their dependences on angle and energy, are insufficient to determine the elements of the density matrix; i.e., to determine the populations and coherences of the excited n=3 state of the H atom as it escapes from the surface.

The contribution of the present work is the extraction of additional information about the nature of the excited state by introducing an applied electric field in the path of the excited particles (Fig. 1). Excited state coherences produced at the surface subsequently evolve in the electric field. This evolution is manifested in changes in the polarization of light ultimately emitted downstream. We have observed these changes in polarization, and have attempted to deduce from this information the coherent superposition of levels responsible.

We have measured six quantities, I, I', S/I, M/I, C/I, and M'/I', as a function of electric field strength from -800 to 800 V/cm. An example of the data is shown in Fig. 3 for H^+ incident energy of 9 keV and incident grazing angle 4°. At zero applied field, S/I is large (~-60%) and C/I is about -12%, in agreement with previous measurements. M/I at zero applied field was observed to vary between about 0 and 10%, apparently sensitive in some undetermined way to the preparation of the surface. M'/I' was positive, indicating linear polarization along the beam direction. S/I, M/I and C/I all oscillate rapidly as a function of field strength. These oscillations are due to quantum interferences arising from electric

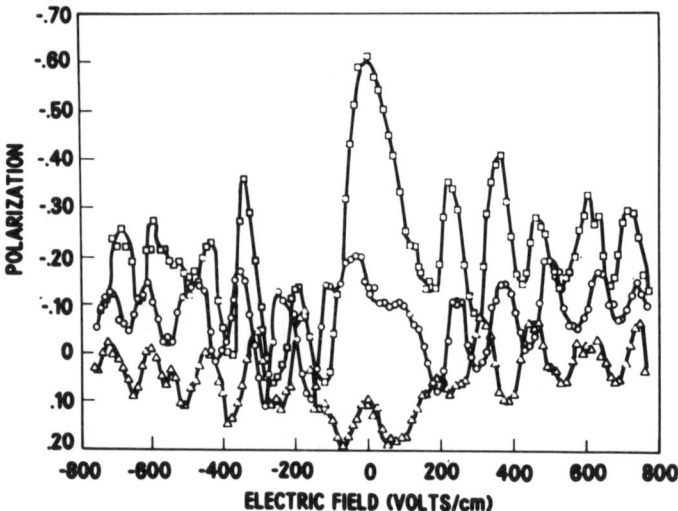

Fig. 3. Experimental measurement of variation of Stokes parameters with strength of applied electric field. Incident proton energy is 9 keV, incident angle is 4 deg., and the target is the (100) face of a Ni single crystal. Squares, S/I. Triangles, M/I. Circles, C/I.

field induced splittings of the near-degenerate H atom n=3 levels. (No detectable changes of polarization with applied electric field were observed for the 3^3D to 2^3P transition of He.) How the analysis of these oscillation may be used to provide information about the nature of the excited state formed at the surface is described in the next section.

Analysis

The analysis proceeds as follows: referring to Fig. 1, we make an initial choice (an intelligent guess) of all elements of the density matrix $\rho(t)$ at time $t=t_o$ corresponding to the instant the n=3 state is formed. The only restriction we impose other than symmetry is absence of spin-dependent interactions during the collision. (We relax this assumption below when we consider possible observation of spin-aligned electron pickup from ferromagnetic targets). Once selected, the density matrix $\rho(t_o)$ is propagated forward in time by numerical integration of the equation

$$\dot{\rho}(t) = i/\hbar^{-1}(H\rho - \rho H). \qquad (4)$$

Fortunately, we know everything necessary about hydrogenic levels and how they evolve in electric fields, so there are no uncertainties about this part of the analysis.

The density matrix is first propogated in region I of Fig. 1, the region of interaction of the escaping particle with the surface, from time t_o to t_1. The Hamiltonian H_I operative in region I describes the interaction of the excited state with the surface. We do not know the precise form of this interaction. We assume that it can be approximated by a constant electric field directed normal to the surface, so H_I takes the form

$$H_I = H_{SI}. \qquad (5)$$

The surface interaction Hamiltonian H_{SI} is assumed to dominate the spin-orbit and Lamb shift terms, so the latter can be neglected in region I [16]. Then the evolution of ρ depends only on the product of the effective surface field and the evolution time t_1-t_0. In the above discussion we have viewed the mechanism as an excited state formation event at time t_0 followed by evolution under the influence of the surface to time t_1. This separation is arbitrary, and experiments of the type described here can only probe the excited state after it leaves the surface region; i.e., after t_1. We employ the above procedure because it provides a convenient way to construct a density matrix at time t_1. Whether $\rho(t_1)$ arose via this two-step mechanism or in some other way can probably be determined only by additional theoretical arguments.

From time t_1 to t_2, region II of Fig. 1, the density matrix evolves without the influence of an external field, subject only to the spin-orbit, H_{SO}, and Lamb shift, H_{LS}, interactions [16]:

$$H_{II} = H_{SO} + H_{LS}. \qquad (6)$$

Hyperfine interactions in the n=3 state are sufficiently weak that they can be neglected here. As seen in Fig. 4, both spin-orbit and Lamb shift interactions are very significant. The time t_2-t_1 that the atom evolves in region II is uncertain, ranging from $1-4 \times 10^{-9}$ sec, depending on where the particle struck the target

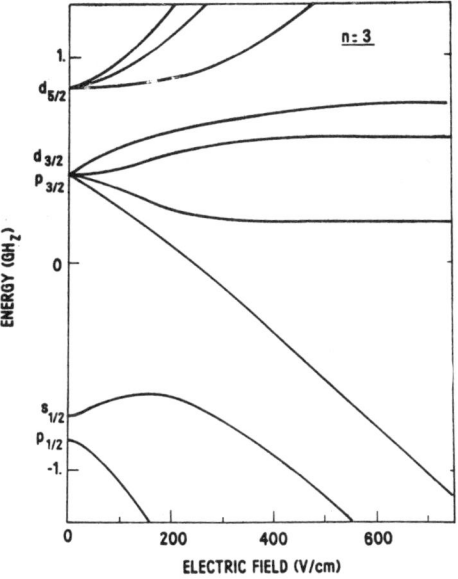

Fig. 4. Energies of n=3 levels of H as a function of applied electric field.

crystal. This uncertainty washes out most (but not all) of the coherence between levels of different zero-field energy in Fig. 4. We have included this loss of coherence as a damping factor in Eq. (4).

In region III the excited state experiences the external electric field, assumed directed along the x axis:

$$H_{III} = H_{SO} + H_{LS} + H_E \tag{7}$$

where H_E is the external field Hamiltonian, and is proportional to the field strength [16]. Entrance into the external field region is not properly described by either the sudden or adiabatic limits. The sudden limit was assumed here, but for quantitative calculations variations of the field strength with position would have to be included explicitly. The field was not sufficiently well defined in the present experiments to permit this.

In region IV the particle evolves free of fields,

$$H_{IV} = H_{SO} + H_{LS}. \tag{8}$$

The time t_4 at which radiation occurs is uncertain by $\pm 0.5 \times 10^{-9}$ sec, as defined by the slit width, so additional damping must be included in this region.

At time t_4 the particle radiates. The relative intensities I and I', and the Stokes parameters S/I, M/I, C/I and M'/I' can be obtained directly from the density matrix $\rho(t_4)$. Explicit expressions are given elsewhere [16].

Using the procedure outlined above, we can compute the Stokes parameters as a function of applied electric field strength to compare with the experimental results of Fig. 3. Examples of calculated results, for different choices of initial density matrix $\rho(t_o)$, are shown in Figs. 5-7. None of them agree quantitatively with experiment, but there are some qualitative similarities. We consider three qualitative features of the experimental results, Fig. 3.

The first feature we consider is the non-zero value of C/I at zero applied electric field. This corresponds to orientation of the major axis of the polarization ellipse at some oblique angle with respect to the surface normal. We suggested previously [4] that this strange result is a consequence of the evolution of the n=3 excited state in the vicinity of the surface. This is consistent with the present studies. If we eliminate the evolution in region I in the calculations, then C/I becomes identically zero at zero applied field for any of the choices of $\rho(t_o)$ we have considered. It is, of course, possible to construct $\rho(t_o)$ for which C/I is non-zero, but there does not appear to be any physical justification for this.

The second qualitative feature of the experimental results we wish to point out is that there is a linear Stark effect even at very low applied electric fields. C/I has non-zero slope at E=0 (Fig. 3), and S/I and M/I change very rapidly at small fields. From Fig. 4 it can be seen that this can only arise from a coherence between the degenerate $3p_{3/2}$ and $3d_{3/2}$ levels. As suggested by Fig. 8, spin-orbit and Lamb interactions remove the degeneracies of all other levels, resulting in small quadratic Stark splittings for applied field strengths less than about 50 V/cm. Thus initial preparation of a pure p state (Fig. 5) or pure d state (Fig. 6) is precluded.

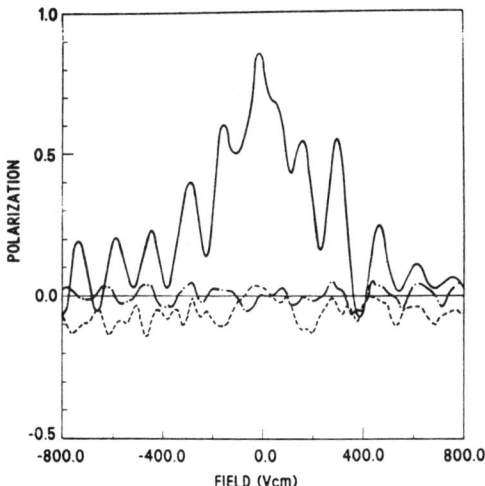

Fig. 5. Variation of Stokes parameters with strength of applied electric field computed by time evolution of density matrix for same conditions as Fig. 3. Solid curve, S/I. Dashed curve, M/I. Dash-dot curve, C/I. Initial state assumed: $p_x + ip_y$.

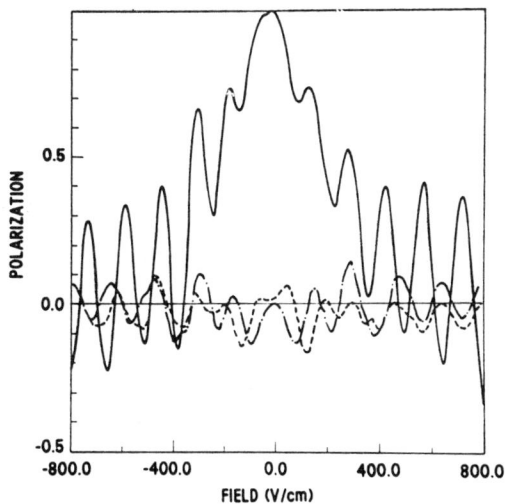

Fig. 6. Same as Fig. 5, but for initial state $d_{x^2-y^2} + id_{xy}$.

The final qualitative feature of Fig. 3 which we wish to consider is its asymmetry with respect to positive and negative electric fields. This can only occur if the wave function itself is asymmetric with respect to reflection through a plane parallel to the surface. Figs. 2g and 2h illustrate how coherent mixtures of hydro-

203

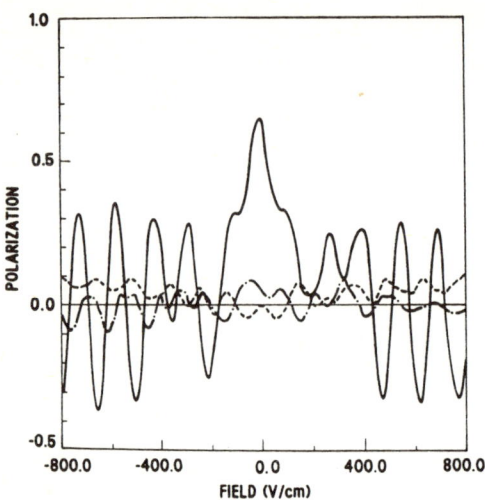

Fig. 7. Same as Fig. 5, but for initial state $\sqrt{1/3}s - \sqrt{1/2}p_x + \sqrt{1/24}d_{z^2} + \sqrt{3/24}d_{x^2-y^2} - i\sqrt{1/2}p_y + i\sqrt{1/2}d_{xy} + \sqrt{1/2}p_z + \sqrt{1/2}d_{xz} + i\sqrt{1/2}d_{yz}$

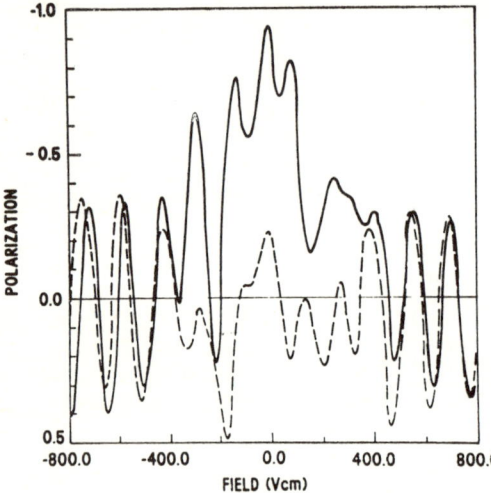

Fig. 8. Variation of S/I with strength of applied electric field computed for spin-aligned electron pickup. Orbital part of initial state same as Fig. 7. Solid curve, spin=+1/2. Dashed curve, spin=-1/2.

genic states can produce wave functions (Stark states) for which the electron density is shifted toward the surface. A reversal of phase of these wave functions would shift the electron density in the opposite direction. The simulation shown in Fig. 7 was produced by choosing $\rho(t_0)$ to be a linear combination of the two Stark states shown in Figs. 2g and 2h, as well as contributions from the similarly shifted

Stark states (not shown in Fig. 2) with odd reflection symmetry with respect to the collision plane. The asymmetric features of Fig. 7 are quite similar to those of Fig. 3. This leads us to conclude, tentatively, that the coherent mixture of states produced at the surface exhibits a shift of electron density toward the surface. (A shift in the opposite direction would produce the reflection of Fig. 7 through E=0). We hope to make this conclusion definitive by achieving a quantitative match with experiments.

Discussion

Our results are consistent with an anisotropic electron pickup mechanism [6]. This mechanism can be understood in terms of a simple classical picture. Consider a Gallilean transformation of our experiment, so that the proton is at rest and the surface travelling 1×10^8 cm/sec in the negative x direction in Fig. 1. At the instant of its pickup, the electron will most probably be located in the region of high electron density between the proton and the surface, and will be moving in the negative x direction. This will produce positive angular momentum about the z axis of Fig. 1, and negative S/I. It is also consistent with the previously discussed asymmetric shift in atomic electron density toward the surface.

If, alternatively, the n=3 level were populated by some incoherent mechanism and then subsequently selectively destroyed by a de-excitation or resonance ionization process, we would expect a different result. Those atomic electrons with smallest overlap with surface states are those on the far side of the atoms and moving in the +x direction. If these were to survive predominantly, then S/I would be the same sign as observed, but the shift in electron density would be in the opposite direction.

Our original goal in this work was to achieve a complete quantitative determination of the density matrix describing the coherent superposition of states initially populated during the neutralization event at the surface. This would provide a very demanding test for theories of ion neutralization at surfaces. Although we have not yet achieved this goal, we have uncovered interesting qualitative information about the neutralization mechanism. Furthermore, we have hope that additional work will allow us to reach the goal.

One of the major difficulties we encountered was the computational complication in dealing with a 18×18 density matrix. We are currently carrying out experiments with vacuum uv detection capabilities so that we will be able to observe Lyman α (n=2 → n=1) radiation. This should greatly simplify the analysis.

We also plan to study effects of applied magnetic fields on the polarization of radiation. For ferromagnetic targets such as Ni, there may be a possibility of producing partially spin-aligned excited atoms from a magnetized sample [14]. Fig. 8 shows predictions for S/I as a function of induced electric field strength for 100% spin alignment in +z and -z directions, assuming the initial density matrix that produced Fig. 7. Spin-alignment effects could possibly be dramatic.

References

1. H. J. Andrä, R. Fröhling, H. J. Plöhn, and D. J. Silver, Phys. Rev. Lett. *37*, 1212 (1976).
2. H. J. Andrä, R. Fröhling, and H. J. Plöhn, in *Inelastic Ion-Surface Collisions*, edited by N. H. Tolk, J. C. Tully, W. Heiland, and C. W. White (Academic, New York, 1977), p. 329.
3. H. G. Berry, G. Gabrielse, A. E. Livingston, R. M. Schectman, and J. Desequelles, Phys. Rev. Lett. *38*, 1473 (1977).
4. N. H. Tolk, J. C. Tully, J. S. Kraus, W. Heiland and S. H. Neff, Phys. Rev. Lett. *41*, 643 (1978); *42*, 1475 (1979).
5. N. H. Tolk, J. C. Tully, J. S. Kraus, W. Heiland and S. H. Neff, Surf. Sci. *90*, 447 (1979).
6. H. Schröder and E. Kupfer, Z. Physik *A279*, 13 (1976).
7. J. Burgdörfer and H. Gabriel, J. Physique *40*-C1, 315 (1979).
8. J. Burgdörfer, H. Gabriel and H. Schröder, Z. Physik A*295*, 7 (1980).
9. T. G. Eck, Phys. Rev. Lett. *33*, 1055 (1974).
10. M. Lombardi, Phys. Rev. Lett. *35*, 1172 (1975).
11. Y. B. Band, Phys. Rev. *A13*, 2061 (1976).
12. R. Herman, Phys. Rev. Lett. *35*, 1626 (1975).
13. A. A. Lucas, Phys. Rev. *B20*, 4990 (1979).
14. C. Rau and R. Sizmann, Phys. Lett. *43A*, 317 (1973).
15. N. Born and E. Wolf, *Principles of Optics*, 4th ed. (Pergamon, Oxford, 1979), p. 30.
16. Explicit prescriptions for the density matrix and Hamiltonian matrices are given in N. H. Tolk, J. C. Tully, J. S. Kraus, C. Rau and R. Morris, to be submitted to Phys. Rev.

LYα Polarization After Electron Capture in Ion-Surface Interaction

Heinz Schröder and Joachim Burgdörfer
Institut für Atom- und Festkörperphysik, Freie Universität Berlin
D-1000 Berlin, West Germany

Abstract. The captures of target electrons into excited states of the projectile are investigated for projectile velocities $v > v_{Bohr}$ and grazing incidence. The polarization of the subsequently emitted light is determined by evaluating the Stokes's parameters. The contribution of electron capture from localized target-electron states is compared with capture from conduction electrons. The role of electron-loss processes is discussed and the influence of the surface-electric field on the evolution of the projectile state (after the capture) is calculated.

Introduction. Fast ion beams reflected at grazing incidence from a surface emit light with high circular polarization [1,2,3], which can be used e.g. for precision measurements of atomic data [4] or for production of beams with polarized nuclei [5]. The theoretical interpretation of the ion surface interaction is not yet clear. Among the possible interactions and effects at least the following ones could be important for production of polarization: electron capture by the ion, excitation or loss of a projectile electron, screening effects inside the solid, and effects of the surface-electric field. We discuss here in more detail only the electron capture and in a crude approximation the electron loss and the surface-electric field.

Capture model. As projectiles we consider only protons in a velocity regime $v \gtrsim 1 a.u.$. The spin orbit interaction is neglected in the collision process, because the collision time is small as compared to the spin orbit interaction time. If not otherwise stated, we give our results for singlet p states, but it is only a matter of angular momentum algebra to extend them to multiplet states. The capture process is treated in Oppenheimer-Brinkman-Kramers approximation which gives for the capture amplitude P_m into a final hydrogen state $\Phi_{H,m}$ (with n=2, l=1, m=±1,0):

$$P_m(\binom{k}{b}, x_{min}) = \frac{1}{i} \int_{x_{min}/v_x}^{\infty} dt \int d^3r \, \Phi_{H,m}(\vec{r}-\vec{R}(t)) \, |\vec{r}-\vec{R}(t)|^{-1}$$

$$\Phi_{T,\binom{k}{b}}(\vec{r}) \exp[i(-\vec{v}\cdot\vec{r} + t(\varepsilon_H + \frac{v^2}{2} - \varepsilon_T)] \, . \tag{1}$$

A straight line trajectory is assumed for the proton $\vec{R}(t) = \vec{v}t$. The energy of the hydrogen or target state is denoted by ε_H respectively ε_T. As target wave function we investigate two different cases: 1.) a conduction electron state $\Phi_{T,k}(r)$ with momentum vector $\vec{k}$ and 2.) a localized target electron state $\Phi_{T,b}$ whose center relative to the projectile path is characterized by the impact parameter b. The parameter x_{min} describes the minimum distance from the surface ($x=0$), at which the capture may become effective; it can be used to simulate shielding inside the metal (total shielding means $x_{min} \geq 0$) or to simulate loss processes, as is discussed later.

For the capture from conduction electrons - in the approximation of a free electron gas inside a step potential - all integrations in (1) can be done analytically [6] for all angles Θ between the velocity $\vec{v}$ of the outgoing projectile and the surface. For electron capture from a localized state - a 1s hydrogenic state is taken for simplicity - an analytic expression can only be given for grazing exit ($\Theta \approx 0$) from the surface [7]. The Stokes parameters of linear polarization M/I (alignment) and C/I and of circular polarization S/I in a subsequent photon emission can be expressed in the usual way in terms of the density matrix [8]. The latter is calculated by numerical integration over momentum space k or the impact parameter b:

$$\rho_{mm'}(x_{min}) = \int \frac{d^3k}{d^2b} P_m P_{m'} (\frac{k}{b}, x_{min}) . \quad (2)$$

<u>Approximation of electron loss</u>. If the capture takes place in the immediate vicinity of the surface or even inside the bulk ($x_{min} < 0$), the formed state will be destroyed with great probability during the outward passage. We therefore weight the capture density matrix for different depth with the survival probability w

$$\rho_{mm'} = \int dx_{min} w(x_{min}) \rho_{mm'}(x_{min}) . \quad (3)$$

The probability $w(x)$ that the excited $n=2$ state can escape from depth x to infinity is estimated from a mean free path concept

$$w(x) = \exp(-\frac{1}{\lambda_{L,0}} \int_x^\infty dx \frac{n(x)}{n_0} \frac{1}{\sin\Theta}) . \quad (4)$$

Here the mean free path λ_L for the loss or deexcitation of the excited electron is scaled with the inverse of the local electron density $n(x)$ which is integrated over the outgoing path (bulk properties are marked by a subscript 0). Thus we neglected electron loss by collisions with the bulk ions. At grazing incidence however, this is not a severe approximation because the collisions with electrons are frequent enough to destroy all excited states formed at the surface. For the mean free path at the surface we tested two values $\lambda_{L,0} = 2$ and 5 a.u. and did not find much difference for model 1. For model 2 (capture from localized states) we investigated only a simplified version with a step function for $w(x_{min})$. As $x_{min} \approx b$ for $\Theta \approx 0$ this means that the density matrix was integrated with equal weight over all impact parameters exceeding $b_c = x_{min}$. A best agreement with experimental values was obtained for $b_c = 7$ a.u. which is of the same order of magnitude as a proposed value in [3].

<u>Results</u> are shown in Fig. 1 for $\Theta = 1°$. The calculated value of C/I

is less than 1% for all target states. The main features of the calculated polarizations (large values of S/I, small M/I, and very small C/I values) agree with several experimental data [1,2] at v ≈ 1a.u.. The dominance of S/I over M/I is caused by the effective restriction to large x_{min} respectively large b. For a quantitative comparison to the experimental data [2] also the polarization of the transition $p^2F_{7/2}-s^2D_{5/2}$ was calculated by coupling an isotropic ^{1}D-core and an isotropic spin to the initial density matrix of the optical electron [8].

At higher velocities v ≳ 2a.u. the calculated polarization for the two target models diverge drastically. Measurements in this regime [9] support the smooth results for the localized target state. This is not a failure of the capture model for conduction electrons, however, but it is a consequence of the fact that capture from conduction electrons becomes unimportant for v≳2a.u., because capture from inner shells dominates at higher velocities.

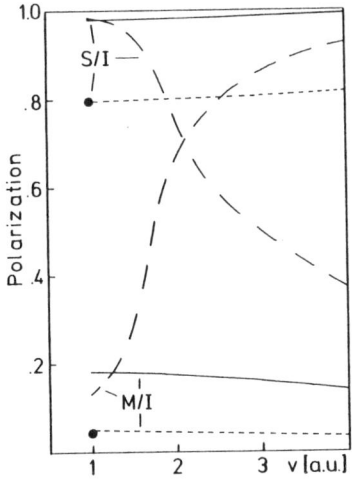

Fig. 1 Polarization due to electron capture at grazing incidence (θ=1°) for transition ^{1}s-^{1}p:

— — model 1 (Cu), $\lambda_{L,o}$=5a.u.

——— model 2 b_c=7a.u.

for transition s $^2D_{5/2}-^2F_{7/2}$

----- from curve ———

● from experiment [2] for Ar

Only if the distance of closest approach to the surface can be kept larger than a few a.u. (by using clean, flat, monocrystalline targets at still smaller θ) could capture from conduction electrons become relevant at high velocities.

Electric field effect. For non-hydrogenic projectiles the surface electric field F should be of minor importance because outside the surface the Stark mixing between excited states is small. In hydrogenic systems however, even small fields (on an atomic scale) lead to a strong perturbation of the nearly degenerate levels via linear Stark effect [10]. Up to now, there is much uncertainty on the value of the electric field near the surface. Therefore we studied the influence of the Stark mixing on the Lyα polarization. Some results are given in Fig.2. As initial value we used the capture density matrix from a localized target electron for θ≈0 and at v=1 and x_{min}=5 a.u.. The Stokes's parameters are given as function of the evolution phase

$$\Phi = \Phi_o/v \sin \theta \quad \text{with} \quad \Phi_o = \int_{x_{min}}^{\infty} dx\, 3F(x) \quad (5)$$

209

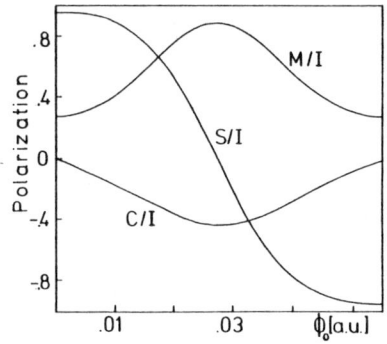

Fig. 2 Dependence of ^1s-^1p polarization in n=2 hydrogen on the evolution phase Φ_0 due to the electric surface field. Initial density matrix is from capture of localized target electrons with b_c=5a.u. v=1a.u., Θ=1°.

Notice that for small grazing angles Θ the dependence of the initial capture density matrix on Θ is smooth (const.+ $O(\Theta)$). The evolution due to the electric field however, behaves as $1/\Theta$. Thus even for small values of F and Φ the polarization can drastically change at small exit angles. In contrast to the zero-field case large values of C/I can be found and a change of sign of S/I becomes possible. A direct comparison to the H_α data of [3] is not meanigful, because the Stark mixing in the n=3 level is much more involved. Nevertheless, assuming a small phase Φ <0.01 (in a.u.) the polarization shows similarities with the data. The main discrepancy arises from M/I which is as large as C/I in our case.

In conclusion we have shown that the electron-capture and loss model can explain several features of the observed polarization phenomena. The electric surface field has been shown to be important for hydrogenic projectiles only.

This work is supported by the Deutsche Forschungsgemeinschaft. Many valuable discussions with Prof. Gabriel, Prof. Andrä, Dr. Winter and R. Fröhling are gratefully acknowledged.

References

1. H.J. Andrä,R.Fröhling,H.J.Plöhn in: Inelastic Ion-Surface Collisions, p. 329, N.H. Tolk, J.C. Tully, W. Heiland, C.W. White(eds) New York: Academic Press 1977
2. H.G. Berry,G.Gabrielse,A.E. Livingston: Nucl. Instr.Meth. 149, 517 (1978)
3. N.H.Tolk,J.C.Tully,J.C.Kraus,W.Heiland,S.H.Neff: Phys. Rev. Lett. 41, 643 (1978)
4. H.Winter,H.J.Andrä: Phys. Rev. A21, 581 (1980)
5. H.Winter,H.J.Andrä: Z.Physik A291, 5 (1979)
6. H.Schröder,E.Kupfer: Z.Physik A279, 13 (1976)
7. J.Burgdörfer,H.Gabriel,H. Schröder: Z.Physik A295, 7 (1980)
8. H.Schröder: Z. Physik A284, 125 (1978)
9. H. Winter,P.H.Heckmann,B.Raith: submitted to Hyperf. Interact.
10. E.Kupfer,H.Winter: Z. Physik A285, 3 (1978)

Light Emission and Circular Polarization by Ion Surface Scattering at Grazing Incidence

W. Graser and C. Varelas

Sektion Physik, Universität München
D-8000 München 40, Fed. Rep. of Germany

1. Introduction

Studies of optical radiation from particles reflected from surfaces provide important information on particle-surface electronic processes. Of particular interest have been measurements of radiation arising from grazing incidence collisions of ions with single crystalline surfaces [1-4]. Here we report on the influence of the oxygen coverage and the crystallographic orientation on the intensity and circular polarization of the light which is emitted from 200 keV He projectiles after scattering at a (110) surface of a (60/40)NiFe single crystal at grazing incidence.

2. Experimental

Figure 1 shows schematically the experimental arrangement. A 200 keV He^+ ion beam with 0.1° divergence is incident with angle ϕ normal and θ azimuthal to the (110)NiFe-surface. The light intensity $I = I_{rh} + I_{lh}$ and the normalized Stokes parameter $S/I =$

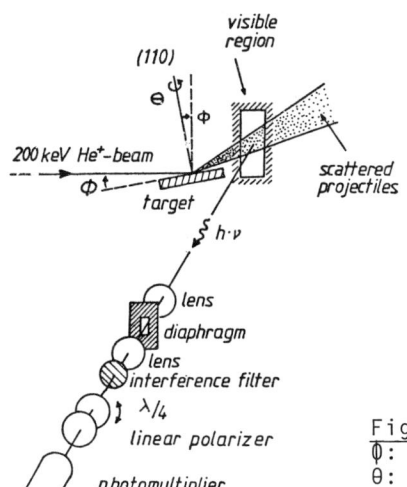

Fig.1 Experimental arrangement
ϕ: angle normal to the surface
θ: azimuthal angle of incidence
interference filter for $\lambda = 468.6$ nm or for $\lambda = 587.6$ nm.

$(I_{rh}-I_{lh})/(I_{rh}+I_{lh})$, where I_{rh} and I_{lh} are the intensities of the right and the left hand circularly polarized light, have been measured by photon counting. The total pressure in the scattering chamber during the experiment was 1×10^{-10} mbar. The crystal surface is first mechanically and then electrochemically polished. After mounting on a 3 axes goniometer and baking in the vacuum chamber the surface is cleaned by 3 keV Ne ion sputtering and annealed at 1000°C by electron bombardment. The quality of the surface was routinely inspected by Auger Electron Spectroscopy (AES) which showed no detectable contamination of S, C, and O, and by Low Energy Electron Diffraction (LEED) which indicated a high contrast p(1x1) structure. For the investigation of the influence of the chemisorbed oxygen atoms on the intensity and the circular polarization of the emitted light high purity oxygen was admitted through a leak valve.

3. Results and Discussion

The first stage of oxidation of the (110)-NiFe-surface is shown in Fig.2a. The calibrated AES shows a Langmuir type chemisorption with saturation of the surface at a coverage of 0.5 monolayers after an exposure of about 5 L (= 500s x 10^{-8} Torr O_2). Under these conditions LEED indicates a p(2x1) overlayer structure. The increasing oxygen coverage leads to a corresponding decrease of the circular polarization S/I both of the ionic He^+ line and of the atomic He^o line, as shown in Fig.2b. The measurements had been done for an angle of incidence $\phi = 5°$, which allows most of the particles to penetrate through the surface into the bulk

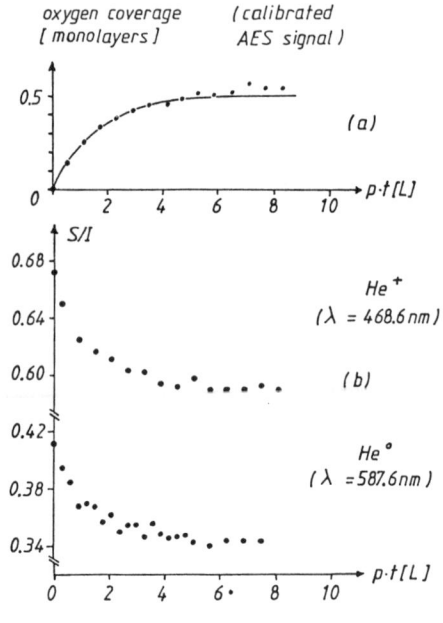

Fig.2
(a) The first stage of oxidation of a (110)-NiFe(60/40)-surface measured by AES as function of exposure (=pressure·time) 1L =10^{-6} Torr·s
(b) Circular polarization S/I of the 468.6 nm (n=4→3) ionic and the 587.6 nm ($3d^3D \to 2p^3P$) atomic line of the scattered He vs exposure of oxygen. The angles of incidence were $\phi=5°$ and θ random to low crystallographic directions. The ion energy was 200keV.

of the crystal. The azimuthal angle θ had been chosen random to low index directions to avoid crystallographic effects. The corresponding decrease of S/I with increasing oxygen coverage of the surface underscores the importance of clean surfaces.

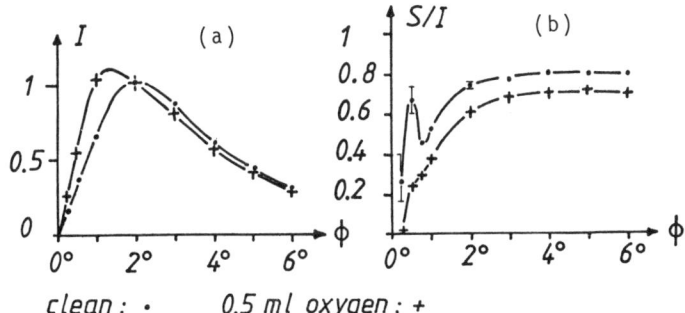

Fig.3 Angular dependence of (a) the intensity I and (b) the circular polarization S/I of the He⁺-468.6 nm transition on the angle ϕ normal to the surface. θ is random.

Figure 3a shows the dependence of I and S/I of the 468.6 nm line of He⁺ on the angle ϕ normal to the surface. The azimuthal angle θ is random to low index lattice directions. The intensity was normalized to a constant number of particles incident on the surface. With increasing ϕ (ϕ > 2°) the light intensity decreases since more ions can penetrate the surface and are lost into the bulk of the crystal. For ϕ < 2° the intensity decreases strongly with decreasing ϕ although the fraction of scattered ions increases [5]. For ϕ ≤ 1° all the ions are specularly reflected at the surface. Thus, the decrease of the light intensity could be caused by a less efficient excitation process with increasing distance of closest approach to the surface. For ϕ ≥ 2° an oxygen coverage of 0.5 monolayers does not influence the light intensity. However, at specular reflection the intensity is increased by a factor of about 1.5. In this case most of the ions are scattered by the chemisorbed oxygen layer.

The normalized Stokes parameter S/I is shown in Fig.3b. S/I starts out at low values and increases rapidly to a value of 0.8 at ϕ = 2° and then remains constant with increasing angle ϕ up to 6°. The strong variation of S/I at specular reflection may be caused by the Stark effect since the surface electric field acts on the hydrogen-like He⁺ ions. For a surface covered by 0.5 monolayers of oxygen the S/I values are lower (according to the previous results in Fig.2).

In order to investigate directional scattering effects, grazing incidence experiments at different azimuthal angles θ were performed. Fig.4a shows the dependence on θ to the ($\bar{1}10$) plane of the intensity and the circular polarization of the He⁺-468.6

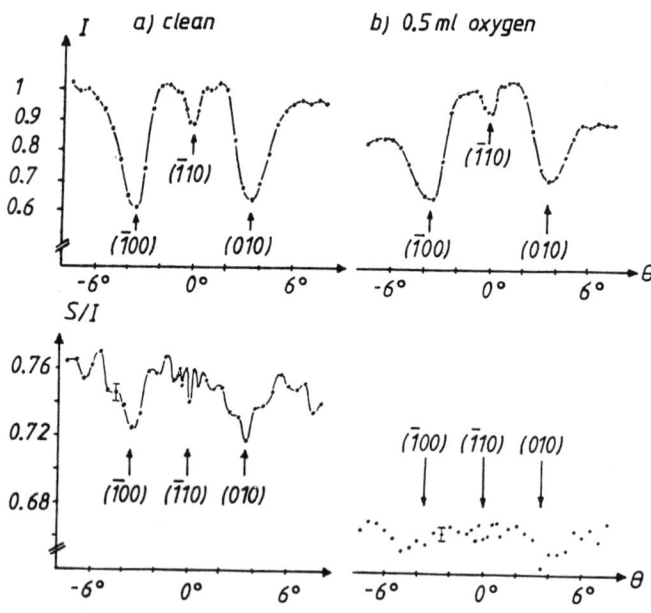

Fig.4 Angular dependence of the intensity I and of the circular polarization S/I of the He$^+$-468.6 nm transition vs θ for planar channeling conditions (a) for a clean surface and (b) for the same surface covered by 0.5 monolayers of oxygen

nm radiation . The grazing angle φ is kept at 3.5°. The intensity minima originate from He$^+$ ions lost by planar channeling into the bulk of the crystal [1]. S/I also shows minima for ion incidence parallel to planar channeling cirections in agreement with previous results [2,3]. The shape of the intensity curves changes if the surface is covered by 0.5 monolayers of oxygen (cf. Fig.4b). We have used this behaviour to locate the adsorbed atoms on the surface [6]. Furthermore, Fig.4b shows that the oxygen coverage decreases the mean value of S/I, as discussed in reference to Fig.2, and smooths the curve drastically. Planar directions are hardly to be seen.

With decreasing angle φ normal to the surface a transition from planar to axial channeling takes place [1]. At an incidence angle φ=1° the ions are axially channeled along the $\langle 1\bar{1}0 \rangle$ atomic rows. Both the He$^+$ and the He^o radiation intensities show deep minima due to ions lost into the bulk of the crystal by axial channeling (cf. Fig. 5a). S/I also shows minima under axial channeling conditions. However, the minima of S/I of the He^o - 587.6 nm radiation is small, while the minimum of S/I of the

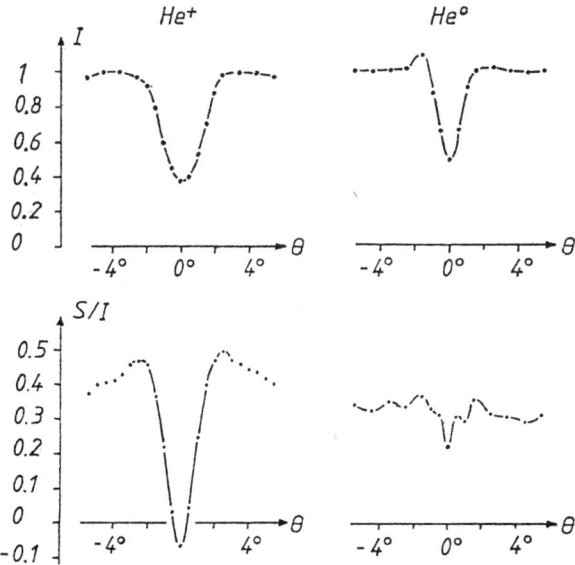

Fig.5 Angular dependence of the intensity and the circular polarization of the He$^+$-468.6 nm and He^o-587.6 nm transition vs. θ for axial channeling along the $\langle 1\bar{1}0 \rangle$ atomic rows

He$^+$-468.6 nm radiation is deep with even a change of sign. This stong variation of the He$^+$ polarization with azimuthal angle θ may again be caused by the various trajectories of the scattered He$^+$ ions at the surface with θ and therefore by the different electric field along the ion paths, which results in a quantum mechanical phase evolution by Stark effect [2,3]. Future experiments should therefore measure, in addition, the Stokes parameter C/I to clear the influence of the Stark effect.

This work was supported by the Bundesministerium für Forschung und Technologie, Bonn, Germany.

[1] C. Varelas and R. Sizmann, Surface Sci. 71 (1978)51
[2] N.H. Tolk, J.C. Tully, J.S. Kraus, W. Heiland, and N.H. Neff, Phys. Rev. Letters 42(1979)1475; Surface Sci. 90(1979)447
[3] H.J. Andrä, H. Winter, R. Fröhling, N. Kirchner, N.J. Plöhn, W. Wittmann, W. Graser, and C. Varelas,Nucl. Instr. Meth. 170(1980)527
[4] P.J. Martin, L. Berzins, and R.J. MacDonald,Surface Sci. 95(1980)L277
[5] R. Sizmann and C.Varelas, Nucl. Instr. Meth. 132(1976)633
[6] W. Graser and C. Varelas, Proc. Europ. Conf. on " Nuclear Physics Methods in Material Research ", Darmstadt, 1980, in press

Orientation of Target and Projectile States After N^+-He-Collisions with Impact Parameter Selection

H. Winter

Fachbereich Physik der Freien Universität Berlin, Boltzmannstraße 20
D-1000 Berlin 33, West Germany

1. Introduction

The ion beam surface interaction at grazing incidence (IBSIGI) (1-5) leaves the atoms or ions in highly oriented angular momentum states which results in a large degree of circular polarization in the emitted light. Even though the first report on this phenomenon dates back only a few years (1) the number of communications related to this new technique is considerable. At present the field seems to have split into two directions: the use of the high orientation produced by IBSIGI and the study of the IBSIGI-interaction process itself.

The first approach takes advantage of the production of oriented levels for high resolution spectroscopy as demonstrated in quantum-beat (6) and level-crossing experiments (7). Another application is based on the transfer of orientation from the electronic shell to the nuclear spins via hyperfine interaction which leads to vector polarized nuclei (8,9); recent data of IBSIGI with N-projectiles of energies up to 15 MeV imply that this technique may serve as a universal tool for spin polarization in nuclear spectroscopy (10).

A second aspect of IBSIGI is the investigation of the interaction itself in order to understand the processes which underlie the production of oriented angular momentum states and to probably get via IBSIGI a deeper general understanding of the excitation mechanism of atoms in ion-solid interactions.

The work reported here is on impact parameter selected gas collisions and is related to the study of the ion-solid interaction since it supports a widely accepted model for the production of orientation by IBSIGI (11). In the following section we briefly outline the main features of this model with respect to this work and describe in Section 3 our experiment and the results.

2. Electron density gradient model

An instructive model for the origin of orientation in the ion-solid interaction is given by SCHRÖDER and KUPFER (11) which applies equally well for IBSIGI and beam-(tilted) foil excitation. In Fig.1 we consider the interaction of an ion of velocity v with a plane surface. In its rest frame the ion

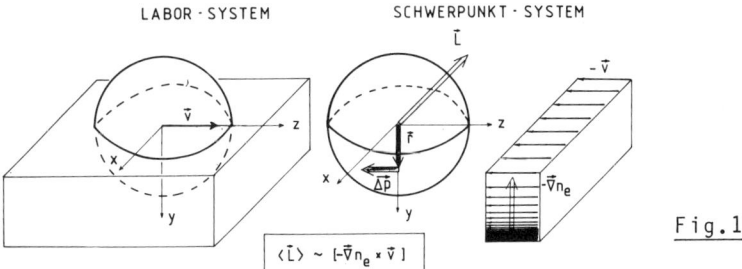

Fig.1

experiences a beam of electrons with $-\vec{v}$ and since the decrease in electron density at a realistic surface is considerable within the typical dimension of an atom, the flux of electrons colliding with the ion in the lower hemisphere is higher than in the upper one. Assuming a transfer of momentum $\Delta\vec{p}$ by the interaction of ion and electron, we end up with an expectation value of the orbital angular momentum $<\vec{l}>$ pointing into the image plane because more vectors r pointing into the lower than into the upper hemisphere of the ion.

As shown by SCHRÖDER and KUPFER $<\vec{l}>$ is proportional to the vector product of the electron density gradient $\vec{\nabla}n_e$ and the velocity $\vec{v}$ and this anisotropic (oriented) population of atomic states can be probed by the circular polarization in the fluorescence light. In all IBSIGI-experiments undertaken so far the sense of rotation in the circular polarization agrees with the model, but for more adequate theoretical descriptions further details on the interaction process have to be taken into account (12).

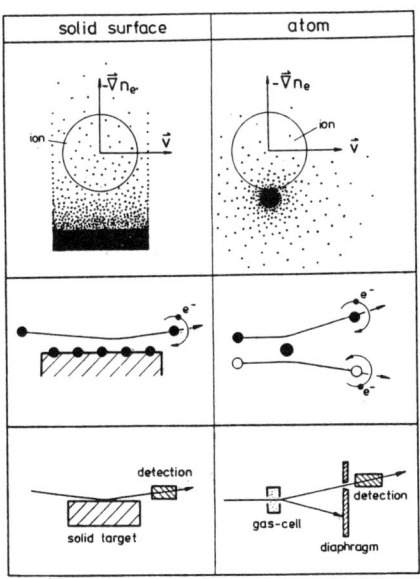

Fig.2 Comparison of ion-surface and ion-atom collisions

217

On the other hand the experimental conditions were rather uncontrolled so that the data are not suited for a direct comparison with theory. An experimental approach with respect to higher reliability and better controlled conditions is the "ideal experiment" described by ANDRÄ and coworkers (13) who pointed out that the preparation of solid targets remains one of the main problems.

A first step in the direction of such an experiment was performed by WITTMANN and ANDRÄ (14) who replaced the solid target by the "ideal surface" of a single atom. The correspondence between ion-solid and ion-atom collision is outlined in Fig.2: as well as a solid surface the atomic shell has a density gradient of electrons, but the difference between both cases is the selection of impact parameters inherent in IBSIGI. This impact parameter selection has to be introduced in a gas collision experiment by putting behind the target area a diaphragm so that due to the repulsive potential between ion and atom only those atoms are detected which have interacted with the upper (or lower) hemispheres of the target atoms. In He^+-Ne collisions (14,13) large fractions of circular polarization are found and as expected from the model above the circular polarization is opposite in sign for upward and downwards scattered atoms.

Because of the need for a sufficient angular resolution after the collision this experimental technique is restricted to levels with long radiative lifetimes (several 10 ns) and no information on the target atoms is obtained. This last point is of some interest with respect to basic statements derived from the model of (11). If we consider in Fig.3 an ion-atom collision in the center of mass frame with projectile and target velocities $\vec{v}_1$ and $\vec{v}_2$, we predict on the base of the density gradient model for both particles of a collision the same sense of orientation. Since projectile and target atom are scattered due to the repulsive potential into opposite hemispheres, we expect for the same scattering angle an opposite sense in the circular polarization of the emitted light from excited projectiles and target atoms.

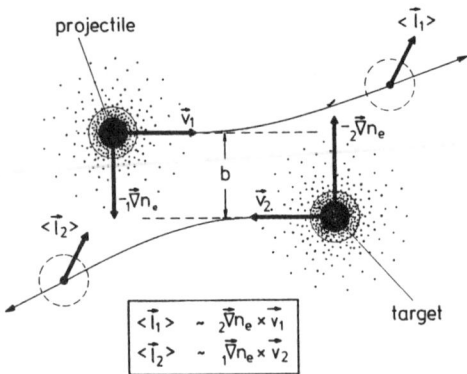

Fig. 3

Interaction of projectile and target atom in the center of mass frame

3. Experiment and results

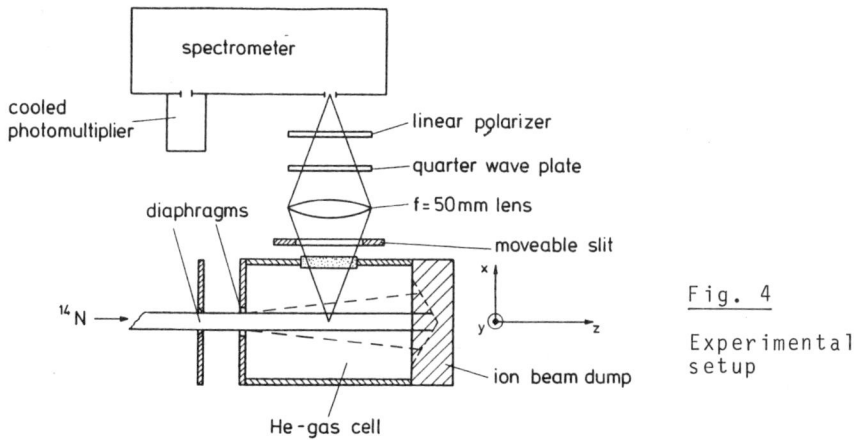

Fig. 4

Experimental setup

The experimental setup is shown in Fig.4. A well collimated beam of 18 keV N^+-ions enters a gas cell with about $5 \cdot 10^{-4}$ Torr He and hits an ion beam dump which is used as a Faraday cup for normalization purposes. The emission within the cell is observed through a window by a detection system which allows to determine the state of polarization of the emitted light. The degree of circular polarization is measured with the help of a COMMODORE CBM3032-minicomputer by scanning the wavelength of the spectrometer and rotating a quarter wave plate in front of a fixed Glan-Thomson polarizer. As outlined in (5) this technique allows a proper background correction even at low signal to noise ratios and a reliable determination of the orientation given by the Stokes-parameter $S/I = (I(r)-I(l))/(I(r)+I(l))$ where $I(r)$, $I(l)$ are the intensities of the components of right- and left-handed circular polarization.

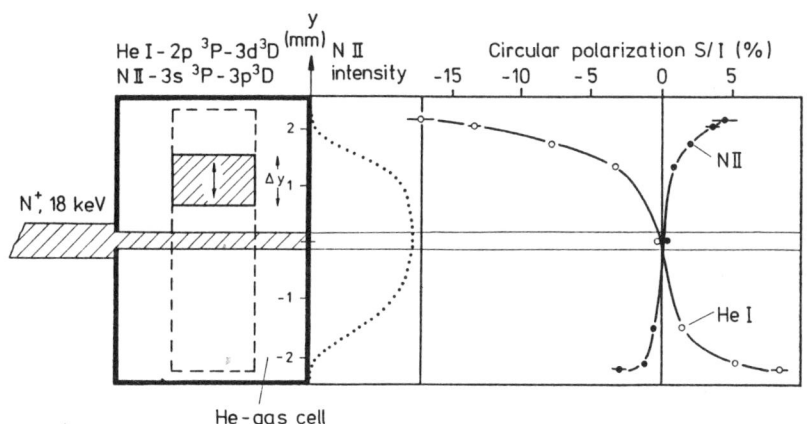

Fig. 5

The scattering angle selection is performed by a slit of Δy = 0.8mm x 5mm in front of the gas target window which is parallel with respect to the z-axis and can be moved parallel to the y-axis. Thus a variable detection zone in the gas target is achieved as indicated in the left part of Fig.5. The distinction of the light emission between projectiles and target atoms is managed by detecting the HeI-$2p\,^3P$-$3d\,^3D$ (588nm) and the NII-$3s\,^3P$-$3p\,^3D$ (568nm) transitions which can be well isolated in the spectrum of both atoms. The intensity of the NII-line while scanning the detection slit is shown in the middle of Fig.5. The resulting curve is strongly influenced by the finite width of ion beam and slit and by the geometrical sensitivity of the detection channel.

In the right of Fig.5 S/I of the HeI- and NII-line is displayed as a function of the detection slit position. The observed effect is in full agreement with the conclusions drawn from the density gradient model: for the same scattering angles the polarization of the light from projectiles and target-atoms are opposite in sign and when changing from the upper to the lower hemisphere of the interaction zone the polarization for both lines changes sign. On the beam axis (y=0) S/I vanishes, because ions having interacted with the lower or upper hemisphere of the target-atoms are detected with equal efficiency (we thus end up with a rotation symmetry around the beam axis and no orientation can occur because of arguments of symmetry).

4. Conclusion

Despite the fact that one cannot expect a quantitative theoretical treatment for the data presented here, the experiment confirms the concept of the electron density model. The technique of the selection of scattering angles within the interaction zone allows in addition to the work of (14) the observation of the target atoms and of short lived levels of the projectiles. Since in the center of mass frame the distinction between projectile and target is irrelevant, the observation of the target atoms is equivalent to the study of a collision of neutral atoms with an ion target. A drawback of this type of experiment is, of course, the bad angular resolution. Coincidence experiments may overcome that problem (15), but with a loss in signal of several orders in magnitude.

Discussions with H.J.Andrä, H.Schröder and J.Burgdörfer and the help of P.Dell'Anna and G.Sedatis for processing the manuscript and J.Thiel and R.Burghardt for preparing the experiment are gratefully acknowledged. This work was supported by the Sonderforschungsbereich 161 of the DFG.

References

(1) H.J.Andrä, Phys.Letters 54A(1975)315
(2) H.J.Andrä, R.Fröhling, H.J.Plöhn and J.D.Silver, Phys. Rev.Lett. 37(1976)1213
(3) H.J.Andrä, R.Fröhling and H.J.Plöhn, in "Inelastic Ion-Surface Collisions", eds. N.H.Tolk, J.C.Tully, W.Heiland and C.W.White, Academic Press, New York (1977) p. 329
(4) H.G.Berry, G.Gabrielse, A.E.Livingston, R.M.Schectman and J.Desesquelles, Phys.Rev.Lett. 38(1977)1473
(5) N.H.Tolk, J.C.Tully, J.S.Kraus, W.Heiland and S.H.Neff, Phys.Rev.Lett. 42(1979)1475
(6) H.Winter and H.J.Andrä, Phys.Rev. A21(1980)581
(7) H.J.Plöhn and H.J.Andrä, private communication
(8) H.J.Andrä, R.Fröhling, A.Gaupp and H.J.Plöhn, Z.Physik A281(1977)15
(9) H.Winter and H.J.Andrä, Z.Physik A291(1979)5
(10) H.Winter, B.Raith and P.H.Heckmann, Hyp.Interactions (1980) in print
(11) H.Schröder and E.Kupfer, Z.Physik A279(1976)13
(12) J.Burgdörfer, H.Gabriel and H.Schröder, Z.Physik A295(1980)7
(13) H.J.Andrä, H.Winter, R.Fröhling, N.Kirchner, H.J.Plöhn, W.Wittmann, W.Graser and C.Varelas, Nucl.Instr.Meth. 170(1980)527
(14) W.Wittmann and H.J.Andrä, Z.Physik A288(1978)335
(15) L.Zehnle, E.Clemens, P.J.Martin, W.Schäuble and V. Kempter, in: "Coherence and Correlation in Atomic Collisions", Plenum Press, New York and London 1980 p.457

Excited Particle Emission

Outer Shell Excitation During Sputtering and Low Energy Ion Scattering

R.J. MacDonald*, C.M. Loxton and P.J. Martin
Department of Physics, The Australian National University
Canberra, ACT 2600, Australia

1. Introduction

The general area of photon emission resulting from the excitation of atomic particles during ion surface interaction was reviewed at the first of these meetings [1]. The more specific aspects of excitation leading to photon emission from sputtered particles were reviewed by Thomas at the second workshop in this series [2]. It is not therefore our intention to once more review the field in general but rather to concentrate on developments since this last review. This means that we will be concerned primarily with the work of only a few groups, since these have been most active in this area in the last two years. We will attempt to present this work in such a way that it highlights our knowledge and understanding of the excitation and other processes contributing to the observed decay by photon emission. The aim of this review will be primarily to highlight what we need to know, and what we ought to measure, rather than to describe in detail what has been measured in the past.

As we will see progress in our understanding of the excitation and de-excitation mechanisms operative in the case of outer shell excitation of sputtered particles has not been remarkable. We have become increasingly aware of the need to accurately define the experimental situation in which measurements are made - further we must define more clearly what we are trying to do experimentally. A lot of the early work in this field has been of a survey nature. This has usefully contributed to our current awareness of what we must now be studying. The initial enthusiasm for ion produced photon emission as a means of quantitative analysis (SCANIIR[3], IBSCA[4]) has abated somewhat as the complexities of the processes affecting the yield of photons have become apparent.

If the excitation and de-excitation processes are known then we should be able to predict such parameters as the yield and yield dependence on energy of the photon emission, the distribution of electrons over excited states, and hence the spectral distribution of intensity of emitted radiation, the spatial distribution of emitted photons, the influence of the energy band structure of the solid on the emitted radiation, and the effect of adsorption of gases on the surface of the target on the emitted radiation (both yield and spectral distribution). These are all parameters we can measure, so they provide an outline for the arrangement of this review. By examining what we know of these experimentally measurable quantities we can assess the degree of understanding of the basic excitation mechanism involved.

---

* Department of Physics, University of Newcastle, 2308, N.S.W. Australia

2. Photon Yield and Energy Dependence of Photon Yield (The Excitation Function)

The original measurements of excitation functions, i.e. the dependence of the photon yield on the energy of the incident ion, were described by White and Tolk [5] and by Tolk et al [6]. This work was done under conditions which we now know to be experimentally undesirable, particularly as the vacuum conditions were bad. The results were fitted by a model incorporating non-radiative decay processes of the type suggested by Hagstrum [7] (van der Weg and Rol [8] first applied this model to photon emission due to sputtering). Since these initial experiments there have been relatively few similar experiments under conditions in which the vacuum has been more acceptable.

Bhattacharya et al [9] have measured photon intensities as a function of incident ion energy for Ar^+ incident on Si in the energy range 50-500 keV. They studied both visible and UV emission but again their vacuum system could provide only a 1 x 10^{-7} torr background. Their current density was high (100-300µA cm^{-2}) but the surface was still likely to be contaminated. By assuming that only the high energy direct and near-direct recoil particles are excited and that the Hagstrum-type de-excitation [7] applies, i.e.

$$P(s) = \exp[-A/av_\perp] \tag{1}$$

where P(s) is the probability that the atom (or ion) in a given excited state survives to decay away from the surface, A/a are constants characteristic of the exchange and $v_\perp$ is the perpendicular component of velocity away from the target, they showed agreement between the model and the experimental results, for A/a values of about 2 x 10^7 cms^{-1}.

Similar experiments and similar interpretations have been described by Heiland et al [10]. Here the incident ions were in the range 400-8000eV, the vacuum 10^{-8} torr. Again photon yields as a function of incident energy were analysed by a model assuming (1) applied to the high energy recoil part of the energy spectrum of sputtered ions. This analysis was also applied to incident ions scattered into excited neutral states, from which photons were subsequently detected. In the latter case $v_\perp$ was taken as the perpendicular component of the incident velocity, while for the sputtering situation a value between 0.5 and 1.0 times the maximum energy transfer in a head-on collision was chosen. Again the model fitted the experimental data, using A/a as a fitting parameter. We will return to these experiments and their interpretation later in this paper.

The most recent and experimentally the most satisfactory measurements of excitation functions are those due to Wright and Gruen [11]. These measurements have been performed under a base pressure of 2.10^{-9} torr, but the system required back-filling to run the ion gun - the pressure during irradiation was of order 5.10^{-5} torr inert gas. Their results for several lines from Be I and Be II bombarded with 50-3 x 10^3 eV He^+, Ne^+, Ar^+, Kr^+ ions, are reproduced in Figure 1. Data for O^+, N^+, D^+ and H^+ bombardment is also available but the authors themselves caution its interpretation as arising from a clean surface. They note, for example, a build-up of oxygen when He^+, H^+ and D^+ are used, presumably because of the lower sputtering yield of these ions.

These results have been analysed in terms of a model developed by Wright and Gruen [12]. This model assumes the dominating effect in the production of secondary ions or of excited atoms or ions is the electron exchange mechanism

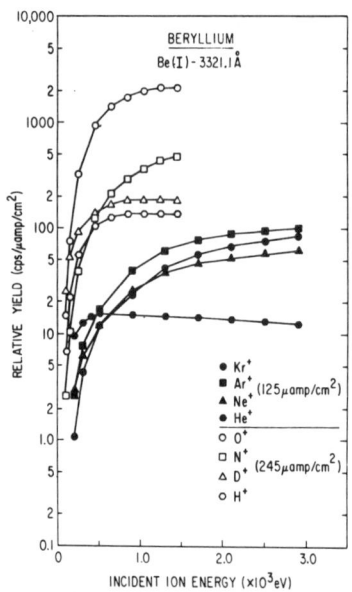 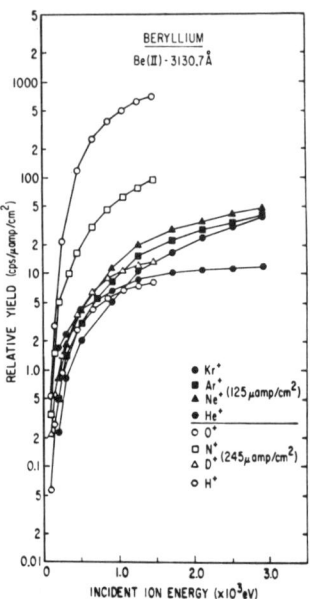

<u>FIG. 1</u>: The excitation functions obtained by Wright and Gruen (11) for the bombardment of Be with various active and inert gas ions. Results are for BeI (332.11nm) and BeII (313.07nm).

between particle and surface. Wright and Gruen [12] then consider a large number of possible events, deriving various excitation probability functions for these events, and then developing expressions for the total yield, and for the energy and velocity distributions of excited and ionised secondary particles. They show their model for the total secondary photon emission yield reduces to

$$Y_i^{ex} \propto JS \exp - (A_i'/v_m) \qquad (2)$$

where J is the incident ion beam current density and S the neutral atom sputtering yield. The term A_i' depends on the form of the excitation function p_i^{ex} chosen - if P_i^{ex} = constant, or $p_i^{ex} \propto v^m \cos \theta^m$, then $A_i' = \Sigma_j (A/a)_j$ where the j represent the possible non-radiative de-excitation paths - if p_i^{ex} is taken to be of the form $\exp[- B/v \cos]$ then $A_i' = \Sigma_j (A/a)_j + B_i$.

Values of A_i' from (2) have been obtained by fitting (2) to the experimental data of Fig.1. The values of A_i' obtained apparently yield very good fits to the data but for a given level, the values of A' due to excitation following sputtering by the different ions are not the same. Taking the simplistic view of the observed photon yield being the result of competition between excitation and non-radiative de-excitation, one would expect that the A_i' values for a given excited state should be the same for the same level independent of the ion inducing the excitation. This is implied by using the cascade model of the flux of sputtered particles. The energy spectrum is the same independent of the incident ion, only the integral over the spectrum (i.e. the sputtering yield) changes. One source of this discrepancy may be that $A' = \Sigma_j (A/a)_j$ for each of the j possible channels of de-excitation - if the distribution of possible channels of de-excitation depends in some way on the incident ion then the sum A' will also do so.

Wright and Gruen (12) provide a model which accounts for Heiland's *et al* [10] result involving the dependence of relative intensity or the reciprocal of the velocity normal to the surface. One problem with this work and with that of Wright and Gruen [11] is that this relationship is observed almost irrespective of the velocity component used. Thus Heiland *et al* [10] choose values (arbitrarily) between the perpendicular component of the maximum recoil energy and half this value. One can verify that similar results are obtained even if one uses say 1/10th the maximum recoil energy. The value of (A/a) or A' is then determined solely by the choice of the scale on the velocity axis, and, as such, is not a reliable parameter to compare with any theoretical result.

While there have been the above attempts at measurement of the excitation function, these attempts have all involved measurements of the relative photon yield. Thus the results have been normalised by dividing by the sputtering yield, and obtained at constant current density but the assumption has been made that the angular distribution is constant so the spectrometer slit always sees the same portion of ejected flux and so collects the same proportion of the emitted radiation. This may not be true particularly if the recoil part of the energy spectrum is the part excited and hence responsible for the radiative emission.

There has only been one attempt to measure absolute photon yields - this has been reported by Tsong [13]. The yields were measured in a system whose detection capabilities were calibrated with a standard lamp. The atoms involved were sputtered from a standard glass target, to reduce the possible influence of non-radiative transitions. The yields of a given excited state were roughly of magnitude 10^{-3} per sputtered atom, while the integrated yields over the excited states of a given atom were of order 10^{-2} per sputtered atom. No other measurements of total yields are available.

There have been some suggestions that secondary ion and photon yields are related. Thus Wright and Gruen [12] treat the probability of ion and photon excitation and survival similarly. Blaise [14] has gathered comparative data on ion and photon reactions to adsorption. Martin and MacDonald [15], Shimizu *et al* [16] have deliberately carried out simultaneous studies of ion and photon emission. Williams *et al* [17] have compared absolute yields of excited neutrals and positive ions from Si and Ni. The latter measurements were performed in an atmosphere of O_2, sufficient to saturate the ion and photon yield. These results show that the secondary ion yield is consistently higher than the secondary photon yield. If the degree of excitation (or ionisation) were simply related to the excitation energy, the experimental observation is difficult to explain. The interpretation of the results is difficult and the extrapolation to clean surfaces not possible because, as the authors point out, the energy spectrum of the secondary ions changes with the state of surface oxidation.

It is obvious from the above survey that there is a serious lack of fundamental data on which a theory can be based. The excitation function and absolute photon yields of more targets, particularly the metallic elements, must be measured. Ideally, secondary ion measurements should also be made to compare with the photon measurements. It is important that the surface be characterised by some technique to determine the extent of the impurity level - the measurements which are made will be of little use unless the extent of enhancement due to contamination is assessable. Effects from the edge of the bombarded spot must also be eliminated from photon measurements.

Since enhancement factors of 50 or more are possible as a result of contamination, the contribution from the edge of a spot may easily exceed the "clean" area contribution.

3. The Spectral Distribution of Emitted Photon Intensity

The study of the intensity distribution of emitted photons and the interpretation of that intensity distribution in terms of the excitation source has produced a lot of controversy. Most of the problems have now been resolved and in most cases result from poorly defined experimental aims and methods. The experimental studies began with attempts to test the model of quantitative secondary ion emission developed by Andersen and Hinthorne [18]. The model is well known, and briefly it assumes that in the region of the surface (though it is not exactly defined as to where this region lies) the ion beam produces a low density plasma in local thermodynamic equilibrium. The secondary ion intensities are then governed by the Saha-Eggert equation and provided either two elemental constituent compositions are known or alternatively some empirical relationships are used, the elemental composition of the specimen can be calculated from measurements of the secondary ion intensities. A vast amount of discussion of this model and its application to quantitative analysis exists in the literature [19]. We will not pursue it further here.

The interest in the model amongst workers involved in photon emission arises from the fact that the region in LTE should also emit photons and further the intensity distribution of these emitted photons should conform to a Maxwell-Boltzmann distribution, provided all the emitting particles are confined within the plasma volume so that excitation and de-excitation may proceed in equilibrium. Kato et al [20] were the first to investigate this possibility using the photon emission. They used a standard plasma physics technique to measure the plasma temperature, assuming LTE did exist at or near the ion-bombarded surface. The derived temperature, obtained from the ratio of intensities of two spectral lines, was roughly equal to the value deduced from the secondary ion measurements.

In later papers Martin and MacDonald [21] studied the LTE plasma concept further. If LTE exists, the intensity of a spectral line from such a source is given by

$$I'_{ij} = A'_{ij}\, h\nu_{ij} n g_i\, Z^{-1} \exp[-E_i/kT] \tag{3}$$

where A'_{ij} is the transition probability, ν_{ij} the photon frequency associated with the transition from the upper state of excitation energy E_i to the lower state of energy E_j, g_i is the degeneracy of the upper level, n the number of atoms, Z the electronic partition function, I'_{ij} the observed intensity and h and k Planck's and Boltzmann's constants respectively. Making adjustments for the detection system and using relative intensity and transition probability, this reduces

$$\ln(N'/g_i A_{ij}) = \ln(N/Z) - E_i/kT \tag{4}$$

where N' is the intensity in counts. Thus a plot of $\ln(N'/g_i A_{ij})$ as a function of E_i should yield straight lines, if we regard the source as a plasma in LTE. Martin and MacDonald [21] used this method as a test for the *possibility* of the existence of LTE. The experimental results indeed yielded straight line fits to experimental points indicating that the possibility of LTE existing was real - the observation of such straight line fits is a necessary but not sufficient condition for the existence of LTE [22]. Similar results have been obtained by others [23], though there has been some disagreement as to whether the so-called "reduced intensity plots" from two components of the same target actually produced plots corresponding to the same plasma temperature [24, 25].

Kelly and his co-workers [26] have examined the problem from a different view. Given that the atoms are excited in the bombardment process, they then decay in free space in front of the target surface. Consequently, the excitation-de-excitation equilibrium assumed in obtaining (3) does not apply. If we assume the distribution of electrons over excited states as a result of the bombardment is describable by a Maxwell-Boltzmann distribution, i.e.

$$P(i) \propto g_i \exp[-E_i/kT] \qquad (5)$$

where $P(i)$ is the probability that the atom is in an excited state of energy E_i, then the total yield from a state i is proportional to $g_i \exp[-E_i/kT]$. This method involves, however, the summation of the intensity in each possible de-excitation transition.

Snowdon et al [17] and Loxton et al [28] have shown that the intensity distribution considering the transition i→j can be written as

$$\ln(N/g_i A_{ij} \tau_i) \propto -E_i/kT \qquad (6)$$

where the term in τ_i results from the branching ratio of the de-excitation of the upper level i. Summation of (6) over all possible transitions leads to the expression used by Kelly et al [26].

There have thus been two approaches to these experiments. One measures the distribution of photon intensity to test the assumption of LTE. The other takes a definite physical model, i.e. excitation and subsequent decay as a free particle and examines the spectral distribution in terms of this model. Both experiments result in straight line fits to the appropriate proportionalities. Both give similar values of the gradient interpreted as a temperature. The most notable difference between the two methods has been demonstrated by Loxton et al [26]. Fig. 2 shows a comparison of the intensity distribution plotted according to (4) and (6). A discontinuity in the plot using (4), which had been interpreted in terms of a possible contribution of non-radiative electron exchange events [29], is removed when the data are plotted according to (6).

Snowdon et al [27,30] have written at length on these problems. Experimentally they have demonstrated the existence of an apparent LTE in gas phase collisions leading to excitation. This result has been demonstrated for metal atoms used as either target or projectile. In addition they have demonstrated the insensitivity of these methods of data presentation to the dependences involved, by fitting functions other than the Maxwell-Boltzmann distribution to the experimental results. From a variety of viewpoints, the concept of a plasma in LTE is not physically attractive (e.g. [27]) and is generally not accepted. Current experimental work also shows the excitation probability distribution may be Maxwell-Boltzmann like, but equally might involve other forms of energy dependence. One is led to the conclusion that these experiments have contributed very little to our understanding of excitation processes at a surface.

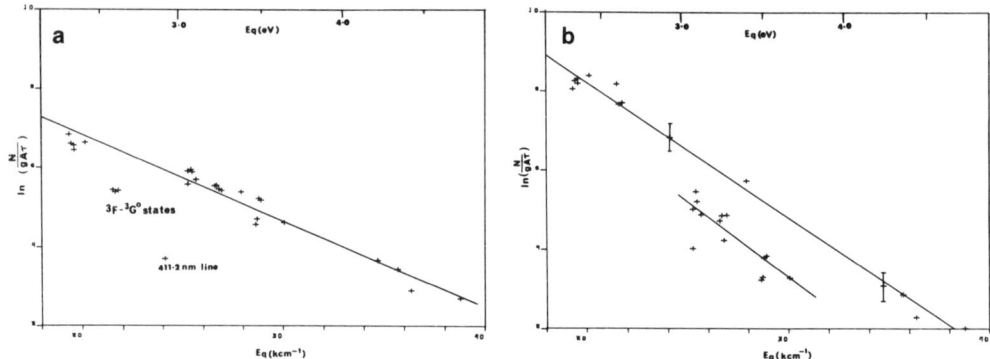

FIG. 2: The spectral distribution of intensity over excited states plotted according to (4) and (6) for the case of Ar$^+$ bombardment of T_i (after Loxton et al (2b)).

It has become equally obvious that geometrical effects must be taken into account in assessing the experimental results. Thus Snowdon [30] attempted to define a condition between observation volume and lifetime under which various methods of analyses were acceptable. If the lifetime of an excited state is such that the characteristic decay distance is long cf. dimensions of the observation volume, then a smaller proportion of excited states will decay within detection range than if the decay length is short w.r.t. the dimensions of the observation volume. Thus Loxton [31] has shown that the slope of the reduced intensity plot of (4) changes as the observation volume moves away from the target surface. Morrow et al [32] have attempted to formulate this problem in terms of the role of the transition probability in the measured photon intensity. However, the conclusion is basically that the decay length relative to the dimensions of the observation volume is the important parameter.

This means that great attention must be paid to the experimental geometry, particularly when comparing experiments. Thus the early experiments of Martin and MacDonald involved bombardment at near 45° incidence with the axis of the monochromator normal to the direction of ion incidence. This geometry was used by Tsong and Yusuf [33]. Under these conditions, since the sputtered atom distribution is usually maximised close to the specular direction, lifetime effects associated with decay of photons outside the observation volume should be minimised. Kelly et al [26,32] have usually used a geometry involving normal bombardment and observation along an axis normal to the incident beam but parallel to the surface. This geometry, also used by Loxton et al [28], will tend to maximise effects associated with the lifetime of excited states on emitted atoms.

These comments plus the observed insensitivity of the measurements to the model assumed in analysis of the experimental results leads us to the conclusion that, at this stage, further experiments on the intensity distribution of emitted photons will not be very rewarding in terms of increasing our basic understanding of the excitation process. The best summary that we can give of the available data would suggest that, as a result of the incident ion bombardment, there is a distribution of electrons over excited states which is approximately Maxwell-Boltzmann. These atoms then decay as free particles in front of the target. The Maxwell-Boltzmann-like distribution could result from cross sections for excitation which vary as the exponential in the excited state energy or alternatively because the energy range accessible to excited states is relatively small, the observed distribution could result from any energy dependence for which an exponential form is a reasonable approximation over a limited energy range. The latter implies virtually any energy dependence is possible. From the work done so far the experiments cannot differentiate between these possibilities.

The photon intensity is enhanced in general by adsorption processes. Experimental studies indicate that as a general rule the relative intensity distribution does not change. This is demonstrated for Ti emission [28]. Further the relative intensity distribution under the same surface conditions depends only very slightly on the incident ion energy and mass. These results also contribute little to our understanding of the excitation process except to suggest that thermal sputtering effects are unlikely to be the source of the excitation [26]. We will discuss the enhancement of photon yields due to adsorption later.

4. Modelling the Excitation Probability Function

Thermal excitation and plasma excitation tend to have been eliminated as the source of the excitation process, mainly by circumstantial and intuitive arguments. There exists no single experiment which definitely excludes the LTE model as the source of excitation. Nevertheless a combination of experimental results and arguments based on number of particles [29], size of excited atoms [27] etc. lead to the conclusion that plasmas in LTE do not exist in the vicinity of the ion bombarded surface. Attention has turned to collisional excitation as the source of excited atoms, though there is some discussion of which part of the collision cascade contributes, i.e. the high energy recoil portion of the spectrum [10] as distinct from the general cascade, which is weighted towards low energy atoms.

Dzioba et al [34] have recently introduced a model which is enjoying some acceptance and which is predicting trends in experimental behaviour, if not numerical agreement. This model owes its origin to experiments aimed at measuring the energy of the excited atom by measuring intensity decay curves away from the target. Borrowing a common technique from beam foil spectroscopy the intensity of photon emission a distance x from the target is given by

$$I = I_0 \exp\left(-\frac{x}{v\tau}\right) \tag{7}$$

where I_0 is the intensity at the target surface, v is velocity of the excited atom away from the target and τ is the lifetime of the excited state. For a sputtering situation the interpretation is complicated by the energy and spatial distribution of particles. However, using a technique suggested by Carter et al [35], it is possible to deconvolute the spatial distribution out of the experimental result.

The first analysis of this type was attempted by Meladenov and Braun [26] but the model has been improved remarkably by Dzioba et al [34]. This is the treatment we will consider here. The yield of photons in front of the target can be written as

$$Y^{int} = K \int_0^{\pi/2} \int_0^0 \frac{dN(E)}{dE} P(E) Q(E) R(E) \sin\theta \cos\theta \, d\theta \, dE$$

where K is a constant combining all terms independent of θ and E, $\frac{dN(E)}{dE}$ is the differential energy spectrum of the sputtered particles, P(E) is the probability of excitation of a sputtered particle of energy E, Q(E) is the probability of the excited state surviving a radiative de-excitation to a perpendicular distance x from the surface and R(E) is the probability of the particle surviving non-radiative de-excitation. The distribution is assumed isotropic, $\frac{dN(E)}{dE}$ is taken as $E/(E+U)^3$, Q(E) is written in the form of (7) as

$$Q(E) = \exp(-bx/E^{\frac{1}{2}} \cos \theta) \qquad (8)$$

while R(E) is usually taken as [7]

$$R(E) = \exp[-a/E^{\frac{1}{2}} \cos \theta]. \qquad (9)$$

Thus integrating over θ

$$Y^{int} = K \int \frac{dN}{dE} \cdot P(E) \cdot E_3 \frac{a+bx}{E^{\frac{1}{2}}} \cdot dE \qquad (9)$$

where $E_3 \frac{a+bx}{E^{\frac{1}{2}}}$ is the exponential integral function.

Dzioba et al used a variety of forms of the function P(E). These are illustrated schematically in Fig.3. The resultant forms of (9) can be

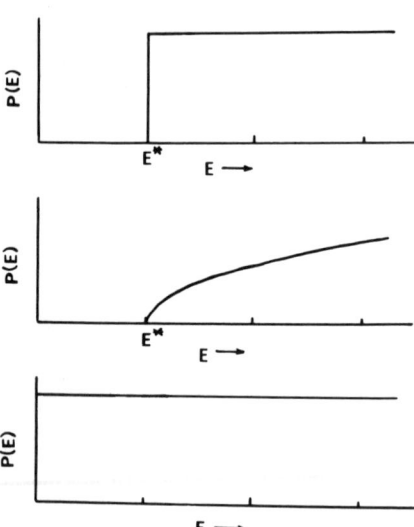

FIG. 3: Forms of the excitation probability function used by Dzioba et al (34, 37).

integrated to give expressions for the total intensity of photons in front of the target Y^{int} and the differential of this Y^{diff} which will give the number of photons seen by a slit in front of the target. The latter forms represent corrections to the intensity decay curves measured experimentally, the correction arising from the angular and energy distribution of sputtered particles. In general the approximation $\frac{dN}{dE} = \frac{E}{(E+U)^3} \simeq \frac{1}{E^2}$ for $E \gg U$ is used. Experimentally the differential form γ^{diff} has been used to convert observed values of I/I_0 to z-values, and then the latter are plotted as a function of x the perpendicular distance from the target surface. Recently, Loxton et al [38] have demonstrated that the experimental conditions may be adjusted so that γ^{int} forms of the curves can be used to analyse the experiment. The experimental results are demonstrated in Fig.4.

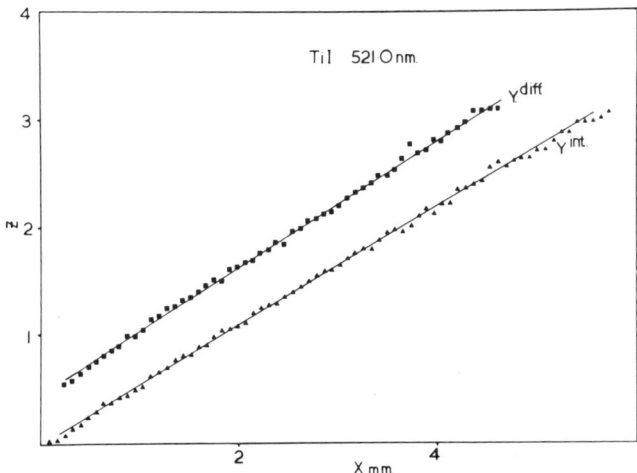

FIG. 4: Comparison of the predictions of Dzioba et al(34,37) for the decay of the intensity of T_i 521.0nm in front of the target. The curves have been shifted for clarity (from Loxton et al (38)) and represent results for the integral and differential yields respectively.

This model has been used to analyse a variety of decay curves obtained from several different targets [27]. The analysis yields a value of E^* which is appropriate to the particular form of $P(E)$ used, as shown in Fig.3. The values of E^* obtained do not show sufficient sensitivity to the form of $P(E)$ to allow one to definitely exclude any one form on the grounds that E^* is too high or too low to be physically acceptable. Consequently most analyses have used the step function form of $P(E)$, Fig.3a, where E^* is then the threshold energy for the excited atom to be removed from the surface. Values of E^* range from 10-100 eV for Li [34], $10-10^3$ eV for Mg, Ca, St, Ba [37,39], 10-500 eV for Ti [38]. These experiments have differed in that the geometry of some is defined by slit systems [37,39] while the geometry of others is defined by lens systems [38]. The latter have been carefully examined to ensure the use of the lens does not affect the experimental result. In Table 1 (Appendix A), we show values for E^* obtained in different experiments.

It is important to appreciate the information which we can obtain from this model. Results so far are not sensitive enough to the forms of $P(E)$ used to allow us to define the form of $P(E)$. This situation might change. The results indicate that a form of threshold function is likely to describe $P(E)$. This is not surprising. The results tell us nothing about the actual mechanism of excitation leading to the form of $P(E)$, but conversely the form of $P(E)$ may give us a clue to the mechanism of excitation. We derive E* from a slope, the slope being independent of the non-radiative de-excitation function R(E). In fact, we can put a = o in equation (9) without influencing the result.

A comparison of the values of E* and other measurements of energies associated with the photon emitters is perhaps meaningful. Other methods of obtaining a value indicative of the energy of the emitting atom have involved the use of measurements of line widths to obtain Doppler broadening contributions [40] and the use of decay curves as a function of distance in front of the target, without corrections inherent in the model of Dzioba et al [41]. Tsong and Yusuf [39] have compared values of E* with values of E obtained from the decay curves. The curve of $\ln[I/I_0]$ as a function of x is usually curved and it has been interpreted as consisting of a sum of decay curves for different particle velocities [36] or, alternatively, as indicating two groups of atoms one with a high average energy, the other with a low average energy [41]. It is the latter interpretation favoured by Tsong and Yusuf [39]. An alternative approach is to compare the values of E* with the value of E obtained from line shape measurements [38,42]. These latter measurements were made with the monochromator axis parallel to the plane of the target surface and so the velocity component parallel to the surface is responsible for broadening. The model assumes all broadening is due to Doppler contributions. The velocity components parallel to the surface derived from the line width are then expressed as an energy and in Fig.5 the results are compared with values of E* obtained from Dzioba's et al model for photons emitted from three different excited states of Ti, where the Ti is sputtered from TiC, TiO_2, TiN, TiB and Ti metal. Most importantly, the behaviour of both E* and $E_{//}$ with respect to energy of the excited state and with source of the Ti* atom are the same. The magnitudes of the E* and $E_{//}$ values are also similar. In Table 2, we compare these values with values of

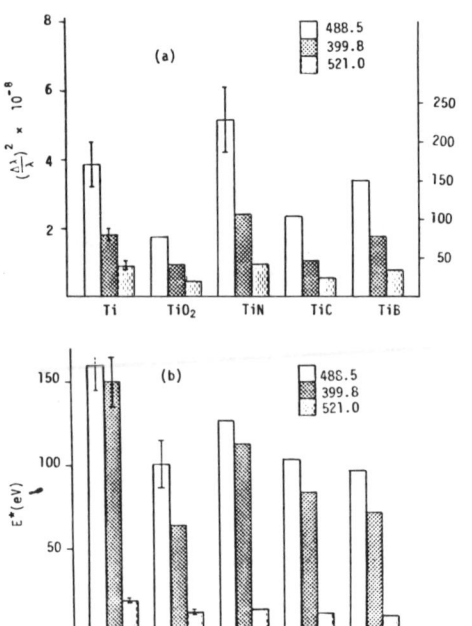

FIG. 5: A comparison between the energies associated with excited atoms emitting photons using the measurement of line width (a) and the Dzioba model (b). (From Loxton et al (42)).

TABLE 2

Comparison of Energy Parameters derive from: $E_\perp$-decay curves using equation 1, E^*-decay curves using the model of Dzioba *et al* and $E_{//}$ line profiles.

λ(nm)	Target	$E_\perp$(eV)	E^*(eV)	$E_{//}$(eV)
521.0	Ti	250	19	40
	TiO$_2$	150	12.5	20
399.8	Ti	3000	150	85
	TiO$_2$	1800	64	40
488.5	Ti	2500	160	185
	TiO$_2$	1700	101	80

E which were obtained by analysing the decay of intensity with distance in front of the target, according to (6) and converting the velocity obtained from the slope to an energy representative of the perpendicular component away from the target. There is an obvious discrepancy between this value and the $E^*, E_{//}$. The discrepancy is opposite to that which one would expect if non-radiative electron exchange processes were contributing to the intensity. Further the similarity of the values of E^* and $E_{//}$ suggests non-radiative de-excitation processes are not very important in determining the photon yield. We will discuss this point further later in this paper.

There are two experimental observations that any model of the excitation function must explain. When a surface is contaminated, the photon yield is in general enhanced, and the distribution of photon intensity with wavelength indicates that the distribution of excited electrons over states is approx-

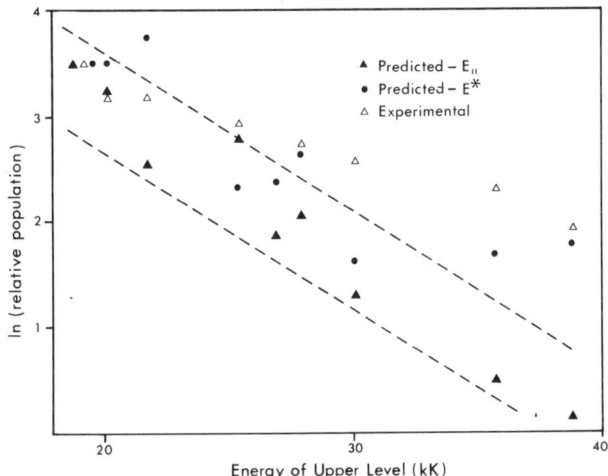

FIG. 6: A comparison between the experimental reduced intensity values of Loxton *et al* (26) with values deduced from measurements of the average energy of sputtered atoms emitting photons (Loxton *et al* (38)). The dashed lines represent the bounds of experimentally observed points, for the case here of Ar$^+$ bombarded T$_i$ and T$_i$ 521.0nm emission.

imately Maxwellian. Using the model of Dzioba et al [34] and a threshold form of the excitation probability function, the threshold energy E* for excitation in the presence of oxygen adsorption or when the sputtered Ti* is ejected from a compound, is observed to decrease. Since the energy spectrum of sputtered neutrals is of the form $1/E^2$, this implies the number of excited atoms and hence the photon intensity may increase significantly. The predictions of the excitation function are thus qualitatively in agreement with observation. The quantitative agreement with intensity ratios obtained by integrating over the energy spectrum is not good [38].

The results shown in Fig.5 give excitation thresholds for three well separated states of Ti*. From these values one can calculate the proportion of atoms excited to these states and hence produce "reduced intensity plots" equivalent to those obtained earlier, from measurements of the spectral distribution over excited states [28,26,23,24]. These are reproduced in Fig.6. The agreement is fair, and not much worse than the scatter in the experimental observations. The indications are that the threshold excitation function could yield an exponential dependence of the probability of occupation of a given level on its excitation energy, but more experimental observations are required.

5. The Effect of Adsorption on Photon Emission from Sputtered and Scattered Low Energy Atoms

We have mentioned already the enhancement of secondary ion and photon yields as a result of adsorption or contamination of the surface. One of the most popular descriptions of this effect has involved the comparison of band structures with atomic levels combined with speculation regarding the contribution of non-radiative de-excitation by electron exchange processes [7]. Alternative models of the enhancement of photon yields have suggested curve crossing type processes involving molecules of the target atom and the adsorbate [14,43]. The situation, however, is not understood and has been hampered by lack of data on the quantitative dependence of photon intensity on the degree of surface coverage.

A typical curve of the photon yield as a function of oxygen adsorption is shown in Fig.7 for Ti emission from atoms sputtered from Ti [28]. These curves indicate that the decrease of the photon intensity at high oxygen exposure rates is closely related to the decrease in the sputtering yield of the Ti at similar high oxygen exposures. Tsong observed similar effects [44]. The numerical values of the enhancement factors vary from element to element and often from observer to observer indicating that the experimental arrangement probably also plays a part in the extent of enhancement observed.

There have recently been a number of interesting studies in which surface adsorption has been monitored by different analytical techniques. Thus Dawson et al [45] and Barber et al [46] have combined SIMS and AES measurements while Fleish et al [47] have combined SIMS and XPS measurements. Of more interest to this paper are the experiments in which MacDonald et al [48] have combined ISS, IPP and SIMS measurements. The SIMS signal was a component of the energy spectrum of the ISS apparatus, centred in the region of 100-200 eV using 1 keV incident ion beams. No mass analysis was used but other experiments [49] indicate that secondary ions of this energy are in the main monatomic ions, the cluster ions almost invariably occurring at lower energies (some 10's of eV or less). In static exposure experiments, in which the surface is exposed to a contaminant and then the gas removed, the changes in

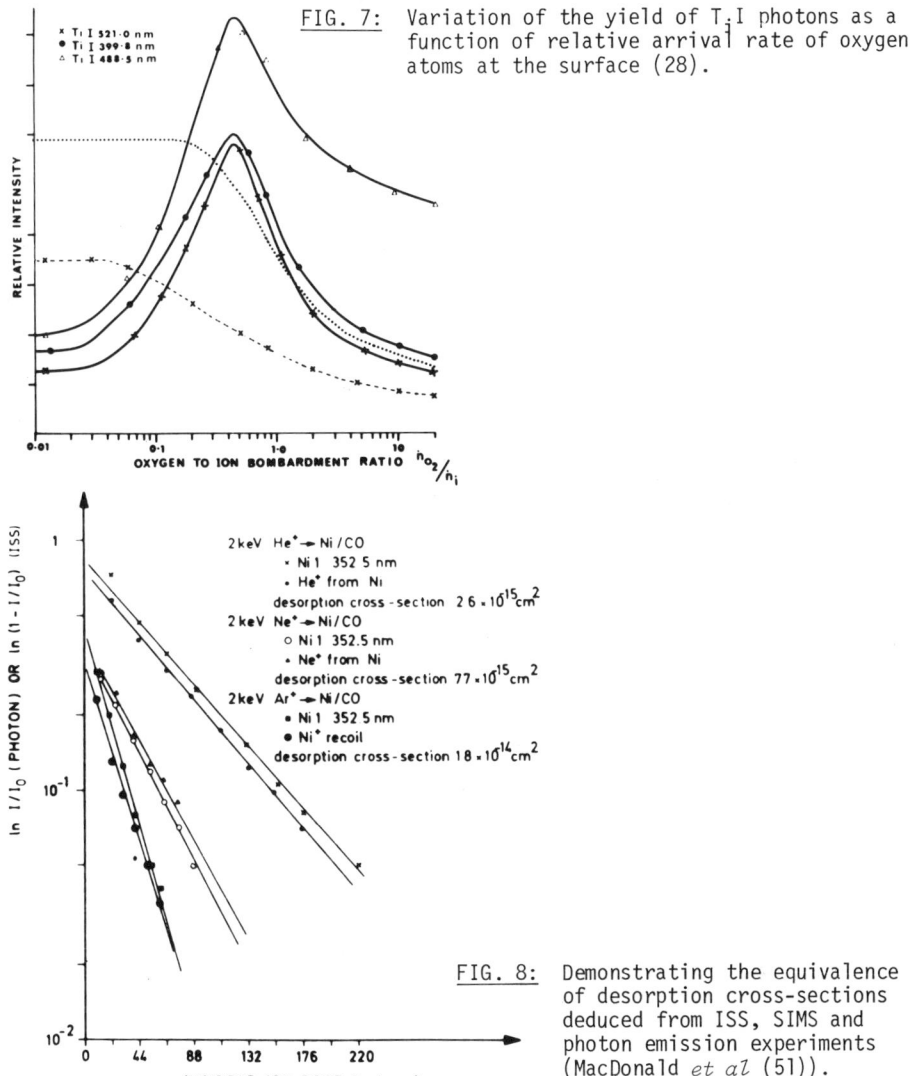

FIG. 7: Variation of the yield of T$_i$I photons as a function of relative arrival rate of oxygen atoms at the surface (28).

FIG. 8: Demonstrating the equivalence of desorption cross-sections deduced from ISS, SIMS and photon emission experiments (MacDonald et al (51)).

signal strength were monitored as a function of fluence. These are of course similar to the desorption measurements of Taglauer and Heiland [50]. MacDonald et al [51] showed that the desorption cross section obtained from measurements of the decay of secondary ion and photon signals with decreasing coverage (increasing fluence) yield the same value for the desorption cross section obtained from ISS measurements. This is demonstrated in Fig.8. The analysis assumes that the relative signal strength is proportional to the coverage of the surface with adsorbate. This implies that the maximum signal enhancement occurs at monolayer coverage.

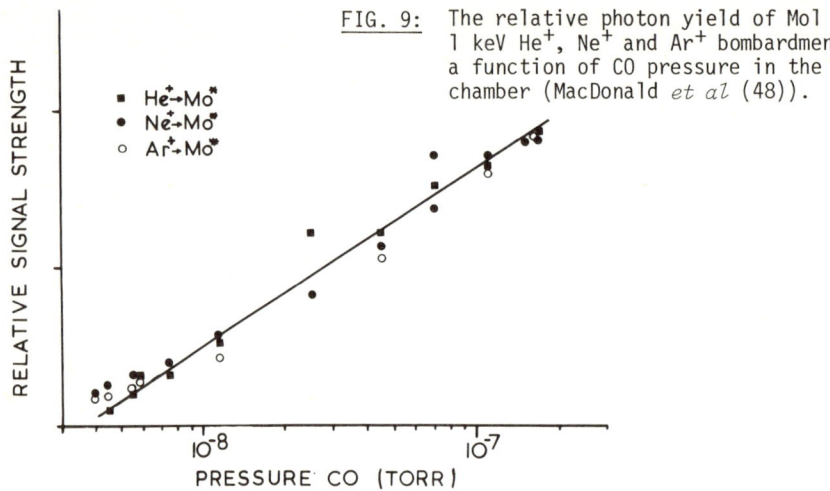

FIG. 9: The relative photon yield of MoI due to 1 keV He^+, Ne^+ and Ar^+ bombardment, as a function of CO pressure in the target chamber (MacDonald et al (48)).

Similar measurements have been made in a dynamic system, i.e. one in which the gas pressure is held constant during the irradiation, the coverage being an equilibrium between rate of arrival and rate of sputtering. The analysis of this followed a model due to Rausch and Thomas [52] and is very similar to an analysis given by Blaise [49]. Again the experimental results are similar to the model predictions but a full comparison was not possible because the rate of arrival of adsorbate atoms relative to the incident ions had to cover at least four orders of magnitude.

One of the most important results of this work, however, is shown in Fig.9. There the relative yield of MoI 386.4 nm photons is presented as a function of the CO adsorbate pressure for 1 keV He^+, Ne^+ and Ar^+ irradiations. There is significant agreement between the relative yields, which are expressed in terms of the value at 2×10^{-7} torr CO pressure. This would indicate that though the sputtering yield changes for the different irradiation conditions, at any given equilibrium coverage the *proportion* of atoms sputtered in excited states is the same, independent of the incident ion. This result favours excitation affecting the collision cascade rather than being restricted to the higher energy atoms resulting from recoil from the incident ion-surface atom collision. The analysis of the static and dynamic coverage dependence provides reasonable agreement to a model relating enhancement linearly to surface coverage and indeed localising the enhancement effect to the vicinity of the adsorbed atom. Such results favour bond-breaking or curve crossing models of the excitation process [14,43].

It is interesting to compare these results with some measurements of the polarised component of the emission resulting from excitation of ions scattered after grazing angle incidence to a metal surface [53]. Andra et al [54] have established general behaviour of the polarised emission following excitation of scattered particles. Many experiments have studied emission from atoms formed by neutralising the incident ion, as well as from ions scattered in the grazing incidence collision event. It was recognised very early in these studies that the extent of polarisation depended on the state of the surface. Martin et al [53] have performed these experiments beginning with a surface prepared by ion bombardment under UHV conditions and studying the polarisation of Balmer H_α, H_β, H_γ and H_δ from hydrogen atoms formed by neutralising and exciting incident H_2^+ molecules scattered in glancing angle collision with a Nb surface. The surface was contaminated under dynamic equilibrium conditions by the admission of oxygen, the degree of circular polarisation as indicated by the Stokes S/I coefficient, was measured using a rotating $\lambda/4$-plate and analyser system [Berry (55)]. The results are shown in Fig.10. The S/I coefficient as a function of background oxygen pressure (dynamic coverage) is shown and compared with the predictions of a model again based on the observed effect being proportional to the surface coverage. It seems possible that the degree of circular polarisation observed is a weighed coverage between the polarisation induced by scattering off the substrate atom and off the adsorbed atom, the weighing being directly proportional to surface coverage of adsorbate.

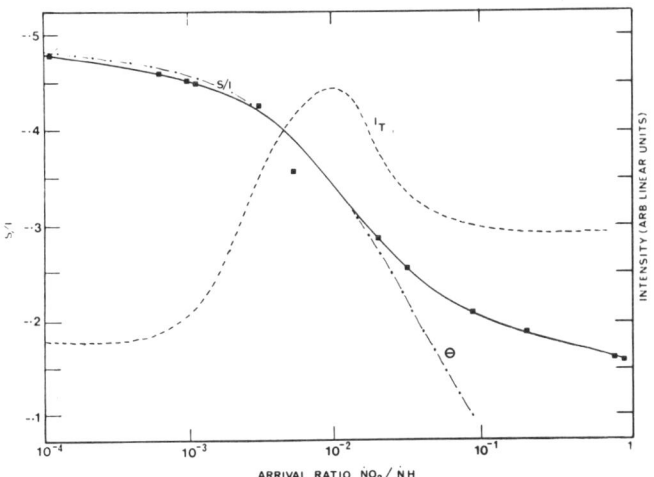

FIG. 10: The variation of degree of circular polarisation of Balmer H_β emission from H atoms scattered off a Nb surface, as a function of relative arrival rate of oxygen atoms at the surface. The 50keV H_2^+ beam was incident at glancing incidence.

These results indicate the enhancement effects are directly related to the surface coverage of adsorbate and are probably localised in the vicinity of the adsorbate atom. This has elements of a band structure picture in that it suggests the perturbation of electron distributions in the solid as a result of the adsorption might be responsible for the enhancement. It is obviously wrong though to extrapolate this idea to the point of comparing band structures of metal, oxide and atom in an adsorption situation. The suggestion of localisation of the enhancement to the vicinity of the adsorbate also could be interpreted in terms of a bond-breaking model, particularly of the type suggested by Blaise [14] and involving curve crossing effects close to the surface.

It is obvious though that further experiments on this enhancement effect must be done under circumstances in which the surface composition is fully characterised. We require quantitative details of the dependence of enhancement on surface coverage. An ideal experiment would include photo-electron spectrometry as well, to characterise the band structure of the solid, and the changes in such structure resulting from the adsorption.

5. Continuum Emission

The observation of continuum spectra from certain targets due to ion bombardment excited a lot of interest in the early years of these experiments. It was generally observed that the continua were very sensitive to the existence of surface contamination. Certain band emissions could be identified with excitation of such bonds as the CH bond or some other known molecular bonds [56], but the continua of interest were those not identifiable with excitation of simple molecules. It became accepted (but not proved) that the continua were the result of excitation of oxides or other molecular species involving the target atom. Recently there has been more activity in this field. Martin and Loxton [57] have studied the continuum emission from Ta. The spectra show several features which correlate with listed band heads of TaO. These features are broad band but discrete and disappear under contaminated conditions. The experiments can be summarised by Fig.11. Here the yields of atomic, molecular and continuuum radiation are compared with the yield of TaO^+ ions obtained with a simple SIMS analyser. These experiments strongly support the idea that the continuum is related to contamination at the surface.

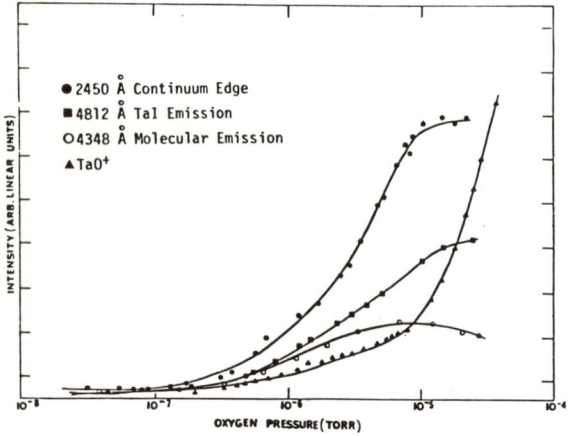

FIG. 11: The dependence of Ta continuum emission, TaI line emission and TaO^+ secondary ion emission on oxygen pressure (Martin and Loxton (57)).

A more detailed study of atomic, molecular and continuum radiation from ion bombarded B or B_2O_3 has been described by Kelly *et al* [58]. They observe features associated with BH molecular emission in the presence of H_2 or CH_4 and features associated with the molecular spectrum of BO from B_2O_3. Braun and Emmoth have observed strong molecular emission from BeH. [59]. Both Braun and Emmoth [59] and Bhattacharya *et al* [60] have observed continuum radiation associated with the excitation of molecules by the incident beam - thus Braun and Emmoth [58] identify Ar_2 radiation and Bhattacharya *et al* [60] identify continua associated with He_2, Ne_2 and Ar_2. The continua arise from the combination of a ground state and excited gas atoms, with the emission due to the transition to the unstable ground state of the molecule. The radiation has been shown to arise from sputtering of unplanted gas molecules but the data is not good enough to indicate whether the combination occurs at the surface or in front of it. The observation of Ar_2 radiation from a previously Ar implanted target bombarded with Kr^+ would suggest the combination involves two sputtered atoms and not one of the incident scattered neutralised ions.

Kelly *et al* [58] also report the observation of an apparent thermodynamic equilibrium distribution of rotational states in the BH molecule. Thomas *et al* [50] reported a similar apparent thermodynamic equilibrium in CH molecules sputtered in excited states from adsorbed species on Si. Snowdon and MacDonald [62] have shown that these apparent equilibrium distributions depend on the ion-molecule pair involved and on the energy, and conclude the observation should not be interpreted in terms of the existence of equilibrium. Kelly *et al* [58] reach the same conclusion.

The origin of the continuum radiation is not clear but Kelly *et al* [58] point out that in those systems where molecular radiation has been identified, e.g. BH, BO, CH and AlH, the states observed correlate adiabatically with ground state atoms or with low lying metastable states. Similarly, Bhattacharya *et al* [60] and Braun and Emmoth [59] favour a molecular formation for the inert gas molecule involving at least one atom in the ground state. This may give some clues to the formation of the excited molecules in general and to identification of other unidentified continuum features.

6. Electron Exchange Processes

In this review we have said little about the role that non-radiative exchange events play in the yield of photons from a bombarded surface. We have mentioned the Hagstrum [7] model, and introduced the survival coefficient by way of the A/a value in (1), but have said little beyond this. Wright and Gruen [12] rely heavily on the non-radiative decay process in their discussion of the excitation and ionisation process. In the Dzioba *et al* [34] treatment, the non-radiative event is of minor importance to the model. Phenomenological explanations incorporating the non-radiative transition and band-models work well for discussion of differences between metals and their oxides [63] but it is generally agreed they have limited use in discussing submonolayer adsorption effects.

There are a number of instances where the predictions of a band structure model do not agree with observation, e.g. Meriaux [64], Morrow *et al* [32], Taglauer *et al* [65], Loxton *et al* [28]. In particular, there is the work of Thomas and de Kluizenaar [66]. These authors deliberately modified the work function of Cu and Al targets by adsorbing Cs on the surface. The work function is thereby lowered. This would have been expected to enhance the photon emission by blocking off the resonant exchange process but experimentally no such enhancement was observed. Yu [67] has disagreed with this interpretation, maintaining that the effects occur very close to the surface where image potentials would cause the excited levels of the neutral atom to be raised relative to the Fermi level of the metal, to positions above the Fermi level, so that the non-radiative transition will always be possible, even in the presence of the Cs layer.

The main quantitative data in support of the electron exchange model in the case of photon emission, comes from studies of the dependence of yield on incident ion energy, as in the experiments of Heiland *et al* [10] and Wright and Gruen [12]. Further evidence comes from the fitting of theoretical line shapes to experimentally determined ones, using the non-radiative process as a fitting parameter. In all cases, one is fitting data with an exponential factor which is very strong term mathematically. Experiments of the type described by Heiland et al [10], for example, are independent of the actual values of the perpendicular component of the velocity used in the expression

$$\ln \frac{J}{JS} \propto - \frac{A}{a v_\perp}$$

where I is the photon intensity, J the current density, S the sputtering yield and v the perpendicular component of velocity away from the surface. Heiland et al [10] used values between v_{max} and $v_{max/2}$ where v_{max} is derived from a head-on collision. Wright and Gruen [34] use a value of v_{max}. The fit is unaffected if we were to assume the appropriate velocity were any fraction of the velocity transferred in a collision. Experimentally, the spectrometer integrates over the angular distribution of sputtered excited atoms, so any value of the velocity might be considered to be similarly appropriate.

The exponential dependence on velocity is a very powerful fitting parameter to be using and it may be that the ability of this factor to reduce a set of data points to an approximate straight line on a semi-logarithm plot is what is reflected in the experimental results. van der Weg et al [68] included such a factor in a model of line shape emitted by sputtered Cu, the model being based on excitation of the higher energy recoil atoms and obtained good agreement with the experiment. Delaney et al [69] showed that a model which included non-radiative transitions and the sputtered atom spectrum could also produce good fits to experimental measurements of the line shape.

It is necessary to develop a new line of experiments to test the role of non-radiative exchange effects. One way might be to look for the emitted Auger electrons predicted by the type of non-radiative transition which could occur. Resonance exchange processes are not likely to be a final determinate in the experimental results because their transition rate is so high that atoms affected by resonant exchanges are likely to be affected a second time by Auger exchange effects since the resonance transition is likely to involve an excited state of an ionised state of the atom. Alternatively, more defined experimental situations of known angle of emission and velocity should be studied. It is interesting that the most recent studies in neutralisation of scattered ions [Overbury et al [70]] suggest exchange processes are almost insignificant compared with collisional processes.

7. Conclusion

We have concentrated on the more recent work done in the general area of outer shell excitation in sputtered particles. It is possible to reach some conclusions. Firstly, the available evidence, while not absolute, is not in favour of the existence of a low density plasma in local thermodynamic equilibrium. The observed Maxwell-Boltzmann-like distribution of excited electrons over available states could arise from a number of physically reasonable situations. The energy dependence of the threshold for excitation derived from the experiments suggested by the work of Dzioba et al [34] provides an approximate explanation for the observed distributions. The observation of similar apparent LTE in the molecular spectra has been shown to be linked to the projectile-molecule combination - indeed it would be physically very difficult to explain both an equilibrium electron temperature in the atoms and the molecules, reached in the duration of the cascade.

The mechanisms leading to excitation are still undefined. Current work would tend to favour a model which involves the cascade as distinct from the higher energy recoils and near recoils. The threshold type behaviour of the excitation function used in the Dzioba et al [34] model is not unexpected and could be related to curve crossing in compound molecules existing for the duration of the collision, along the lines of the Fano-Lichten [63] model. Enhancement effects would then be related to different thresholds for the possible excitation channels.

All the current analysable data on the effect of adsorption on the observed intensities points to the effect being directly related to the adsorbate coverage and implies a localisation of the effect around the adsorbed atom. This observation further enhances suggestions that the excitation event involves molecular effects in the collision and that the excitation involves the cascade rather than the higher energy recoil particles.

Continuum emission is finally yielding some data which is not completely qualitative - identification with a few molecular bands has recently been possible. The direct correlation with oxygen on the surface has also been established using the SIMS measurements, but there has not been any real attempt to compare molecular emission in the present of different gases. Nitrogen is known to induce enhancement but little has been done on the molecular spectra. Similarly, CO adsorption is known to introduce continuum features into the spectrum of Mo, but there has not been an explicit comparison with the effect of oxygen for example. In the case of Mo, the continuum is in the same place as that introduced by oxygen adsorption. This might well be evidence for the dissociated adsorption of CO.

One conclusion which is obvious from this survey is that future experiments must be carefully planned. There is a great need for basic elementary data, such as the dependence of photon yield on incident ion energy, but the experimental situation must be well defined. The lifetime of the level being studied must be taken into account in ensuring that similar experimental situations are being investigated. The target surface must be characterised by some means, so the degree and type of contaminant is known. Total yields should be measured, the method of Tsong and Yusuf [15] being extended to alternative target-incident ion combinations. In studies of the effect of adsorption and of continuum emission, the surface coverage must be quantified. Alternative information such as the change in energy distributions of electrons in the solid is required, preferably concurrently with the photon data. We can gain little further insight into the mechanisms involved by yet another transient study, or another intensity distribution experiment.

From the work cited in this review, the single collision is emerging as the source of the excitation. This collision may involve homonuclear species, as in a cascade in a clean metal, or heteronuclear species, when adsorption contributes to the experiment. It may be necessary for researchers in this field to begin experiments in gas phase collisions, and compare the results directly with ion-solid experiments. The work of Snowdon et al [27,30] on comparisons of intensity distributions derived from ion-solid and ion-surface collisions points to the success of these experiments. The experiments involved will entail low energy ion-atom collisions involving metal ions - metal atom beams for example. This is a new and difficult experimental regime for many researchers in this field. Experiments involving laser-excitation of molecular beams and the subsequent behaviour of this excitation during ion- or atom-surface interaction may be tailored to provide answers to some of the questions still being asked in this field.

APPENDIX 1

Table 1 A Summary of Available Measurements of Energy of Excited Sputtered Atoms.

TARGET	LINE OBSERVED (inM)	ION AND ENERGY	MEASUREMENT	VALUE	REF
Al	AlI 309	300 keV Xe+	1	350eV + high	1
	AlI 396	300 keV Xe+	1	1.1keV+ high	1
Al$_2$O$_3$	AlI 396	20 keV Ar+	1	54eV + 8keV	2
	AlI 670	20 keV Ar+	1	12keV	2
Al	AlII 359	300 keV Xe+	1	3.2keV+ high	1
	AlIII360	300 keV Xe+	1	3.6keV+ high	1
Ca	CaI 411	30 keV Ar+	1	16keV	3
	CaI 423	30 keV Ar+	1	very high	3
CaF$_2$	CaI 423	12 keV Ar+	2	1580	4
	CaI 424, 436, 443, 453, 469, 504, 428, 432, 535, 586	30 keV Ar+	1	1.4-20keV	3
CaF$_2$	CaII 393.4	12 keV Kr+	2	1060	4
	CaII 396.8	12 keV Kr+	2	1280	4
Cr	CrI 427	8 keV Cr+	1	100eV	7
Glass	CrI 485.4		2	318	8
Glass	CrI 425.4		1	165,1494	8
Cu	CuI 511,515,522	20 keV Ar+	1	~3keV	2
Cu	CuI 325	60 keV Ar+	1	1keV	9
Cu	CuI 325	300 keV Xe+	1	1keV	1
Cu	CuI 327	300 keV Xe+	1	1keV	1
GaAs	GaI 403, 417	25 keV Ar+	1	2-3 keV	10
Glass	SiI 288.2		2	570	8
Glass	SiI 288.2		1	227, 11867	8
Si	SiII	20 keV Ar+	1	Slow	2
	SiIII	20 keV Ar+	1	Slow	2
	SiIV	20 keV Ar+	1	Slow	2
LiF	LiI 392,399, 413,427, 460,457, 610,671	20 keV Ar+	1	80eV	6
Li,LiF	LiI 671,323, 274,497, 427,399, 610,460,413	12 keV Kr+	2	31,9,7,28, 24,40,44,24,53	5
NaCl	NaI 330,569,498	12 keV Kr+	2	97,83,306	5
NaCl	NaI 498,515,569, 589,590,616	20 keV Ar+	1	1.5 - 3keV	6

Table 1 (continued)

TARGET	LINE OBSERVED (inM)	ION AND ENERGY	MEASUREMENT	VALUE	REF
Glass	NaI 589		2	141	8
Glass	NaI 589		1	118,4192	8
MgF$_2$	MgI 285.2, 383.5 517.8	12keV Kr+	2	40,640,1050	4
SrF$_2$	SrI 466.7	12keV Kr+	2	9600*	4
	SrII 407.8	12keV Kr+	2	2270	4
	SrII 421.6	12keV Kr+	2	1300	4
BaF$_2$	BaI 553.5	12keV Kr+	2	2700	4
Glass	BaI 553.5		2	1047	8
Glass	BeI 553.5		1	235, 16418	8
BaF$_2$	BaII 455.4	12keV Kr+	2	1950	4
BaF$_2$	BaII 493.4	12keV Kr+	2	1380	4
Glass	BeII 493.4		2	2258	8
Glass	BaII 493.4		1	1176, 10352	8
Glass	NiI 346.1		2	357	8
Glass	NiI 346.1		1	129, 831	8
Ti	TiI 468,506,521 453,486	20keV Ar+	1	10-100eV	6
Ti	TiI 398 521,399.8	20keV Ar+	1	~5keV	6
TiO$_2$	TiI 488.5 521,399.8	55keV Ar+	3	40,85,185	12
Ti	TiI 488.5 521,399.8	55keV Ar+	3	20,40,80	12
TiO$_2$	TiI 488.5 521,399.8,	55keV Ar+	2	19,150,160	12
Ti	TiI 488.5 521,399.8,	55keV Ar+	2	12.5, 64,100	12
TiO$_2$	TiI 488.5 521,399.8	55keV Ar+	1	250,3000,2500	12
TiB	TiI 488.5 521,399.8	55keV Ar+	1	150,1800,1700	12
TiC	TiI 488.5 521,399.8	55keV Ar+	3	10.5, 72, 97	12
TiN	TiI 488.5 521,399.8, 488.5	55keV Ar+	3	11.5, 84,104	12
		55keV Ar+	3	14, 113, 127	12

Note.

(a) Method 1 - exponential decay $\exp -\frac{\alpha_c x}{V}$

    2 - Dzioba et al [34] with step function for excitation probability.

    3 - Doppler broadening.

(b) The results of Tsong and Yusuf [8] using method 1 involve the separation of the decay curve into two regions corresponding to high and low energies. Both values are shown.

(c) *the authors Dzioba et al [4] have doubts about this value

Acknowledgement

The authors wish to acknowledge the financial assistance of the Australian Research Grants Committee, which have made our contributions to this field possible. We thank the Australian Institute of Nuclear Science and Engineering for the loan of one of the spectrometers. We thank the workshop of the Physics Department, and the Australian National University in general for providing the facilities. We acknowledge the many discussions with colleagues all over the world which have helped us in the research programmes in which we have been involved.

REFERENCES

1. C.W. White, E.W. Thomas, W.F. van der Weg and N.H. Tolk in "Inelastic Ion Surface Collisions", edited by N.H. Tolk, J.C. Tully, W.Heiland and C.W. White, Academic Press, N.Y. 1977.

2. G.E. Thomas, Surface Science, 90 (1979) 381-416.

3. C.W. White, D.L. Sims and N.H. Tolk, Science 17 (1972) 481.

4. I.S.T. Tsong and A.C.McLaren, Anal. Chem. 48 (1976) 699.

5. C.W. White and N.H. Tolk, Phys. Rev. Letters 26 (1971) 486.

6. N.H. Tolk, D.L. Sims, E.B. Foley and C.W. White, Radiation Effects 18 (1973) 221.

7. H.D. Hagstrum, Phys. Rev. 96 (1954) 336; see also H.D. Hagstrum in ref. 1.

8. W.F. van der Weg and P.K. Rol, Nucl. Instr. and Meth. 38 (1965) 274.

9. R.S. Bhattacharya, D. Hasselhamp and K.H. Schartner, J. Phys. D. Appl. Phys. 12 (1979) 255.

10. W. Heiland, J. Kraus, S. Leung and N.H. Tolk, Surface Science 67 (1977) 437-50.

11. R.B. Wright, Ming-Biarm Liu, and D.M. Gruen, J. Nucl. Mat. 76 (1978) 205; R.B. Wright and D.M. Gruen, Nucl. Instr. and Methods 170 (1980) 577; R.B. Wright and D.M. Gruen, J. Chem. Phys. (1980).

12. R.B. Wright and D. Gruen, J. Chem. Phys. 72 (1980) 147-171.

13. I.S.T.Tsong and N.A. Yusuf, Appl. Phys. Letts. 33 (1978) 999.

14. E. Blaise, Surface Science 60 (1976) 65-75.

15. P.J. Martin and R.J. MacDonald, Radiation Effects 32 (1977) 177-185.

16. R. Shimizu, T. Okutani, T. Ishitani and H. Tamara, Surface Science 69 (1977) 249.

17. P. Williams, I.S.T. Tsong and S. Tsuji, Nucl. Instrum. and Methods 170 (1980) 591.

18. C.A. Andersen and J.R. Hinthorne, Anal. Chem. 45 (1973) 1421.

19. See, for example, references in R.J. MacDonald and R.F. Garrett, Surface Science 78 (1978) 371-385.

20. M. Kato, P. Shimizu and T. Ishitani, Tech. Reports Osaka University 24 (1974) #1199.

21. P.J. Martin and R.J. MacDonald, Surface Science 62 (1977) 551-566.

22. F. Cabannes and J. Chapelle, Chapter 7 of M. Venugapolas "Reactions Under Plasma Conditions", Wiley Interscience 1971.

23. I.S.T. Tsong, Surface Science 69 (1979) 609 - see also ref. 2.

24. I.S.T. Tsong, Surface Science 75 (1978) 159L.

25. R.J. MacDonald, R.F. Garrett and P.J. Martin, Surface Science 75 (1978) 155L.

26. C.J. Good-Zamin, M.T. Shekata, D.B. Squires and R. Kelly, Rad. Effects 35 (1978), 139.

27. K. Snowdon, Rad. Effects 42 (1979) 185.

28. C. Loxton, R.J. MacDonald and P.J. Martin, Surface Science 93 (1980) 84.

29. C. Loxton and R.J. MacDonald, 7th International Conference on Atomic Collisions in Solids, Moscow USSR, September 1977.

30. K.J. Snowdon, G. Carter, D.G. Armour, B. Andressen and E. Veje, Surface Science 90 (1979) 429-441.
K.J. Snowdon, B. Andressen and E. Veje, Radiation Effects 40 (1979) 19.
G. Carter, D.G. Armour and K.J. Snowdon, Radiation Effects 35 (1978) 175.

31. C.Loxton, unpublished work.

32. M.R. Morrow, O. Auciello, S. Dzioba and R. Kelly, Surface Science, in press.

33. I.S.T. Tsong and N.A. Yusuf, Surface Science 90 (1979) 417.

34. S. Dzioba, O. Auciello and R. Kelly, Rad. Effects 45 (1980) 235.

35. G. Carter, G. Fisher, R. Webb, S. Dzioba, R. Kelly and O. Auciello, Radiation Effects 45 (1979) 45.

36. G.M. Mladenov and M. Braun, Phys. Stat Sol. (a) 53 (1979) 631.

37. S. Dzioba and R. Kelly, Surface Science, in press.

38. C. Loxton and R.J. MacDonald, Surface Science, to be published.

39. I.S.T. Tsong and N.A. Yusuf, Nucl. Inst. and Methods, in press.

40. C. Snoek, W.F. van der Weg and P.K. Rol, Physica 30 (1964) 341.
    W.F. van der Weg and D.J. Bierman, Physica 44 (1969) 206.
    C.S. White, D.L. Sims, N.H. Tolk, D.V. McCaughan, Surface Science 49 (1975) 657.
    R. Hippler, W. Kruger, A. Scharmann, K.H. Schartner, Nucl. Instrum. and Methods 132 (1976) 439.
    C.M. Loxton, R.J. MacDonald and P.J. Martin, Surface Science 93 (1980) 84.

41. C. Snoek, W.F. van der Weg and P.K. Rol, Physica 30 (1964) 341.
    V.V. Gritsyna, T.S. Kiyan, R. Goutte, A.G. Koval and Ya.M. Fogel, Izv Akad. Nauk. SSSR 35 (1971) 578.
    V.V. Gritsyna, T.S. Kiyan, A.G. Koval and Ya. M. Fogel, Radiation Effects 14 (1972) 77.
    M. Braun, B. Emmoth and I. Martinson, Phys. Scripta 10 (1974) 133.
    T.S. Kiyan, V.V. Gritsyna, Ya. M. Fogel, Nucl. Inst. and Methods 132 (1976) 543.
    M. Braun, Phys. Scripta 19 (1979) 33.
    C. Loxton, R.J. MacDonald and P. Martin, Surface Science 93 (1980) 84.
    G.M. Mladenov and M. Braun, Phys. Stat. Sol.(a) 53 (1979) 631.

42. C. Loxton, R.J. MacDonald and E. Taglauer - to be published in Surface Science Letters.

43. G.E. Thomas, Radiation Effects 31 (1977) 185.

44. I.S.T. Tsong and S. Tsuji, Surface Science 94 (1980) 269.

45. P. Dawson, Surface Science 57 (1976) 229; 65 (1977) 41; 71 (1979) 247.
    P. Dawson and Wing-Cheung Tam, Surface Science 81 (1979) 164.

46. M. Barker, R.S. Bardoli, J.C. Vickerman and J. Wolstenholme, Proc. 7th International Vacuum Congress and 3rd Int. Conf. on Solid Surfaces (Vienna 1977) 893.

47. T. Fleisch, G.L. Oh, W.N. Delgass and N. Winograd, Surface Science 81 (1979) 1.
    T. Fleisch, N. Winograd and W.N. Delgass, Surface Science 78 (1978) 141.

48. R.J. MacDonald, E. Taglauer and W. Heiland, Applications of Surface Science 5 (1980) 197-211.

49. G. Blaise, Proceedings NATO Congress on Ion Beam Characterisation of Materials, Corscia 1977.

50. E. Taglauer, U. Beitat, G. Marin and W. Heiland, J. Nucl. Mat. 63 (1976) 193; Surface Science 63 (1977) 507;
    E. Taglauer, U. Beitat and W. Heiland, Nucl. Inst. and Methods 149 (1978) 605.

51. R.J. MacDonald, W. Heiland and E. Taglauer, Appl. Phys. Letters 37 (1978) 516.

52. E.O. Rausch and E.W. Thomas, Nucl. Inst. and Methods 149 (1978) 511.

53. P.J. Martin, L. Berzins and R.J. MacDonald, Surface Science Letters, in press.

54. A.J. Andrä, H. Winter, F. Fröhling, N. Kirchner, H.J. Plöhn, W. Wittmann, W. Graser and C. Varelas, Nucl. Instr. and Methods 170 (1980) 527.

55. H.G. Berry, G. Gabrielse and A.E. Livingston, Appl. Optics 16 (1977) 3200.

56. N. Andersen, G.W. Carriveau, K. Jensen and E. Veje, Phys. Letts. 35A (1971) 19.

57. P.J. Martin and C. Loxton, Radiation Effects Letts, in press.

58. R. Kelly, S. Dzioba, N.H. Tolk and J.C. Tully, Surface Science, in press.

59. M. Braun and B. Emmoth, Nucl. Inst. and Methods 170 (1980) 585.

60. R.S. Bhattacharya, K.G. Lang, A. Scharmann and K.H. Schartner, J. Phys. D: Appl. Phys. 11 (1978) 1935.

61. G.E. Thomas, E.E. De Kluizenaar and H.P. Palenius, Chem. Phys. 7 (1975) 303.

62. K.J. Snowdon and R.J. MacDonald, Nucl. Instrum. and Methods 170 (1980) 351.

63. P.J. Martin, A.R. Bayly, R.J. MacDonald, N.H. Tol, G.C. Clark and J.C. Kelly, Surface Science 60 (1976) 349.

64. J.P. Mériaux, App. Phys. 7 (1978) 313.

65. E.Taglauer, W. Heiland and R.J. MacDonald, Surface Science 90 (1979) 661.

66. G.E. Thomas and E.E. De Kluizenaar, Nucl. Inst. and Methods 132 (1976) 449.

67. Ming L. Yu, Surface Science 90 (1979) 442.

68. W.F. van der Weg and D.J. Berman, Physica 44 (1969) 206.

69. P. Delaney and R.J. MacDonald, 7th International Conference on Atomic Collisions in Solids, Moscow, Sept. 1977.

70. S.H. Overbury, P.F. Dittner and S. Datz, Nucl. Inst. Meth. 1980, in press.

REFERENCES FOR APPENDIX 1.

1. M. Braun, B. Emmoth and I. Martinson, *Physica Scripta* 10 (1974) 133.

2. V.V. Gritsyna, T.S. Kijan, R. Goutte, A.F. Koval and Ya. M. Fogel, *Bull. Akad. Sci.* U.S.S.R. 35 (1971) 530.

3. T.S. Kijan, V.V. Gritsyna, Ya. M. Fogel, *Nuc. Instr. and Methods* 132 (1976) 435.

4. S. Dzioba and R. Kelly, *Surface Science* (in Press).

5. S. Dzioba, O. Anciello and R. Kelly, Rad. Effects **45** (1980) 235.

6. V.V. Gritsyna, T.S. Kijan, A.G. Koval and Ya. M. Fogel, Rad. Effects **14** (1972) 77.

7. H. Kerkow, Phys. Stat. Sol. (a) **10** (1972) 501.

8. I.S.T. Tsong and N.A. Yusuf, Nucl. Inst. and Methods. **170**, (1980) 357.

9. C. Snoek, W.F. van der Weg and P.K. Rol. Physica, **30**, (1964) 341.

10. A.G. Koval, G.I. Vyagin, V.V. Bobkov, Yu A. Klumovskii, S.S. Strel'chenko and Ya. M. Fogel. Sov. Phys - Tech Phys. **18** (1974) 1105.

11. C. Loxton, R.J. MacDonald, P.J. Martin, Surface Sci. **93** (1980) 84.
    C. Loxton, R.J. MacDonald and E. Taglauer, Surface Sci. Letters - to be published.
    C. Loxton and R.J. MacDonald, Surface Science, to be published.

Comment on the Energy Distribution of Excited Recoil Atoms

Peter Sigmund
Fysisk Institut, Odense Universitet
DK-5230 Odense M., Denmark

1. Introduction

In studies of the charge and excitation state of sputtered atoms [1-4], some knowledge is required about the energy spectrum of the flux of excited recoil atoms. It is the purpose of this note to demonstrate, on the basis of a simple model, that this energy spectrum may differ substantially from the equivalent spectrum of all (excited and unexcited) recoil atoms, and to provide qualitative insight into where the main differences should be expected.

2. The Model

KESSEL's model [5,6] assumes a threshold reaction governed by a probability $P(r_o)$ for excitation of a prescribed type,

$$P(r_o) = \theta(r_c - r_o) \quad , \tag{1}$$

where r_o is the distance of closest approach between colliding nuclei, r_c the internuclear distance at which the reaction sets in (e.g., an appropriate curve-crossing), and θ a step function,

$$\theta(x) = \begin{cases} 1 \\ 0 \end{cases} \text{for} \quad \begin{matrix} x > 0 \\ x < 0 \end{matrix} \quad . \tag{2}$$

The distance of closest approach may be related to the impact parameter p by means of the relation [7]

$$1 - \frac{V(r_o)}{E_r} - \frac{p^2}{r_o^2} = 0 \quad , \tag{3}$$

where $V(r)$ is the interatomic potential and E_r the relative energy in the center-of-mass system. Eq. (3) holds for elastic collisions, i.e., it is assumed that inelasticity is weak enough in order not to influence the collision dynamics.

Eliminating r_o from (1) and (3), one finds an excitation probability

$$P = \theta(1 - \frac{V(r_c)}{E_r} - \frac{p^2}{r_c^2}) \quad , \tag{4}$$

i.e., excitation takes place for $p < r_c \sqrt{1 - V(r_c)/E_r}$.

3. Total Excitation Cross Section

The total excitation cross section σ_e follows immediately from (4),

$$\sigma_e = \pi r_c^2 (1 - V(r_c)/E_r) \quad ; \tag{5}$$

it increases monotonically with increasing energy from a threshold defined by $E_r = V(r_c)$. Elastic collision cross sections are known [8] to essentially show the opposite behaviour.

4. Excitation Rate in Collision Cascade versus Energy of Exciting Particle

A large number of recoil atoms may participate in a collision cascade initiated by an atom of energy E. Every atom with a recoil energy exceeding the threshold given by (5) may undergo collisions in which atoms are excited. It is of interest, therefore, to provide an estimate of the relative significance of high- and low-energy atoms in the total excitation rate.

Let $X(E, E_o) dE_o$ be the mean number of excitations caused by atoms with instantaneous energy (E_o, dE_o) in a cascade initiated by an atom of energy $E (> E_o)$. By means of standard arguments [9], one finds the following integral equation for $X(E, E_o)$,

$$\int d\sigma(E,T) \{X(E,E_o) - X(E-T,E_o) - X(T,E_o)\} = \sigma_e(E) \delta(E-E_o) \, , \tag{6}$$

where $d\sigma(E,T)$ is the differential cross section for elastic scattering and T the recoil energy. It is assumed here that the primary atom has the same mass as the target atoms.

The solution to (6) is available [9,10]; it reads

$$X(E, E_o) \sim \Gamma_m \frac{\sigma_e(E_o)}{E_o S(E_o)} \cdot E \quad \text{for } E \gg E_o \, , \tag{7}$$

where $S(E_o) = \int T \, d\sigma(E_o, T)$ is the stopping cross section, and

$$\Gamma_m = \frac{m}{\psi(1) - \psi(1-m)} \tag{8}$$

a constant factor, determined by the exponent in the power cross section for elastic scattering [8],

$$d\sigma(E,T) = C \, E^{-m} \, T^{-1-m} \, dT \quad ; \quad 0 < m < 1. \tag{9}$$

C is a well-defined constant [8]. Since $S(E_o) \propto E_o^{1-2m}$, the dependence of $X(E, E_o)$ on E_o is like

$$X(E_o) \propto E_o^{2m-2} - E_c E_o^{2m-3} \, , \tag{10}$$

where $E_c = 2V(r_c)$ is the threshold energy for excitation in equal-mass collisions. In the relevant range of m-values, i.e., for $m \leq 1/2$, this function peaks toward low values of E_o, with a maximum at

$$(E_o)_{max} = \frac{3-2m}{2-2m} E_c \tag{11}$$

which lies in the interval $1.5\, E_c < (E_o)_{max} \lesssim 2E_c$.

We may conclude from these results that in estimating the energy distribution of excited recoil atoms, one needs to properly include excitation processes at energies quite close to threshold.

5. Differential Excitation Cross Section

The cross section for energy transfer (T,dT) to a recoil atom in a collision leading to excitation can be written in general terms,

$$d\sigma_e(E,T) = dT \int_0^{p_c} 2\pi p\, dp\, \delta(T - T(p,E)), \tag{12}$$

where $p_c = r_c(1-E_c/E)^{1/2}$. Since, for repulsive interaction, T is monotonic in p, (12) can be simplified into

$$d\sigma_e(E,T) = \begin{cases} d\sigma(E,T) \\ 0 \end{cases} \text{for} \quad \begin{matrix} T > T(p_c) \\ T < T(p_c) \end{matrix} \tag{13}$$

where $d\sigma(E,T)$ is given by (9).

The remainder of this section is devoted to the problem of finding a tractable expression for $T(p_c)$ with regard to the dependence on E. This problem has a rather simple solution for $E \gg E_c$ (cf. (20a) below); however, in view of the necessity to properly treat excitations near threshold, a more comprehensive estimate is mandatory.

We note first that (9) is equivalent with a scattering law

$$T = \frac{1}{E}[m\pi p^2/C + E^{-2m}]^{-1/m} \tag{14}$$

for equal masses. Thus,

$$T(p_c) = \frac{1}{E}[\frac{\pi m r_c^2}{C}(1 - \frac{2V(r_c)}{E}) + E^{-2m}]^{-1/m}. \tag{15}$$

From LINDHARD's magic scattering formula [8], one readily derives that

$$\frac{2V(r_c)}{E} = (\frac{C}{\pi m r_c^2})^{1/2m} \cdot \frac{\beta}{E}, \tag{16}$$

where

$$\beta = [\tfrac{1}{8}(\tfrac{3}{m} - 1)]^{-1/2} \quad , \tag{16a}$$

so that

$$T(p_c) = E(1 + \zeta^{-1} - \beta \, \zeta^{-1-1/2m})^{-1/m} \tag{17}$$

with

$$\zeta = \frac{C \, E^{-2m}}{\pi m r_c^2} \quad . \tag{18}$$

In order that excitation take place the condition

$$\zeta < \beta^{-2m} \tag{19}$$

has to be fulfilled.

From (17), one obtains the following boundaries for $d\sigma_e(E,T)$ (13),

$$(\frac{C}{\pi m r_c^2})^{1/m} \cdot \frac{1}{E} < T < E \qquad \text{for } \zeta \ll \beta^{-2m} \, , \tag{20a}$$

and

$$E\{1 - \frac{\beta^{2m}}{m^2}(1 - \frac{C}{\pi m r_c^2}(\frac{\beta}{E})^{2m})\} < T < E \qquad \text{for } \zeta \simeq \beta^{-2m} \, ; \tag{20b}$$

(20a) and (20b) reflect the limits of colliding-atom energy far above and near threshold, respectively.

If, instead of applying (20b), we apply (20a) *even near threshold*, we introduce a modified threshold energy E_t given by

$$E_t = (\frac{C}{\pi m r_c^2})^{1/2m} \tag{21}$$

according to (20a). This differs from the "real" threshold E_c (19),

$$E_c = \beta \, (\frac{C}{\pi m r_c^2})^{1/2m} \tag{22}$$

by the factor β. According to (16a), we have $\beta = 1.26, 1.00, 0.85$, and 0.76 for $m = 1/2, 1/3, 1/4$, and $1/5$, respectively. The error made in utilizing (20a) is, therefore, inappreciable, even with reference to the final remark in sect. 3.

6. Energy Distribution of Excited Recoil Atoms

After these preliminaries, one may introduce the function $F_e(E, E_o)dE_o$, which is the mean number of atoms recoiling from collisions leading to excitation with a recoil energy (E_o, dE_o), as a result of an initial cascade energy E. For E far above threshold, most excitations take place at small energies, hence target-target collisions dominate, and we can first restrict attention to the equal-mass case.

Again by means of standard arguments [9], one arrives at the integral equation

$$\int d\sigma(E,T)\{F_e(E,E_o) - F_e(E-T,E_o) - F(T,E_o)$$
$$- \delta(T-E_o)\theta(ET-E_t^2)\} = 0 \quad , \tag{23}$$

where the last term follows from (13), (20a), and (21).

Since (6) can be interpreted as defining a Green's function, the solution of (23) can be written directly as an integral,

$$F_e(E,E_o) = \int dE' \frac{X(E,E')}{\sigma_e(E')} K(E'E_o)\theta(E'E_o-E_t^2) \quad , \tag{24}$$

where $K(E,T) = d\sigma(E,T)/dT$. Insertion of (7) and (9) and integration yield

$$F_e(E,E_o) \sim \begin{cases} \Gamma_m E E_t^{2m-2} E_o^{-2m} & E_o<E_t<<E \\ \Gamma_m E E_o^{-2} & \text{for} \quad E_t<E_o<<E. \end{cases} \tag{25}$$

It follows that for $E_o > E_t$, $F_e(E,E_o)$ is identical with the recoil density for all atoms, excited or unexcited [9-11], while for $E_o < E_t$, a much flatter energy spectrum is found (Fig.1).

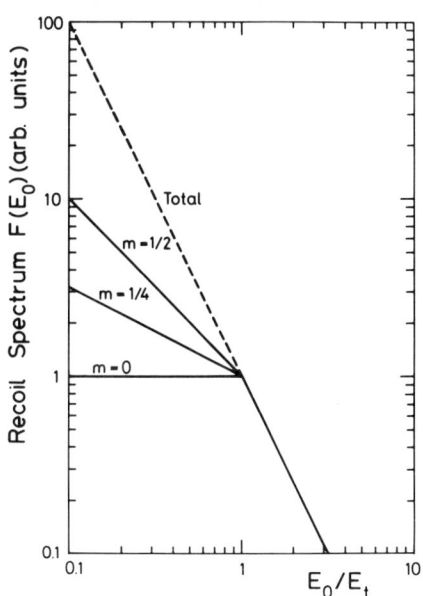

Fig. 1 Energy spectrum $F_e(E_o)$ of recoil atoms in collision cascade, emerging from collisions leading to excitation; according to (25). The corresponding distribution for all recoiling atoms is included for comparison (dotted curve). The spectra coincide for $E_o \geq E_t$

Eq. (25) represents the energy distribution of atoms recoiling from collisions leading to excitation. The energy distribution of excited atoms is, then, found by multiplication with an appropriate branching ratio ω_s.

As is well known [9-11], (25) remains valid for bombardment with an ion of different mass so long as inelastic energy losses do not dominate.

7. Note on the Flux of Excited Sputtered Atoms

An energy distribution of the present type is one of the necessary ingredients in a theory describing the flux of excited sputtered atoms. This is not the place to present a comprehensive theory of excitation phenomena in sputtering. However, an estimate will be given under the following simplifying (and possibly unjustified) assumptions,

- i) all excitations take place *in the bulk* in accordance with (1), essentially in the first recoil,
- ii) excited states survive until the excited atom is either sputtered or coming to rest, and
- iii) cross sections for collisions undergone by excited atoms do not differ from those applying to atoms in their ground states.

Under these assumptions, integration of (25) over E_o from E_o to infinity, and division by the stopping power $NS(E_o)$ [12] leads to the following expression for the flux of sputtered excited atoms,

$$G_e(E_o, \theta_o) dE_o d^2\Omega_o = \frac{\Gamma_m}{4\pi} \frac{F_D(0)}{E_o NS(E_o)} |\cos\theta_o| d^2\Omega_o \cdot g(E_o), \quad (26)$$

where

$$g(E_o) = \begin{cases} \frac{1}{1-2m} \frac{E_o}{E_t} [2(1-m) - (E_o/E_t)^{1-2m}]\omega_s & E_o < E_t \\ & \text{for} \\ \omega_s & E_o > E_t \end{cases} \quad .(26a)$$

For $g(E_o) \equiv 1$, this reduces to the total flux of sputtered atoms, apart from a surface binding condition which can be applied by the familiar procedure [12,13].

Figure 2 shows G_e for the case of $m = 1/4$, in comparison with the total flux of sputtered atoms, both functions being uncorrected for surface binding. It is seen that the low-energy peak is much less pronounced for excited sputtered atoms than that for all sputtered atoms. It is, however, clear that low-energy atoms strongly dominate the sputtered flux, and that there is no evidence on this basis to consider E_t as a cutoff energy *toward low energies*, as was stipulated in [14].

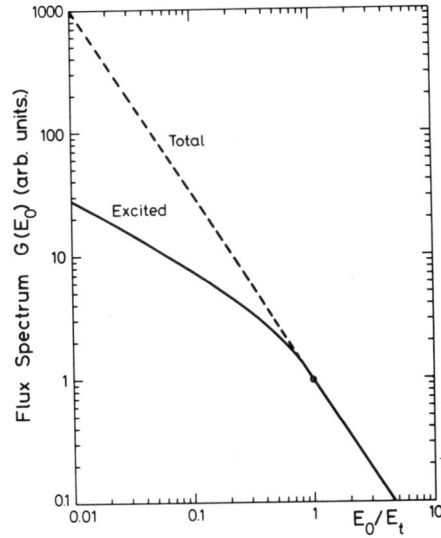

Fig. 2 Energy spectrum $G_e(E_o)$ of the flux of excited sputtered particles, according to (26) and (26a). The flux of all sputtered atoms is included for comparison (dotted curve). Normalizing constants are chosen so that the spectra coincide for $E_o = E_t$ (i.e., $\omega_s = 1$ in (26a)). $m = .0.25$. Simplifying assumptions are specified in the text (sect. 7)

Acknowledgement

This work was stimulated by discussions with K. Wittmaack and P. Williams during the Feldkirchen Workshop. The main result, equation (25), was presented as a discussion remark with appropriate precautions. In fact, a spurius factor $(E_t/E_o)^{1-m}$ had entered into the result for $E_o > E_t$, this overemphasized the dropoff at high energies.

References

1. G.E. Thomas, Surf. Sci. 90, 381 (1979)
2. G. Blaise, A. Nourtier, Surf. Sci. 90, 495 (1979)
3. T.R. Lundquist, Surf. Sci. 90, 548 (1979)
4. P. Williams, Surf. Sci. 90, 588 (1979)
5. Q.C. Kessel, Bull. Am. Phys. Soc. 14, 946 (1969)
6. R.K. Cacak, Q.C. Kessel, M.E. Rudd, Phys. Rev. A 2, 1327 (1970)
7. H. Goldstein, Classical Mechanics, Addison-Wesley (1953)
8. J. Lindhard, V. Nielsen, M. Scharff, Mat. Fys. Medd. Dan. Vid. Selsk. 36, no. 10 (1968)
9. P. Sigmund, Rev. Roum. Phys. 17, 969 (1972)
10. P. Sigmund, Rev. Roum. Phys. 17, 1079 (1972)
11. P. Sigmund, Appl. Phys. Lett. 14, 114 (1969)
12. P. Sigmund, in Inelastic Ion-Surface Collisions, N.H. Tolk et al., eds., Acad. Press (1977), p. 121
13. M.W. Thompson, Phil. Mag. 18, 377 (1968)
14. S. Dzioba, O. Auciello, R. Kelly, Rad. Eff. 45, 235 (1980); S. Dzioba, R. Kelly, Surf. Sci. 100, 119 (1980).

Quantitative Aspects of Outer-Shell Excitation in Ion-Surface Collisions

I.S.T. Tsong

Materials Research Laboratory, The Pennsylvania State University
University Park, PA 16802, USA

Abstract

Outer-shell excitation of sputtered atoms is a well-known inelastic process in ion-solid collisions. Quantities that are readily measurable in this process are the photon yields and the velocities (or energies) of the excited atoms. Useful information on possible basic mechanisms responsible for excitation can be derived from such quantitative data. In this report, the experimental procedures used to perform these measurements are described and the results relevant to the understanding of the excitation process in sputtering are discussed in detail.

1. INTRODUCTION

During the bombardment of solid surfaces by energetic heavy ions, a fraction of the sputtered particles leave the surface in electronically excited states. The subsequent decay of these excited states give rise to the emission of photons which can be observed outside the surface. This phenomenon has been known for some considerable time and the history of the observations of sputter-induced optical emission has been reviewed in detail by Thomas [1] recently.

The purpose of this paper is not to present yet another review on the general phenomenon of outer-shell excitation in sputtered particles since this has been dealt with extensively in the previous two Workshops [1,2], but to present specifically a summary of recent quantitative measurements undertaken in the study of photon emission. Quantitative data can be readily obtained on two observables: the photon yield, i.e., the number of photons produced from a given transition per sputtered atom, and the velocity (or energy) of an ejected atom in a given excited state. It is not difficult to see that measurements on the yield and energy will provide useful hints on the basic mechanisms of the outer-shell excitation process.

2. PHOTON YIELD

In a sputter-induced optical emission experiment, the observed photon yield, $I_{fi}(\lambda)$, of a spectral line of wavelength λ associated with an atomic transition from initial level i to final level f can be expressed as

$$I_{fi}(\lambda) = N\epsilon b p \eta A \text{ photons s}^{-1} \qquad (1)$$

where N is the total number of atoms sputtered per unit area per second; (if there are more than one atomic species present in the target, then N is the number of atoms of any given species sputtered per unit area per second); ε is the excitation efficiency, i.e., the ratio of number of atoms in upper level i to the total number of atoms sputtered; b is the probability that the atom will decay from upper level i to lower level f; (b is usually identified as the branching ratio where $b_{ji} = A_{ji}/\Sigma A_{fi}$); p is the probability that the excited atom will undergo radiative decay; η is the quantum efficiency of the optical detection system, i.e., the number of photons detected per emitted photon at a particular wavelength λ; and A is the area bombarded by the ion beam.

One notices in equation (1) that the problem of how the atom comes to exist in excited states i has not been addressed. Such a process is as yet unspecified. To make matters simple, we assume p = 1 for oxide type insulator targets and for metal surfaces with saturated oxygen coverage. Otherwise the argument on the cleanliness or degree of oxidation of the target surface will arise which of course will affect the determination of absolute photon yields.

The photon yield of a particular transition is then given by

$$Y(\lambda) = \varepsilon b = I_{fi}(\lambda)/N\eta A \text{ photons per sputtered atom} \tag{2}$$

where the quantities on the right-hand side can be experimentally determined.

The quantum efficiency, η, can be determined by standard calibration procedures such as those described by Andersen et al. [3]. A convenient black body standard for calibration is a tungsten lamp with a quartz envelope operated at 3000°K. The spectral output of such a lamp is well documented [4].

The sputtering rate, N, is simply determined by measuring the volume of target material eroded during a given bombardment time. A surface profilometer is very suitable for this purpose.

The absolute photon yields for a number of elements are given in Tables 1 and 2. For the elements listed in Table 1, the photon yields were measured by bombarding insulators such as NBS standard glasses (SRM 611 and RM 30), GdF_3 film and natural obsidian glass. The photon yields in Table 2 were obtained from Si and Ni targets saturated with oxygen by backfilling the target chamber. In both cases the bombarding ions were Ar^+ at 20 keV. Full experimental details for these two Tables are to be found in references [5] and [6]. The uncertainty in the photon yield determination is typically about ±25%, due mainly to the reproducibility of yield measurements and the accuracy and precision of the quantum efficiency calibration of the optical system.

The lines chosen for study all have short lifetimes, i.e., 10^{-8} seconds or less, such that the photons are emitted reasonably close to the target surface and within the field of view of the optical detection region. What has not been allowed for is the repopulation of the initial states by cascades from long-lived upper states. So the photon yield given for each individual transition is actually the sum of the contribution from the de-excitation of state i to state f and the cascade contribution.

Table 1 Absolute photon yield expressed as number of photons emitted per sputtered atom for the prominent lines of 34 elements. The detection limits for these elements are also given.

Element	Line (Å)	Energy of upper level (eV)	Photon per sputtered atom ($\times 10^{-3}$)	Detection limit (ppm by weight)
Ag	3281	3.77	10.14	200
Al	3962	3.14	2.55	50
B	2497	4.96	0.26	100
Ba	5535	2.24	1.44	800
Be	2348	5.27	0.22	50
Ca	4227	2.93	3.00	50
Cd	2288	5.41	0.21	200
Ce	4187	3.51	0.095	2500
Co	3453	4.01	0.41	500
Cr	4254	2.91	1.73	300
Cs	4555	2.72	4.58	100
Cu	3247	3.81	3.83	150
F	6902	14.52	0.007	3000
Fe	3581	4.31	0.75	300
Ge	3039	4.95	0.75	500
H	6563	12.09	0.35	15
In	4511	3.02	2.18	500
K	7665	1.61	2.69	20
Li	6708	1.85	8.37	1
Mg	2852	4.34	6.99	40
Mn	4034	3.07	2.47	200
Mo	3798	3.26	5.33	150
Na	5890	2.10	7.41	1
Ni	3415	3.65	0.71	400
P	2536	7.20	0.003	7500
Pb	4058	4.37	0.082	5000
Re	3460	3.58	0.75	400
Si	2882	5.07	1.28	250
Sr	4607	2.69	4.20	200
Ta	3311	4.43	0.083	3500
Ti	3653	3.44	0.18	700
Tl	3519	4.48	4.65	500
Zn	2138	5.79	0.009	6300
Zr	3601	3.59	0.11	1300

Tables 1 and 2 refer only to yields of one particular spectral line of any given element. Since nearly all elements emit more than one strong line, the integrated photon yield or the total number of excited atoms per sputtered atom is certainly higher than the values given. In Table 3, the integrated photon yields for all the excited states observed in Si, Al, Ca and Na are listed. These numbers are derived from the sum of all yields from all observable lines in the spectrum of each of the four elements.

3. COMPARISON BETWEEN PHOTON AND ION YIELDS

The secondary ion yields for Si^+, Si^{++} and Ni^+ are also given in Table 2. These values were determined by Williams [6] who calibrated the absolute transmission of the mass spectrometer by measuring the total sputtered ion current with a movable probe which could intercept the secondary ions some 5 cm from the sample. The transmission, defined as the ratio of ions detected to ions intercepted by probe, was found to be 0.01, with a reproducibility over many months of better than ±50%.

Table 2 Absolute ion and photon yields for Si and Ni saturated with oxygen

Species	Ionization potential or excitation energy (eV)	Ions or photons per sputtered atom
Si^+	8.15	1×10^{-2}
Si^{++}	16.34	1×10^{-6}
Si I 2882 Å	5.08	1.6×10^{-3}
Si I 2516 Å	4.95	2.2×10^{-4}
Si II 3856 Å	10.07	1.4×10^{-5}
Si II 6347 Å	10.07	1.0×10^{-5}
Ni^+	7.63	6×10^{-4}
Ni I 3415 Å	3.63	1.0×10^{-4}
Ni I 2326 Å	5.49	3.1×10^{-7}

Table 3 Total photon yield of all observable excited states in a given element

Element	Yield (photons/sputtered atom)
Si	1.8×10^{-3}
Al	8.6×10^{-3}
Ca	1.1×10^{-2}
Na	1.2×10^{-2}

The secondary ion yields were determined under the condition of full oxygen coverage of the target surface, identical to that in the photon measurements. It is interesting to see that the singly charged ions of Si and Ni show much higher yields compared to photons. The implications of this observation will be discussed in a later section.

4. VELOCITY DISTRIBUTIONS OF EXCITED ATOMS

One way of determining the velocity distributions of sputtered excited atoms is to measure the Doppler shift and broadening of the spectral lines. Van der Weg and his co-workers [7,8] were first to attempt this and they found a peak shift of 0.6 Å indicating a mean energy of ~ 1 keV for excited Cu atoms. In addition, they obtained reasonable agreement by fitting the line shape of Cu I 3247 Å emission to a model based on binary collisions at the surface combined with non-radiative de-excitation processes by electron tunneling [9]. However, later Doppler line shape measurements by Hippler et al. [10] on a variety of ion-bombarded surfaces could only yield limited agreement with such a model. They conclude that contributions from collision cascades should also be taken into account.

White et al. [11] also observed Doppler broadening of the Si I 2882 Å line from a Si surface and to a lesser extent from an SiO_2 surface, although an estimation of the velocities was not possible because of inadequate instrumental resolution.

A second method of determining velocity distributions is by observing the decay of the photon yield as a function of the perpendicular distance away from the solid surface undergoing bombardment. This intensity decay follows the well-known relation

$$I = I_0 \exp(-t/\tau) \tag{3}$$

which can be rewritten as

$$I = I_0 \exp(-x/v_\perp \tau) \tag{4}$$

where I is the intensity or photon yield from a particular transition,

x is the distance away from the surface,

$v_\perp$ is the velocity component normal to the surface,

and τ is the lifetime of the excited atom.

$\tau = 1/\sum_f A_{fi}$ where A_{fi} is the transition rate or the so-called Einstein coefficient of transition.

By plotting log I versus x, a smooth curve is generally obtained. One standard treatment is to divide up the curve into straight-line segments and assign average velocities to these segments, provided of course that cascading from upper levels is negligible. Following such a treatment of their data, Gritsyna et al. [12,13] and Braun et al. [14] suggest that sputtered excited atoms have two reasonably well-defined velocity distributions, one at medium energy (10^2-10^3 eV) and the other at high energy (10^3-10^4 eV).

Recently, Dzioba et al. [15] proposed an alternative way of interpreting the intensity decay data by assuming, similar to Wright and Gruen [16], that excited atoms originate from those sputtered by a collision cascade process, and that there is a threshold kinetic energy for the excitation to occur. We shall deal with such a model and its application to data interpretation in some detail in the following sections.

4.1 Theory

The theory of velocity (or energy) distributions of excited atoms given below follows closely the derivations of Dzioba et al. [15] and Wright and Gruen [16,17]. We have chosen to use energy as the variable rather than velocity although it does not make any difference either way.

We start with the assumption that outershell excitation occurs in those atoms sputtered by a collision cascade process. The energy spectrum of sputtered atoms in the energy interval between E and E + dE and in the angular interval between θ and θ + dθ with θ the angle of atom's trajectory with the surface normal is given by

$$\Phi(E) dE d\Omega = \frac{CE \cos\theta}{(E+E_b)^{n+1}} dE (2\pi \sin\theta d\theta) \tag{5}$$

where $\Phi(E)$ is the sputter flux density energy distribution and has the units of number per unit area per unit time per unit energy. E_b is the surface binding energy. The solid angle $d\Omega = \sin\theta d\theta d\phi = 2\pi\sin\theta d\theta$; $d\phi$ is the azimuthal angle, about which the sputter flux is assumed to be isotropic. According to Thompson [18] and Carter et al. [19], n = 2. C is a constant for a given primary ion and energy and target material.

The usual expression for the probability of the excited atom to survive non-radiative decay near the surface is given by [9,20]

$$R(E,\theta) = \exp(-a/E^{\frac{1}{2}}\cos\theta) \ . \tag{6}$$

The probability of radiative de-excitation of the excited atom in state i at a time t after it leaves the surface is

$$P_i(t) = \exp(-\gamma_i t) \ . \tag{7}$$

Equation (8) is similar to equation (3) with

$$\gamma_i = \frac{1}{\tau} = \sum_f A_{fi} \ .$$

It can be rewritten as

$$P_i(t) = \exp\left(-\frac{\gamma_i x}{v_\perp}\right) = \exp\left(-\frac{\gamma_i x m^{\frac{1}{2}}}{2^{\frac{1}{2}} E^{\frac{1}{2}} \cos\theta}\right) \ . \tag{8}$$

Finally, we introduce the probability $P_{ex}(E)$ of a sputtered species with kinetic energy E being found in a particular excited state. The integral yield of photons as a function of perpendicular distance from the target, as would be relevant if the target were viewed parallel to the surface, can then be written

$$Y^{int} = K \int_0^{\frac{\pi}{2}} \int_0^\infty \Phi(E) R(E,\theta) P_i(t) P_{ex}(E) dE \sin\theta d\theta$$

$$= K \int_0^{\frac{\pi}{2}} \int_0^\infty E(E+E_b)^{-3} \exp\left(-\frac{a+\gamma_i x m^{\frac{1}{2}} 2^{-\frac{1}{2}}}{E^{\frac{1}{2}}\cos\theta}\right) P_{ex}(E) \sin\theta\cos\theta dE d\theta \ . \tag{9}$$

Dzioba et al. [15] postulated $P_{ex}(E)$ as a step function in which $P_{ex}(E) = 0$ for $E < E^*$ and $P_{ex}(E) = 1$ for $E \geq E^*$ where E^* is defined as the threshold kinetic energy for excitation to take place. We shall discuss later the validity of such a postulate in our model for the excitation mechanism. Equation (9) can now be written as

$$Y^{int} = K \int_0^{\frac{\pi}{2}} \int_{E^*}^\infty E(E+E_b)^{-3} \exp\left(-\frac{a+bx}{E^{\frac{1}{2}}\cos\theta}\right) \sin\theta\cos\theta dE d\theta \tag{10}$$

after putting $b = \gamma_i m^{\frac{1}{2}} 2^{-\frac{1}{2}}$. For $E_b \ll E$, we can approximate $E(E+E_b)^{-3}$ to simply E^{-2}. Substituting z for $(a+bx)/E^{*\frac{1}{2}}$ and integrating, we obtain after normalizing

263

$$Y^{int}(z) = z^{-2}\{1-\exp(-z)-3zE_4(z)\} \qquad (11a)$$

$$Y^{int}(0) = 1 \qquad (11b)$$

where $E_n(z) = \int_1^\infty k^{-n}\exp(-zk)dk \qquad (12)$

k being any variable

and $E_{n+1}(z) = \frac{1}{n}[e^{-z}-zE_n(z)] \qquad . \qquad (13)$

For the case where the light emission is viewed through a narrow slit, i.e., equivalent to observing the emission at distance between x and x+dx from the target surface, we differentiate equation (11a) and normalize again to obtain

$$Y^{diff}(z) = (3/2z^2)[1/z-(1+1/z)\exp(-z)-zE_3(z)] \qquad (14a)$$

$$Y^{diff}(0) = 1 \qquad , \qquad (14b)$$

A plot of equations (11) and (14) is given in Fig. 1. In an experiment, the photon count rate, i.e., Y^{diff}, is converted to z using Fig. 1. A plot of z versus x will yield E*, the threshold kinetic energy required for atoms to appear in excited states.

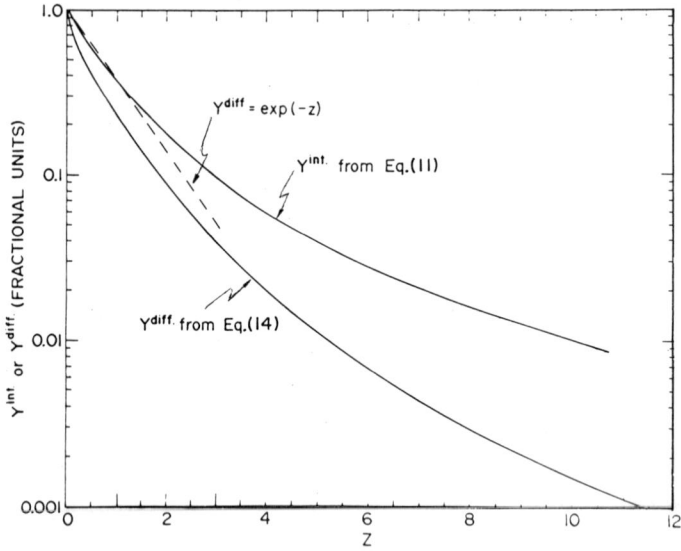

Fig.1 The normalized photon yield functions in integral and differential form versus the dimensionless distance variable, z, given by $(a+\gamma_i m^{\frac{1}{2}} 2^{-\frac{1}{2}} x)/E^{*\frac{1}{2}}$ where E* is the assumed threshold kinetic energy. The curve exp(-z) would apply if there was a unique kinetic energy and no angular distribution.

4.2 Experimental

Once again as in the photon yield experiments we used an oxide glass as our target. One of the main reasons for choosing an oxide target was to try to measure the lowest velocities of the excited atoms since previous Doppler shift and broadening measurements on metal targets yielded exceedingly high energies, i.e., in the keV range [7,8]. Moreover, White et al. [11] observed greatly reduced Doppler broadening from an SiO_2 target compared to from Si. One possible explanation for low-velocity excited atoms originating from oxidized surfaces is the blocking of non-radiative transitions in the electron-tunneling model [2]. We therefore would like to test the validity of such a model, as well as trying to determine a minimum threshold kinetic energy using the theory of Dzioba et al. [15] outlined in the previous section.

The compositions of the two oxide glass targets were:

$2SiO_2 \cdot Na_2O \cdot BaO \cdot \frac{1}{2}NiO_2$

and $2SiO_2 \cdot Na_2O \cdot BaO \cdot \frac{1}{2}Cr_2O_3$.

The polished glass surface was bombarded by a 20 keV Ar^+ ion beam at normal incidence. The beam spot at the target surface was 1 mm in diameter and the current density was about 30 $\mu A\ mm^{-2}$. We used two different experimental approaches to measure the decay in photon yield (intensity) as a function of distance away from the surface in order to determine the velocity (energy) distribution.

The first approach [21] is illustrated in Fig. 2. The emission from the excited particles was dispersed with a Jarrel-Ash 82-410 monochromator and the photons were detected by a cooled Hamamatsu R955 photomultiplier. The region of emission observed by the monochromator was defined by two 150 μm wide slits parallel to the target surface. The first of these was placed inside vacuum close to the target, about 3 cm from the beam spot, while the second slit was the entrance slit of the monochromator. A fused silica window separated the two slits. No lenses were used. The 150 μm slit-width was chosen as the best compromise between detection sensitivity and spatial resolution. When the target was undergoing Ar^+ bombardment, the decay in intensity as a function of distance from the surface was measured by moving the target in small steps along the beam direction and the number of photon counts per second was recorded at each position.

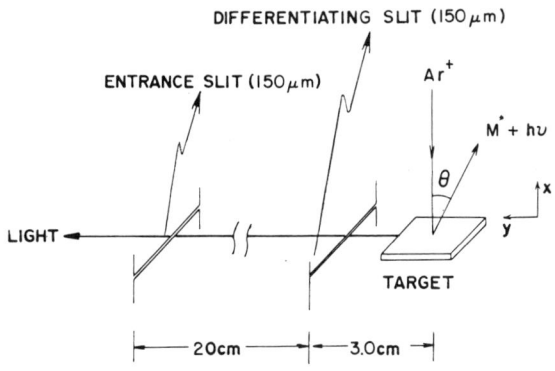

Fig.2 Schematic representation of the defining-slits system used for measuring intensity decay. The target is moved by small increments along the x-axis, i.e., the direction of the beam.

The second experiment [22], shown in Fig. 3, is essentially a slitless approach which involves observing the intensity decay in real-time by direct-imaging of the light emission using an optical multichannel analyzer (OMA). The bombarding condition is identical to the first experiment. The light emitting region in front of the bombarded surface was imaged by the two-lens system (measured magnification 4.65×) directly on to the active

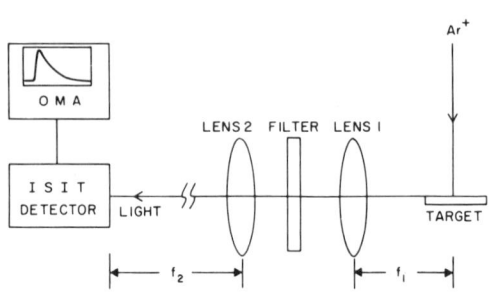

Fig.3 Schematic diagram of the experimental set-up used to directly image the light-emitting region on to an intensified silicon intensifier target (ISIT) detector and observing in real-time the intensity decay on an optical multi-channel analyzer (OMA). The focal lengths of the lenses are f_1 = 2.54 cm, f_2 = 11.43 cm, giving an effective magnification of ∿4.5×.

area of an intensified silicon intensifier target (ISIT) detector which is essentially a vidicon tube with superior spectral response from ultraviolet to infrared. The scan pattern of the electron beam in the ISIT was adjusted such that the light was sampled from a region equivalent to a 300 μm thick slice perpendicular to the surface. The wavelength of the spectral line of interest was selected by a narrow band interference filter. The intensity-versus-distance information was displayed on the optical multichannel analyzer (OMA) with background noise subtracted. In this way, we were able to study the intensity decay over a distance of about 2.6 mm above the surface. Since this distance was displayed on the OMA over 512 channels, the effective spatial resolution was therefore ∿5 μm, a great improvement over the defining-slits approach.

In both experimental configurations, the probability of excitation of a sputtered atom by collision with primary Ar^+ ions during its flight from the target over a distance say ℓ = 1 cm is extremely small. The probability of excitation for the atom over such a distance is simply $n_i \sigma_{exc} \ell$, where n_i is the density of the primary ion beam, and in our case for a 20 keV Ar^+ beam of 30 μA mm^{-2}, n_i = 6 × 10^8 ions cm^{-3}. σ_{exc} is the cross-section for excitation of a sputtered atom from ground state to a given excited state. From Ar-Ar collision studies [23], we have $\sigma_{exc} \approx 10^{-18}$ cm^2. Therefore the probability of collisional excitation = $n_i \sigma_{exc} \ell$ = 6 × 10^{-10} photons per sputtered atom. Compared to the photon yields in Tables 1, 2 and 3, this is at least a factor of 10^5 smaller.

4.3 Results and Discussion

4.3.1 Defining Slits Experiment

The decay in photon yield or intensity, I, as a function of distance away from the target surface is shown in Fig. 4 as log I versus x plots. These decay curves clearly show a smooth velocity distribution. However, following the curve-fitting practice of previous workers [12-14], each curve is

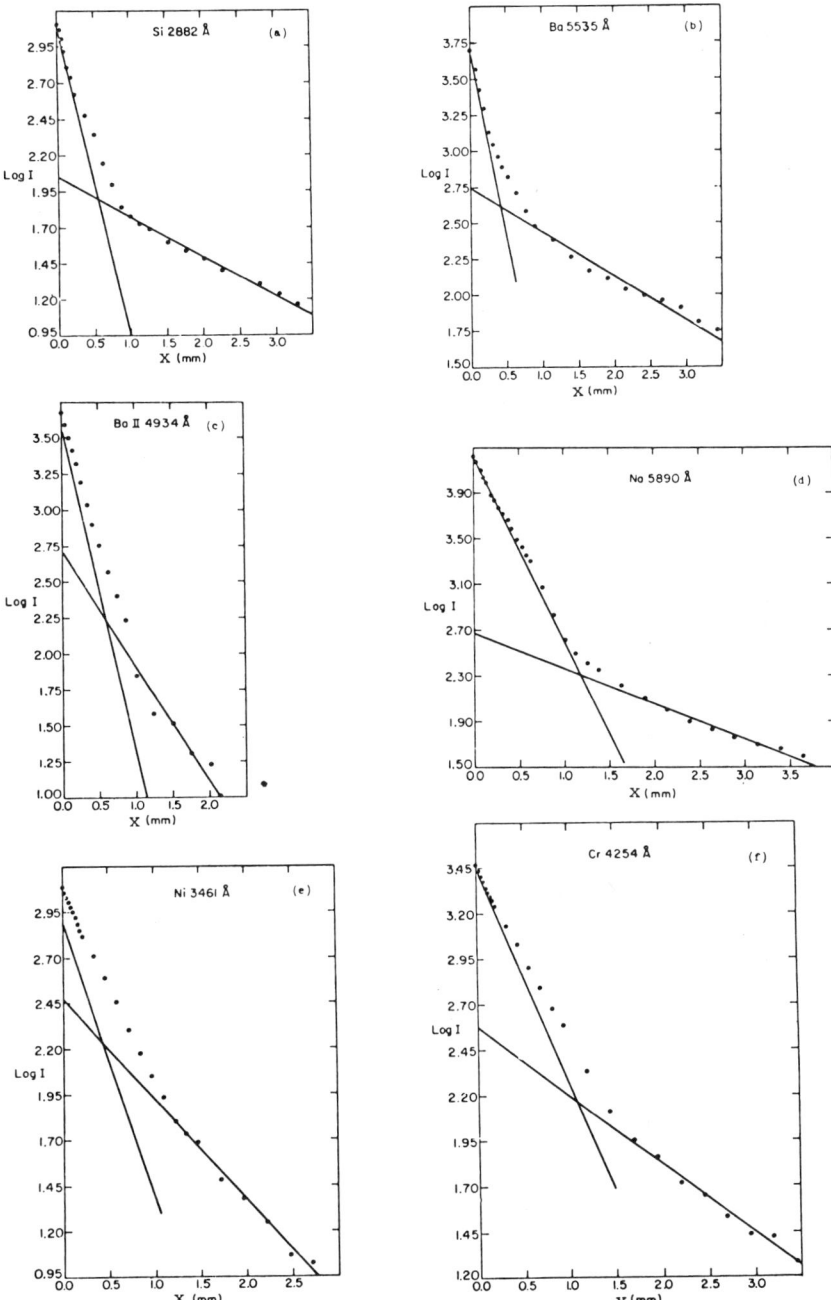

Fig.4 Intensity decay curves expressed as log I versus distance, x, for lines (a) Si I 2882 Å, (b) Ba I 5535 Å, (c) Ba II 4934 Å, (d) Na I 4890 Å, (e) Ni I 3461 Å and (f) Cr I 4254 Å. Note that when x = 0, I - I_0, the maximum intensity. The straight lines are drawn to determine the velocities of the fast and slow particles. I is measured in photon counts per second.

broken up into two straight-line segments to which we can assign average velocity values. Accordingly, the tails of the decay curves in Fig. 4 can be interpreted as due to fast sputtered produced after having undergone only a few binary collisions. Straight line fits to these tails yield, v_f, velocities of fast excited atoms according to equation (4). The decay close to the surface is largely due to excited atoms travelling more slowly. After the intensity contributions due to fast atoms are subtracted, straight-line fits produce values for, v_s, velocities of slow excited atoms. The values of v_f, E_f and v_s, E_s are included in Table 4 for comparison with threshold energies E^* determined from z versus x plots shown in Fig. 5. Attention must be drawn to the fact that v_f, E_f and V_s, E_s refer only to the normal component of the velocity (or energy) of the atoms.

The z versus x plots in Fig. 5 enable us to calculate E^* since by definition

$$z = (a + 2^{-\frac{1}{2}} \gamma_i m^{\frac{1}{2}} x)/E^{*\frac{1}{2}} \quad . \tag{15}$$

One notes the fact that while the plots for Ba II, Ni I and Cr I lines produce continuous straight lines, those for Si I, Ba I and Na I lines show breaks in the lines. Dzioba et al. [15] attribute the discontinuities to cascading from upper levels. The values for E^*, the threshold kinetic energy of excited atoms, are shown in Table 4.

Repopulation of the initial states of the transition studied by cascades from upper levels can seriously affect our results. If the cascade contribution is large, it will tend to make the velocity and energy values appear high. Jensen and Veje [23] have shown that for metallic lithium bombarded by Xe^+ ions, cascade corrections can be as high as 35%. For the lines studied, a spectral scan showed up no lines arising from the cascades. However, this negative result may be due to the poor light collection geometry required for this experiment as well as the weak intensities usually associated with cascade lines. Indeed, a glance through the appropriate spectroscopic tables [25-28] reveals that all six lines can be repopulated by cascades.

In Fig. 4, the tails of the decay curves have rather poor statistics (count rates $\leq 10^2$ c/s). Any cascade contribution will affect the determination of v_f and E_f adversely. Braun et al. [14] have also noted this point in their work. We wish to stress that no great importance should be attached to the v_f and E_f values in Table 4, apart from the fact they are comparable to values reported by Gritsyna et al. [12,13]. The conclusion is therefore that the approach taken by Gritsyna et al. [12,13] to determine the velocity of 'fast' sputtered atoms is invalid because of the cascading problem.

The slower velocities, v_s, determined from Fig. 4, are also affected by cascades although this is less serious because of the high count-rates near the surface. The values of v_s were determined after the contributions due to the tail had been subtracted. The energy component normal to the surface, E_s, is of the order of $10\text{-}10^2$ eV. The ionized Ba II line has higher kinetic energy than the other atomic lines, which agrees with the findings of Braun et al. [14] for Xe^+-bombarded Al. However, one cannot rule out the possibility that a slight charging of the glass target surface could give rise to the higher energy observed in Ba excited ions.

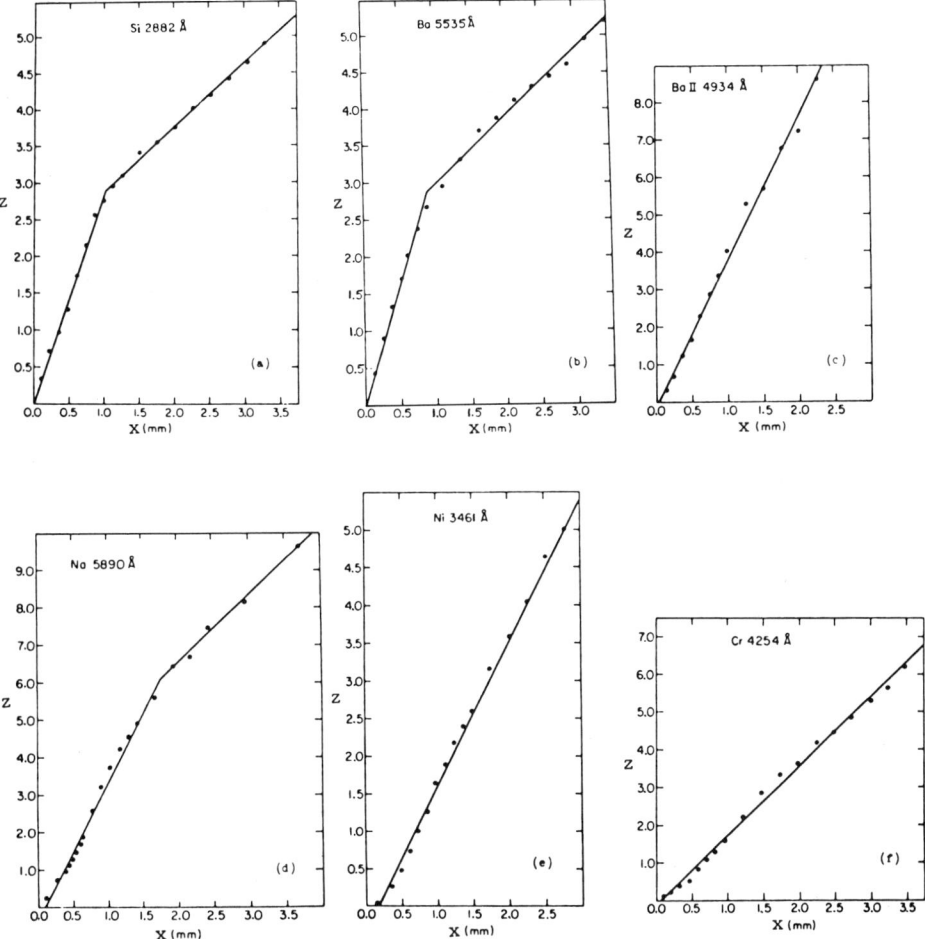

Fig.5 Plots of z versus x according to the equation $z - (a+2^{-\frac{1}{2}}\gamma_i m^{\frac{1}{2}} x)/E^{*\frac{1}{2}}$ to determine E^*, the threshold energy for excitation, for lines (a) Si I 2882 Å, (b) Ba I 5535 Å, (c) Ba II 4934 Å, (d) Na I 5890 Å, (e) Ni I 3461 Å, and (f) Cr I 4254 Å.

The threshold kinetic energies required for excitation to take place, E^*, determined from the z versus x plots in Fig. 5 lie in the $10-10^3$ eV range. This is somewhat surprising as one instinctively expects that E^* would lie near the surface binding energy, typically 1-10 eV, as the energy spectra of sputtered atoms [18,29,30] and secondary ions [31,32] have their maxima. One can argue that the inherently poor spatial resolution ($\geq$ 150 μm) tends to discriminate against measuring excited atoms with low energies, i.e., $\leq$ 10 eV. It is interesting, however, to note that although our experiment had better spatial resolution than that of Dzioba et al. [15] and we could measure short-lived excited states, $\sim 10^{-9}$ s, the E^* values in Table 4 are generally higher than those reported for long-lived Li and Na excited states by Dzioba et al. [15].

Table 4 Results from the defining-slits experiment

Line (Å)	E_i (eV)	Transition	v_s ($\times 10^4 ms^{-1}$)	E_s (eV)	v_f ($\times 10^4 ms^{-1}$)	E_f (eV)	τ ($\times 10^{-8}$)	E^* (eV)
Si I 2882	5.08	$3p^1D-4s^1P^0$	4.34±0.59	272±40	31.28±5.35	14258±2440	0.52	684±140
Ba I 5535	2.24	$6s^1S-6p^1P^0$	1.90±0.44	256±30	15.91±3.03	18038±3430	0.83	1150±130
Ba II 4934	2.51	$6s^2S-6p^3P^0$	2.46±0.28	432±50	7.32±1.19	3805±620	0.78	830±150
Na I 5890	2.11	$3s^2S-3p^2P^0$	1.49±0.19	26±5	8.84±1.76	932±190	1.59	31±7
Ni I 3461	3.60	$4s^3D-4p^5P^0$	1.10±0.15	36±5	2.78±0.44	236±40	2.81	101±20
Cr I 4254	2.91	$4s^7S-4p^7P^0$	0.97±0.10	21±3	2.64±0.46	189±30	4.50	40±8

E_i = energy of the upper level

v_s = velocity component normal to the target surface of the 'slow' group of excited atoms

E_s = kinetic energy associate with v_s

E^* = threshold kinetic energy of excited atoms

v_f = velocity component normal to the target surface of the 'fast' group of excited atoms

E_f = kinetic energy associate with v_f

τ = life-time of the excited states = $1/\sum_f A_{fi}$

In comparing the E_s and E^* data in Table 4, it should be remembered that E_s refers only to the component normal to the surface and therefore, not surprisingly, has lower values.

4.3.2 OMA Experiment

The advantages of this experiment lie on the capability of the OMA to observe the intensity decay in front of the target in real-time and to provide vastly improved spatial resolution, i.e., ∼5 μm. A typical decay curve, this particular one being for the Cr I 5260 Å line, is shown in Fig. 6.

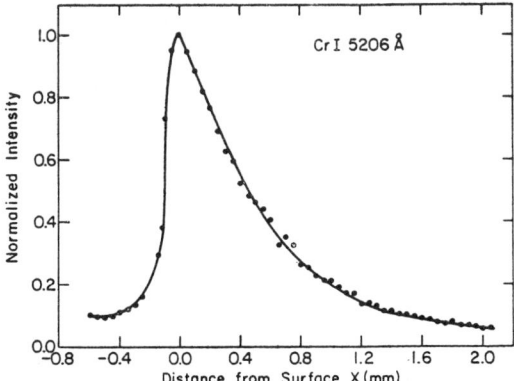

Fig.6 Decay of photon yield as a function of distance for the Cr 5206 Å line determined by the OMA method. The decay curve shown has had the background subtracted and the cross-talk between channels corrected.

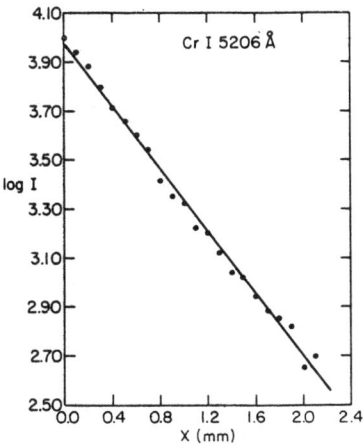

Fig.7 A plot of Log I as a function of the distance form the surface, x, for the Cr I 5206 Å line deduced from the data shown in Fig.6.

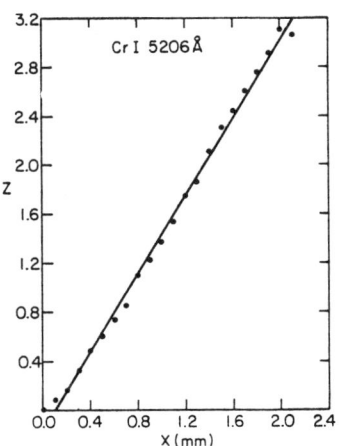

Fig.8 A plot of z versus x for the Cr I 5206 Å to determine E^*, the threshold kinetic energy for excitation.

This curve is not taken directly from the OMA during bombardment of the target, but one which has had the background noise subtracted and the cross-talk between channels corrected. A plot of log I versus x from this curve shows a straight line (Fig. 7) as is also the case for the z versus x plot (Fig. 8). It is obvious that in this region, the decay curve can be approximated by a single exponential.

Values of $v_\perp$ and $E_\perp$ can be calculated from Fig. 7 using equation (4) and values of E* from Fig. 8 using equation (15). The results for four lines of Na I, Cr I, Ba I and Ba II are shown in Table 5. The data presented are actually the average of two separate runs which produced similar results.

Table 5 Results from the OMA experiment

Line (Å)	E_i (eV)	Transition	$v_\perp$ (x10^4 m/sec)	$E_\perp$ (eV)	τ (x10^{-8} sec)	E* (eV)
Na I 5890	2.11	3s - 3p	2.36 ± 0.13	65 ± 3	1.59	58 ± 3
Cr I 5206	3.32	5s - 5p	3.73 ± 0.13	319 ± 11	1.72	324 ± 13
Ba I 7059	2.94	6s - 6p	4.06 ± 0.14	1184 ± 43	1.40	1278 ± 58
Ba II 4554	2.72	6s - 6p	6.87 ± 0.80	3378 ± 150	0.67	3720 ± 144

E_i = energy of the upper level

$v_\perp$ = velocity component perpendicular to surface

$E_\perp$ = kinetic energy associated with $v_\perp$

τ = life-time of the excited states = $1/\Sigma A_{fi}$

E* = threshold kinetic energy of excited atoms

The main purpose of undertaking the OMA experiment is to answer the criticism that poor spatial resolution will discriminate against measuring excited atoms with low energies. Nevertheless, despite the much-improved resolution, the results obtained are very similar to those determined by the defining-slits experiment. Once again cascades from upper levels could influence the results, i.e., produce higher energies. However, the Ba I 7059 Å line is supposedly cascade-free according to spectroscopic tables and yet it has a threshold kinetic energy of $\sim 10^3$ eV. The conclusion of this experiment is therefore, in agreement with the previous experiment, that excited atoms have high kinetic energies and that the threshold kinetic energies of these atoms lie in the range of $10-10^3$ eV.

5. MODEL OF OUTER-SHELL EXCITATION

Two important points emerge from the studies of photon yield and energy distribution of excited atoms: (1) The number of photons produced per sputtered atom is generally quite low $\sim 10^{-3}$-10^{-4}. The secondary ion yield is almost always higher than the photon yield. (2) The kinetic energies of excited atoms are much higher than those of sputtered atoms [18] and secondary ions [15]. An additional point which is widely known is the fact that both secondary ion and photon yields are enhanced by the presence of oxygen in the target either in the form of a bulk oxide or in the form of adsorbed oxygen on the surface. We shall propose a model incorporating these experimental observations to explain the mechanism for producing excited atoms and secondary ions. In doing so, we will not go into the detail of examining the pros and cons of various existing models since this has been done in two recent reviews [1,33], but will simply present our argument in a straightforward manner.

A model which takes into account the influence of oxygen on yields of excited atoms and secondary ions is the level-crossing process for the dissociation of an M-O quasimolecule leaving the surface proposed by Blaise [34] and Thomas [35]. In the scheme put forward by Blaise [34], crossings of excited levels and the ionized level occur in the outward-going direction, i.e., as $r \to \infty$, of the potential energy curves. Since the excited energies lie below the ionization energy, the probability of the M-O system dissociating into excited M* and neutral O is higher than dissociating into ionized M^+ and O^-. This of course predicts a higher excited yield than ion yield which is contrary to experimental observation. In Thomas's scheme [35], it is speculated that if sufficient energy is available in the collision, curve-crossing processes in the receding quasi-molecule can lead to the population of excited molecular states which, via dissociation can yield excited or ionized sputtered atoms. However, details of how the crossings occur are not given.

Our proposed model of excitation and ionization, first put forward in an earlier work [6], features a combination of the concepts advanced by Blaise [34] and Thomas [35] with suitable modifications. Fig. 9 shows a schematic representation of our model. During the final collision in the collision cascade process, an atom pair consisting of a substrate metal atom and an adsorbed oxygen atom is ejected. Since the laboratory scattering angle in such a collision is limited, there is good probability that both atoms are ejected as a quasi-molecule. In so doing, the system first is compressed and moves up the repulsive side (where $r \to 0$) of the potential energy curve before rebounding with enough energy to dissociate. The dissociation time is $\sim 10^{-13}$ seconds, long enough for the system to move more than a few angstroms away from the surface, i.e., out of the range of surface electronic interaction [33].

Curve (a) in Fig. 9 follows the progress of the sputtering and dissociation of the M-O quasi-molecule into ions M^+ and O^-. $V(r_0)$ represents the potential well in which the chemisorbed M-O system is sitting. We note that on the outgoing path, curve (a) crosses curve (b) at C_1 which could lead to the dissociation of the sputtered M-O into M and O ground state neutral atoms. There is also another crossing point C_2 for curves (a) and (b) on the repulsive side, of the potential energy, which could allow the subsequent dissociation to follow curve (b) rather than (a).

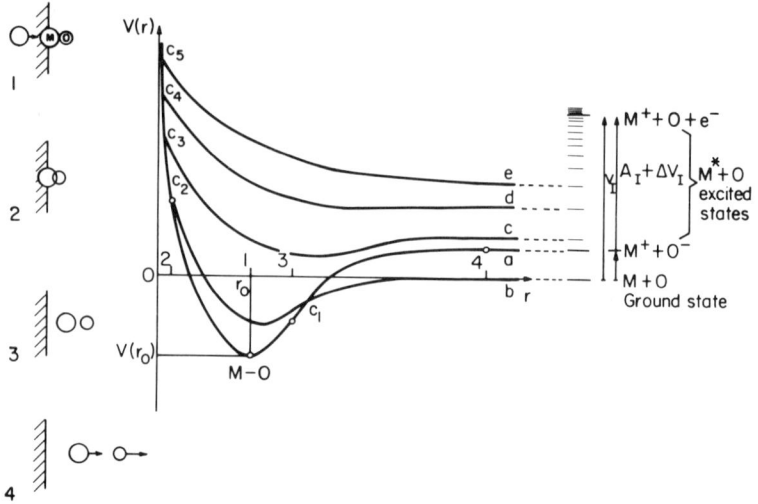

Fig.9 Illustration of the sputtering of an M-O quasi-molecule and a schematic potential energy diagram of the M-O system during its dissociation with increasing distance, r, from the surface. The sequence of events 1, 2, 3 and 4 is marked on the r axis. The crossing points are labeled C_1, C_2, ..., etc. V_I is the ionization energy of M. A_I is the electron affinity of O, and ΔV_I is ionization energy depression due to interaction with the surface.

On the right of the potential energy curves in Fig. 9 are the various energy levels of the two separated atoms. V_I is the ionization potential of the M atom and A_I is the electron affinity of the O atom. According to Blaise [34], because of the electron affinity of oxygen (A_I = 1.46 eV), the energy level of the M^+ and O^- lies below that of $M^+ + O + e^-$. Therefore it is easier for the M-O system to dissociate into $M^+ + O^-$ rather than $M^+ + O + e^-$. In our model (see Fig. 9) we have introduced ΔV_I which further depresses the energy level of M^+ and O^- to below that of the first excited state of M^* and ground state O. This ΔV_I term is due to the Coulomb interaction between the surface of a polar solid and the departing quasi-molecule. The concept of depression of ionization potential is not new. In a gaseous plasma, the well-known Saha-Eggert equation contains a ΔV_I term to take into account the depression of the ionization potential due to Coulomb interaction of the particles. The concept of surface-polarization due to oxygen adsorption is first introduced by Williams and Evans [36] and Williams [33]. Depending on whether the oxygen is adsorbed above or incorporated beneath the substrate surface, a surface dipole is created which modifies the local surface potential barrier, increasing the barrier for electron emission when the oxygen (generally more electronegative than the substrate) is outermost and lowering the barrier for electron emission when the electropositive substrate atom is outermost. Because potentials are relative quantities, the effect of increasing the surface potential barrier is exactly equivalent to reducing the ionization potential of an atom outside the surface. Indirect evidence of oxygenation above and underneath the surface atoms has been provided by transient measurements in secondary ion [36] and photon [37] studies.

By putting the $M^+ + O^-$ energy level below the excited levels of M^* not only means that ionization is now energetically much easier than excitation, but excitation can only occur when the final collision is energetic enough to allow the system to reach crossing points (C_3, C_4, C_5 in Fig. 9) on the repulsive side of the potential energy curve. Such a situation has been discussed in the Fano and Lichten model [38,39] for inelastic collision of atoms in the gas phase.

To form excited states, therefore, the input energy required in the final collision has to be considerably higher in order to move up to the crossing points. The end result of this process is dissociation products of considerable excess energy. This is consistent with the fact that the kinetic energies of excited atoms tend to be high. The concept of threshold kinetic energies for excited atoms is also consistent with specific collisional energy transfer in order to reach the crossing points on the repulsive portion of the potential energy curve. The higher energies involved in excitation suggests that this process occurs in the early part of the collision cascade.

According to our model, excitation and ionization of sputtered atoms result from the same basic process, i.e., transfer of energy during the final collision between a moving atom in the collision cascade and an M-O system on the surface, which accounts for the similarities observed in the oxygen enhancement of photon and ion yields. The lower photon yield compared with ion yield results from the fact that the energetics of ion formation near a surface are more favorable than excited atom formation. In contrast to Blaise's model [34] there is no need for a well-defined upper energy limit of the radiative excited states. The absence of photons from highly excited levels is a consequence of the exceptionally high energy input required in the final collision to move up to the crossing points.

ACKNOWLEDGEMENT

This work was supported by the Office of Naval Research (L. R. Cooper). I wish to thank Peter Williams and Roger Kelly for frequent stimulating discussions which lead to the ideas outlined in this paper.

REFERENCES

1. G. E. Thomas, Surface Sci. 90, 381 (1979).
2. C. W. White, E. W. Thomas, W. F. Van der Weg and N. H. Tolk in "Inelastic Ion-Surface Collisions", eds. N. H. Tolk, J. C. Tully, W. Heiland and C. W. White, Academic Press (1977) p. 201.
3. N. Andersen, K. Jensen, J. Jepsen, J. Melskens and E. Veje, Appl. Optics 13, 1965 (1974).
4. O. C. Jones, National Physical Laboratory Tables of Spectral Power Distribution (1970).
5. I. S. T. Tsong and N. A. Yusuf, Appl. Phys. Letters 33, 999 (1978).
6. P. Williams, I. S. T. Tsong and S. Tsuji, Nucl. Instrum. Meth. 170, 591 (1980).
7. C. Snoek, W. F. Van der Weg and P. K. Rol, Physica 30, 341 (1964).
8. W. F. Van der Weg and D. J. Bierman, Physica 44, 206 (1969).
9. H. D. Hagstrum, Phys. Rev. 96, 336 (1954).
10. R. Hippler, W. Krüger, A. Scharmann and K. H. Schartner, Nucl. Instrum. Meth. 132, 439 (1976).
11. C. W. White, D. L. Simms, N. H. Tolk and D. V. McCaughan, Surface Sci. 49, 657 (1975).

12. V. V. Gritsyna, T. S. Kiyan, R. Goutte, A. G. Koval and Ya. M. Fogel, Bull. Acad. Sci. USSR 35, 530 (1971).
13. T. S. Kiyan, V. V. Gritsyna and Ya. M. Fogel, Nucl. Instrum. Meth. 132, 435 (1976).
14. M. Braun, B. Emmoth and I. Martinson, Phys. Scripta 10, 133 (1974).
15. S. Dzioba, O. Auciello and R. Kelly, Rad. Effects 45, 235 (1980).
16. R. B. Wright and D. M. Gruen, Nucl. Instrum. Meth. 170, 577 (1980).
17. R. B. Wright and D. M. Gruen, J. Chem. Phys. 72, 147 (1980).
18. M. W. Thompson, Phil. Mag. 18, 377 (1968).
19. G. Carter, D. G. Armour and K. J. Snowdon, Rad. Effects 35, 175 (1978).
20. W. F. Van der Weg and P. K. Rol, Nucl. Instrum. Meth. 38, 274 (1965).
21. I. S. T. Tsong and N. A. Yusuf, Nucl. Instrum. Meth. 170, 357 (1980).
22. N. A. Yusuf and I. S. T. Tsong (to be published).
23. Th. J. M. Sluyters and J. Kistemaker, Physica 25, 1389 (1959).
24. K. Jensen and E. Veje, Z. Physik 269, 293 (1974).
25. W. L. Wiese, M. W. Smith and B. M. Glennon, Atomic transition probabilities, Vol. 2 (U.S. Gov't. Printing Office, 1969).
26. B. M. Miles and W. L. Wiese, Critically evaluated transition probabilities for Ba I and II, NBS Technical Note (1969).
27. C. E. Moore, A. multiplet Table of astrophysical interest (revised edition) (Nat. Bureau of Standards, 1972).
28. C. H. Corliss and W. R. Bozman, Experimental transition probabilities for spectral lines of seventy elements, NBS Monograph 53 (1962).
29. R. V. Stuart and G. K. Wehner, J. Appl. Phys. 35, 1819 (1964).
30. R. V. Stuart, G. K. Wehner and G. S. Anderson, J. Appl. Phys. 40, 803 (1969).
31. K. Wittmaack, in "Inelastic Ion-Surface Collisions", eds. N. H. Tolk, J. C. Tully, W. Heiland and C. W. White, Academic Press (1977) p. 153.
32. T. R. Lundquist, J. Vac. Sci. Technol. 15, 684 (1978).
33. P. Williams, Surface Sci. 90, 588 (1979).
34. G. Blaise, Surface Sci. 60, 65 (1976).
35. G. E. Thomas, Rad. Effects 31, 185 (1977).
36. P. Williams and C. A. Evans Jr., Surface Sci. 78, 324 (1978).
37. I. S. T. Tsong and S. Tsuji, Surface Sci. 94, 269 (1980).
38. U. Fano and W. Lichten, Phys. Rev. Lett. 14, 627 (1965).
39. W. Lichten, Phys. Rev. 164, 131 (1967).

Theory of Charge States in Sputtering

Z. Šroubek
Institute of Radio Engineering and Electronics
Czechoslovak Academy of Sciences
Prague 8, Lumumbova 1, Czechoslovakia

1. Introduction

The ionization of particles sputtered from solid state substrates is a complex phenomenon which involves motions and strong electronic perturbations of a large number of atoms and may comprise several independent physical processes. Thus, a theoretical description of charge states of sputtered particles which would be more than a working hypothesis, would probably be a complex one. Fortunately, experimental evidence of recent years indicates that the ionization processes must be governed by certain general rules which might be retained in suitably simplified models. It is hoped that such models will allow the development of a sufficiently transparent theory, i.e. one which is capable of predictions and is helpful in the interpretation of experimental data.

In view of the crucial importance of experimental data for the development of an adequate theory of charge states of sputtered particles we will first briefly review the experimental information on ionization gathered in recent years. In the second part of the paper we will deal in detail with one group of ionization theories, i.e. the theory based on tight-binding description of valence electrons of the sputtered particle and of the substrate. Within the framework of the tight-binding scheme, we will suggest a simple model system and we will show how far theoretical predictions agree with experiments.

2. Experimental Evidence

Most of the relevant experiments have been carried out on Secondary Ion Mass Spectrometers (SIMS). Despite the large activity in this field, only few SIMS experimental results were characterized by other physical and chemical methods to allow further data processing. Nevertheless, there were several exceptions and, particularly in recent years, reliable data have been accumulated which clearly enable extracting certain general relations governing the ionization processes. The subject is well described in the quoted references (the list of references is by no means complete) and we will confine ourselves to a short summary:

2.1 Rather well seems to be documented the observation that the positive ion ionization probability $N^+/N^o = R^+$ of particles sputtered from one substrate depends upon their ionization energy I as

$$R^+ = B^+(T_e^+)\exp(-I/kT_e^+),\qquad(1)$$

where T_e^+ is the parameter called the effective temperature. In the definition of R^+ we use, for simplicity, the assumption that the number of sputtered positive ions N^+ is much smaller than the number of sputtered neutral particles N^o. Similarly, the negative ion ionization probability $N^-/N^o = R^-$ depends upon the electron affinity A of the sputtered particle as

$$R^- = B^-(T_e^-)\exp(A/kT_e^-).\qquad(2)$$

The functions $B^+(T_e^+)$ and $B^-(T_e^-)$ vary only slightly with $T_e^\pm$. Eqs. (1) and (2) have been first introduced and experimentally tested by ANDERSON and HINTHORNE [1]. Controversy arose when these authors gave a specific meaning to the parameter T_e by assuming that a plasma in local thermodynamic equilibrium is formed at the surface. The formation of plasma and the consequent application of the Saha-Eggert equation was accepted with intense scepticism because the assumption of thermal equilibrium at the bombarded surface has contradicted the current understanding of sputtering process. Another statistical approach was suggested by Jurela [2], who used the Dobretsov equation and arrived to relations formally similar to (1) and (2). No matter what is the correct physical interpretation of (1) and (2), the suggested exponential relation between I and R^+ has passed through rather stringent experimental tests. The most systematic experimental verification, so far, of formal correctness of Eq. (1) has been carried out by MORGAN and WERNER [3, 4] with the use of a large number of NBS steel and silicate standards. It should be noted at this point that the commonly used semilog plots are rather insensitive to the exact form of the exponential dependence. The general precision of most ion yield data is such that the ion-yield expressions of the form $R^+ \propto \exp(bI^m)$, where b is a constant, are indistinguishable for $0.5 < m < 1.5$. Thus the exponential dependence (1) with $m = 1$ is simplest and physically quite plausible but, unless confirmed by other tests, still tentative.

2.2 YU [5] has shown in his experiments on well defined metallic surfaces that R^- of sputtered atoms depends on the macroscopic work-function ∅ of the cesiated substrate as

$$R^- \propto \exp(b - \emptyset)^{1/2},$$

where b is a constant. As mentioned in 2.1, semilog fits are rather insensitive to the actual form of the exponent over the limited range in which experimental data are taken. As pointed out by YU, his data can be fitted equally well by exponents linear in ∅. If we denote by Δ∅ the change of the work-function due to externally induced changes of surface properties, the dependence of R^- on Δ∅ can then be well described by

$$R^- \propto \exp(-\Delta\emptyset/kT_e^-),\qquad(3)$$

where we have implicitly assumed that T_e^- does not depend upon the changes of $\emptyset$. It should be stressed that the change of the work-function $\Delta\emptyset$ must occur within the interatomic distance above the surface and thus only $\Delta\emptyset$ due to surface dipoles is relevant. In the reported experiments [5] this does not pose any problem because $\Delta\emptyset$ on metal surfaces is only due to the surface dipole layer. On semiconductor surfaces there is a considerable contribution to $\Delta\emptyset$ which is caused by surface charges and associated space charges and band bending. The contribution to $\Delta\emptyset$ caused by band bending should not influence the ionization probability and should be subtracted from the work function data.

YU has also shown [5] that T_e^- are different for different species sputtered from the same metallic substrate. More experimental work is necessary to prove that the change of T_e^- is entirely due to different electronic structures of the sputtered atoms and not due to possible substrate changes.

A logical extension of (1)-(3) is the relation

$$R^+ \propto \exp(\Delta\emptyset/kT_e^+) , \qquad (4)$$

used and discussed in [6] and [7]. Unfortunately, the general relationship between different T_e in (1)-(4), if it exists, is not yet known. For certain ionic clusters (Ti_nO_m) the effective temperatures T_e^+ and T_e^- seem to be very close [6, 7], but generally T_e^- is lower than T_e^+.

In this context we should mention the work of DELINE et al. [8] in which the exponential dependences (1) and (2) have been used to interpret ionic yields of ions sputtered from pure elements. Basic idea of their work is that heavy bombardment of elemental surfaces by O or Cs ions creates new surfaces which are fully characterized by O or Cs concentrations and are independent of the element itself. It was suggested in Ref.[8] that the concentrations of O or Cs are proportional to the inverse value of the sputtering yield of a given element. The product of the ionic yield and the inverse value of O or Cs concentrations should then be matrix independent and the major remaining yield-determining variables should be simply the parameters of the sputtered atom, i.e. ionization potential I or electron affinity A. The dependence of the product on I or A fits very closely the exponential dependences (1) or (2), respectively. The parameter T_e^+ for oxygen bombarded elements was found to be 7500 K and T_e^- for cesium bombarded elements was found to be 3000 K.

2.3 There are many indications that T_e^+ in (1) is higher in oxidized metal samples [9] than in pure metals. This fact must be taken into account when one wishes to interpret the change of R^+ during oxidation by means of the relation (4) [10]. The change of T_e^+ may actually influence the value of R^+ more significantly than the change of $\Delta\emptyset$ and may explain the seemingly anomalous increase of positive ion yields with the decrease of the work-function in certain materials [10]. It is quite natural to expect that also T_e^- increases with oxidation, in agreement with the observation that ion yields of negative ions in most cases also increase with oxidation.

2.4 The parameter T_e^+ in (1) depends upon the kinetic energy of the sputtered atom. It has been shown convincingly by WERNER and MORGAN [3, 4] that the value of T_e is slightly higher (usually less than by 50%) for secondary ions with the kinetic energy between 40-60 eV than for ions with kinetic energy between 0-20 eV. Preliminary experiments on Ga^+ sputtered from cesiated GaAs indicate that also the effective temperature in (4) slightly increases with the increase of the secondary ion energy [11].

2.5 The energy spectra of secondary ions have been measured by many authors. Most extensive studies have been published recently by RUDAT and MORRISON [12]. The consensus is that majority of spectra peaked at energies below 10 eV and that the dependence of R^+ on the secondary ion kinetic energy E_s is rather small (i.e. $R^+ \propto E_s^n$ where $n < 1$). Further studies of the R^+ energy dependence, particularly in context with the precise measurement of neutral particle energy distribution [13] and with the computer simulation of the sputtering [14], are highly desirable.

2.6 Another important effect which should be studied in more detail is the primary ion energy dependence of the ionization probability R^+. An extensive work regarding this problem was performed by WITTMAACK [15]. According to his study all sputtered ions can be divided into two groups. The intensity of ions of the first group depends only insignificantly upon the change of the primary ion energy between 0.5-5 keV. The intensity of the ions of the second group changes by several orders of magnitude within the same interval of primary ion energies. The second group includes multicharged and some single charged ions like H^+ sputtered from Si:H samples [16] or Al^+ at higher energies [17]. The mechanism responsible for their ionization involves violent collisions and deep level hole creations and is, apparently, quite different from the ionization mechanism of the first group. Undoubtedly, also the change of the work-function should influence quite differently the ionization probabilities of the ions of the first and of the second group.

In the conclusion of this section we can summarize that many ionic species, particularly those which belong to the first group discussed in 2.6, are ionized by a process which is well described by the phenomenological expressions (1)-(4). The range of validity of these formulas and the relations between the parameters T_e must be carefully investigated. For example, there is no convincing proof yet of the conjecture that T_e^- in (2) is the same as T_e^- in (3), or that T_e^+ in (1) is the same as T_e^+ in (4). As there seems to be no physical objection against this conjecture, we will assume in the next chapter that the expressions (1), (3) and (2), (4) can be combined and written as

$$R^+ = \exp[(\emptyset^+ + \Delta\emptyset - I)/kT_e^+] , \quad (5)$$

$$R^- = \exp[(\emptyset^- - \Delta\emptyset + A)/kT_e^-] , \quad (6)$$

where $\emptyset^+$, $\emptyset^-$ and T_e^+ and T_e^- are phenomenological parameters which, in general, may be different for different particles sputtered from the same substrate. It is the purpose of the following theoretical description of the ionization process to give justification of (5) and (6) and to interpret, eventually, the phenomenological parameters T and $\emptyset$ in terms of microscopical parameters of the system.

3. Theory

As mentioned under 2.6 of the preceding chapter, ions produced during sputtering can be divided in two groups. Ions of one group are produced by violent, usually binary collisions with core-hole excitations followed by Auger decay. A thorough theoretical treatment of this mechanism has been carried out by JOYES [18] and subsequently discussed and reviewed by many authors (e.g. Ref.[19]). In this paper we deal with processes which are believed to be responsible for the production of majority of low energy ions and which involve many smaller collisions in collision cascades rather than a violent collision. The theory of such processes should encompass detailed description of the transfer of electron excitations in the collision cascade to the outgoing atom or atomic cluster and is, by its nature, rather complex. To make such complicated problem tractable, simplified physical models must be used. Ultimate check of the correctness of these models can be made only by comparing theoretical results with the experimental results described in the previous chapter.

3.1 Tight Binding Method

The first simplifying assumption is that the electronic structure of the substrate and the sputtered atom is described in the one--electron tight-binding (TB) approximation. The TB scheme, despite its crudeness, has the advantage that even complicated physical situations can be modeled by an appropriate choice of few parameters and is currently the only scheme which allows analyzing the ionization process in more detail. The tight-binding Hamiltonian, suitable for ionization studies, has the following form:

$$H = \varepsilon_a(t) c_a^+ c_a + \sum_i \varepsilon_i(t) c_i^+ c_i + \sum_{i \neq j} V_{ij}(t) c_i^+ c_j + \sum_j [(V_{ja}(t) c_j^+ c_a + h.c.)] ,\qquad(7)$$

where subscripts i and j refer to the electron orbitals $|i\rangle$ and $|j\rangle$ of the substrate atoms with diagonal energies ε_i and ε_j, and a refers to the assumed single valence orbital $|a\rangle$ of the sputtered atom. The $|a\rangle$ orbital can exchange electrons with substrate orbitals via the hopping integrals V_{aj}. The hopping integrals between substrate orbitals $|i\rangle$ and $|j\rangle$ are denoted by V_{ij}. The time-dependence of ε and V parameters in (7) is of fundamental importance in the theory of ionization and distinguishes the Hamiltonian (7) from its statical counterpart.

If the substrate is semiinfinite and with translational symmetry, the Hamiltonian (7) can be conveniently rewritten in the k-representation. One gets essentially the Anderson Hamiltonian [20, 21] with two-electron interaction term averaged in the Hartree approximation and with the one-electron interaction term depending on time [22]. The Hamiltonian has then the form

$$H = \varepsilon_a(t) c_a^+ c_a + \sum_k \varepsilon_k(t) c_k^+ c_k + \sum_k [(V_{ak}(t) c_a^+ c_k + h.c.)] , \quad (8)$$

where the subscript k refers to the substrate $|k\rangle$ orbitals and a refers to the single valence orbital $|a\rangle$ of the sputtered atom. The $|a\rangle$ orbital exchanges electrons with the substrate via the hopping integral V_{ak}. The diagonal energies of $|a\rangle$ and $|k\rangle$ orbitals are denoted by ε_a and ε_k, respectively. Models for the static adsorption studies are usually chosen such that k is a good quantum number and the Hamiltonian (8) can be used. This approach is generally not acceptable in dynamical problems. Firstly, if the atoms of the substrate are in motion, wavevectors k are not good quantum numbers. But even if we neglect atomic motions in the substrate, any local electron excitation in such systems spreads immediately through the semiinfinite substrate and the sputtering spot is immediately deexcited.

As it will be discussed later, the sputtering process takes place in the interval of the order of $\tau = 2 \times 10^{-14}$ sec. If one assumes that the cascade region resembles, in the short time interval, a highly disordered solid, the excitations may spread by a hopping process and travel in the time τ the distance d given by

$$d = (\tau v_e \ell)^{1/2} , \quad (9)$$

where ℓ is the length of the hop and v_e is the electron group velocity. For the electronic bandwidth of 3 eV and ℓ equal to the interatomic separation 3 Å, one gets d equal to 10 Å, i.e. the distance of three interatomic distances. The implication is that for a description of the ionization process it may be sufficient to assume a substrate with relatively few atoms. Just a two-atomic, two-level model [23] is clearly not enough but the other extreme, a semiinfinite substrate described by (8) is also incorrect (unless, of course, one takes special precautions to include the finite speed of electron excitations which makes the mathematics untractable). It has been shown [22] that the semiinfinite substrate gives even less reasonable results, i.e. leads to smaller values of R^+, independent of the interaction term V, and to a larger velocity dependence of R^+ than a comparable two-level system. Therefore, the study of two-level systems has its merits and is, as shown below, instructive.

3.2 Two-level Model

An appropriate Hamiltonian, for a two-orbital two-atomic system consisting of the orbital $|a\rangle$ on one atom and of the orbital $|i\rangle$ on the other atom, has the form

$$H = \varepsilon_a c_a^+ c_a + \varepsilon_i c_i^+ c_i + [V_{ia}(t) c_i^+ c_a + h.c.],$$

where we use the same notation as in the Hamiltonian (7). For simplicity, we assume that ε_a and ε_i are time independent and that ε_a is always smaller than ε_i. The system (we neglect spin degeneracy) is occupied by one electron which is, at the time $t = -\infty$, in the orbital $|a\rangle$ and the orbital $|i\rangle$ is unoccupied. The interaction $V = V_{ij}$ is then slowly turned on as $V(t) = V \exp(-\lambda_0 t)$ where $\lambda_0 \to 0$. At the time $t = 0$ it starts to fall off from the value V to zero, according to the chosen time dependence $V(t)$ ($t > 0$). The ionization probability

$$R^+ = 1 - \langle c_a^+(+\infty) c_a(+\infty) \rangle \tag{10}$$

can be calculated straightforwardly, provided R^+ is much less than one [24]. Then

$$R^+ = \left| \frac{1}{\hbar} \int_{-\infty}^{+\infty} V(t) \exp\left[\frac{i}{\hbar}(\varepsilon_i - \varepsilon_a)t\right] dt \right|^2. \tag{11}$$

We have chosen four different forms of $V(t)$ for $t \geq 0$, namely

$$V(t) = V\exp(-\lambda_1 t), \tag{12}$$

$$V(t) = V/2 \, [1 + \cos(\pi \lambda_2 t)] \tag{13}$$

for $1/\lambda_2 \geq t \geq 0$ and zero for $t > 1/\lambda_2$,

$$V(t) = V(1 + t^2 \lambda_3^2)^{-1}, \tag{14}$$

$$V(t) = V[\text{ch}(\lambda_4 t)]^{-1}. \tag{15}$$

The corresponding values of R^+ are, according to the formula (11), equal to

$$R_1^+ = \hbar^2 V^2 \lambda_1^2 / (\Delta \varepsilon)^4, \tag{12'}$$

$$R_2^+ = \hbar^4 V^2 \lambda_2^4 \pi^4 / 2(\Delta \varepsilon)^6, \tag{13'}$$

$$R_3^+ = \pi^2 V^2 \lambda_3^{-2} 4^{-1} \hbar^{-2} \exp(-2\Delta \varepsilon / \lambda_3 \hbar), \tag{14'}$$

$$R_4^+ = \pi^2 V^2 \lambda_4^{-2} 4^{-1} \hbar^{-2} \exp(-\pi \Delta \varepsilon / \lambda_4 \hbar), \tag{15'}$$

where we have denoted the difference $\varepsilon_i - \varepsilon_a$ by $\Delta \varepsilon$. To be able to compare (12')-(15') quantitatively, we have normalized the dependences $V(t)$ in (12)-(15) by assuming that all $V(t)$ must decay from the value $V(0) = V$ to the value of V/e in the same time interval $t = \lambda_1^{-1}$. The normalization yields: $\lambda_2 = 0,568 \, \lambda_1$; $\lambda_3 = 1,3 \, \lambda_1$; $\lambda_4 = \lambda_1$. We have used (12) as the reference formula because it has a clear physical meaning. If the spatial dependence of V in the direction z parallel to the motion is described by $V\exp[-\gamma(z-d)]$, where d is the equilibrium distance, then (12) describes the process in which the two atoms start to separate at t = 0 with the constant speed $v = \lambda_1/\gamma$. Assuming a typical value of $\gamma = 1.5$ Å^{-1} and $v = 1.95 \times 10^{14}$ (Å·sec^{-1}) corresponding to a particle with the mass equal to 20 proton masses and the kinetic energy E_S equal to 40 eV, one gets $\lambda_1 = 3\times10^{14}$ sec^{-1}. Substituting into the normalized expressions (12')-(15'), and assuming V = 1 eV, $\Delta\varepsilon = 2$ eV, one obtains:

$$R_1^+ = 2.4\times10^{-3}, \quad R_2^+ = 1.6\times10^{-4}, \quad R_3^+ = 3\times10^{-6}, \quad R_4^+ = 1.7\times10^{-13}.$$

Similarly, for $E_S = 20$ eV one gets

$$R_1^+ = 1.2\times10^{-3}, \quad R_2^+ = 4\times10^{-5}, \quad R_3^+ = 8\times10^{-9}, \quad R_4^+ = 5.2\times10^{-19}.$$

The results demonstrate how sharply the ionization probability depends upon the choice of fine details of the V(t) function. The V(t) functions with discontinuities in the derivatives (Eq. (12)) or in the space (Eq. (13)) lead to the power-law dependences (12') and (13') on $\Delta\varepsilon$ and λ, and to reasonably large values of R^+. The realistic, perfectly regular V(t) dependences, either Lorentzian (14) or exponential (15), lead to the exponential laws (14') and (15'), to extremely small values of R^+, and to unreasonably large $\Delta\varepsilon$ and λ dependences.

3.3 Model with a Substrate Consisting of an Infinite Number of Energy Levels

As mentioned above and explained in Ref [22] one cannot expect anything better by introducing instead of one, infinite number of levels above and below the Fermi energy. To demonstrate it, we will first generalize (11) for a continuum of levels above the Fermi energy ε_F by rewriting (11) as

$$R^+ = \frac{1}{\hbar^2} \int_{\varepsilon_F}^{\infty} \left| \int_{-\infty}^{+\infty} V(t)\exp[\frac{i}{\hbar}(\varepsilon_i - \varepsilon_a)t]dt \right|^2 \varrho(\varepsilon_i)d\varepsilon_i, \quad (16)$$

where $\varrho(\varepsilon_i)$ is the density of empty electronic states. In general, V depends upon ε_i but we will further neglect this dependence. To include the neutralization process, due to the tunneling of electrons from the occupied states below the Fermi energy or due to the Auger process, we employ the procedure described by NØRSKOV and LUNDQVIST [25]. These authors have shown that an excitation (i.e. the transfer of the electron from the sputtered atom into empty states above ε_F), which is made at the time t, survives until the time $t = \infty$ with the probability

$$P(t) = \exp[-\int_t^{\infty} W(t')dt'], \quad (17)$$

where W(t') is the neutralization probability caused either by resonance tunneling or by Auger process. Accordingly, the phase factor in Eq. (16) must be damped, i.e. multiplied by P(t). The resulting formula is

$$R^+ = \frac{1}{\hbar^2} \int_{\varepsilon_F}^{\infty} \left| \int_{-\infty}^{+\infty} V(t) \exp[\frac{i}{\hbar}(\varepsilon_i - \varepsilon_a)t - \int_t^{\infty} W(t')dt'] dt \right|^2 \varrho(\varepsilon_i) d\varepsilon_i . \quad (18)$$

It has been shown by BLANDIN et al. [26] using the Green's function technique and by BLOSS and HONE [27] using the equation of motion technique that (18) can be obtained from the Hamiltonian (8) quite generally, in any order of V.

To facilitate the analytical solution of the integral (18) we shall assume that the neutralization is only due to the resonance tunneling and that the transfer integral V describing the tunneling is the same as that in (16). Furthermore, we neglect any fine structure of the band, i.e. we put $\varrho(\varepsilon_i) = \varrho$. Then, the neutralization probability is given by

$$W(t) = \frac{2\pi}{\hbar} V^2(t)\varrho = \frac{2}{\hbar} \Delta(t) . \quad (19)$$

When the sputtered atom is at $t = 0$ at the surface or very close to it, the one-electron width $\Delta(0)$ is several electronvolts and is thus much larger than $\hbar\lambda$ ($\hbar\lambda \doteq 0.1$ eV). Also the difference $\varepsilon_F - \varepsilon_a$ is in all practical cases much larger than $\hbar\lambda$. In these limits the integral (18) has an analytical exponential solution

$$R^+ = \frac{2}{\pi} \exp[-\frac{\pi(\varepsilon_F - \varepsilon_a)}{\lambda_1 \hbar}] = \frac{2}{\pi} \exp[-\frac{\pi(I - \emptyset)}{\lambda_1 \hbar}], \quad (20)$$

where I is the ionization energy of the sputtered atom and $\emptyset$ is the work-function of the substrate. To compare (20) quantitatively with the two-level system, we again assume that $\varepsilon_F - \varepsilon_a = 2$ eV and $\lambda_1 = 3\times 10^{-14}$ sec^{-1} (for $E_S = 40$ eV). For the kinetic energies $E_S = 20$ eV and $E_S = 40$ eV we obtain from (20) the values of R^+ equal to 6×10^{-21} and 4.2×10^{-15}, respectively. Eq. (20) differs from the corresponding two-level expression (12') because the discontinuity of V(t) at $t = 0$ has been cut off in the integral (18) by the large damping near the surface. Without a discontinuity in V(t) the ionization probability (20) must and indeed has the same exponential form as (15'). The only difference between (15') and (20) is the different prefactor (Eq. (20) is independent of V). As expected, the actual numerical values of R^+ are smaller for the continuous level system than for the two-level system [22].

It has been pointed by NØRSKOV and LUNDQVIST [25] that the exponential energy dependences (20) or (15') allow to introduce an effective temperature $T_{eff} = \lambda_1 \hbar / \pi k$, where k is the Boltzman factor. The calculated values of T_{eff} are 700 K and 500 K for $E_S = 40$ eV and $E_S = 20$ eV, respectively. They are, compared to the values observed experimentally (2000-15000 K), rather small.

Moreover, the E_S dependence of T_{eff} is unrealistically large (see 2.5 in the experimental part of the paper). These shortcomings of (20) led NØRSKOV and LUNDQVIST [25] to invoke the possibility that the energy difference $\varepsilon_a - \varepsilon_F$ may be substantially reduced near the surface due to the effect of the classical image potential. When, for example, the energy difference $\varepsilon_a - \varepsilon_F$ is reduced in the critical region to $(\varepsilon_a - \varepsilon_F)_{eff} = C_1(\varepsilon_a - \varepsilon_F)$, where C_1 is a constant smaller than one, the parameter T_{eff} effectively increases to $\lambda_1 \hbar / \pi k C_1$. There are several difficulties with the model. Firstly, because of the small value of $(\varepsilon_a - \varepsilon_F)_{eff}$, the parameter T_{eff} and therefore the value of R^+ would be very sensitive to the details of the electronic potential near the surface. Secondly, there is no firm experimental or theoretical evidence that the large reduction of $\varepsilon_a - \varepsilon_F$ near the surface exists in all materials, metals or semiconductors. On the other hand, experiments show a great resemblance of ionization processes in all studied materials. Furthermore, to achieve high effective temperatures the energy difference $(\varepsilon_a - \varepsilon_F)_{eff}$ must reach a very small value, comparable to kT at room temperature. One would expect then that the external heating of the substrate to higher temperatures would increase the value of R^+ considerably but such effect has not been observed experimentally. Nevertheless, despite these difficulties, the model is very interesting and needs experimental testing on suitable systems.

3.4 Cluster Model

It has been demonstrated in [28] that the rather critical sensitivity of the ionization process to details of the interaction mechanism is largely due to the neglect of atomic motions in the substrate. When atomic motions are included one has to describe the ionization process by means of the Hamiltonian (7) and the problem becomes extremly complicated. Fortunately, according to the estimate (9), only relatively small number of atoms in the collision cascade contributes significantly to the ionization. Therefore, it is physically not unreasonable to study the ionization process with the help of models which consist of only few atoms. Even then the finding of R^+ is a complex mathematical problem and computers must be used. Details of the calculation are described in [28] and can be summerized as follows. Firstly, one must define the dependence of matrix element V_{ij}, V_{ia}, ε_i and ε_a on the atomic coordinates x_i, y_i, z_i of all atoms in the system. The atomic coordinates x_i, y_i, z_i depend in turn on time. The time dependence of the coordinates is described by the classical equations of motion

$$M_i \ddot{x}_i = F_i(x_i, \dot{y}_i, z_i, x_j, y_j \ldots\ldots\ldots) , \qquad (21)$$

where M_i is the mass of the atom i and F_i is the total classical interatomic force acting on this atom. Secondly, we must solve the set of the operator equations of motion, obtained from the Hamiltonian (7) in the form

$$i\hbar \frac{dC_a}{dt} = \varepsilon_a C_a + \sum_j V_{aj} C_j \qquad (22)$$

$$i\hbar \frac{dC_i}{dt} = \varepsilon_i C_i + \sum_{j \neq i} V_{ij} C_j + V_{ia} C_a ,$$

simultaneously with the particle equations of motions (21). The initial values of C_i and C_a are found by finding stationary solutions of (22) at $t = -\infty$. The results of such calculation are molecular orbitals $|m\rangle$ which are characterized by certain specific combinations of the coefficients $C_i^m(-\infty)$. Conveniently, we choose from the whole set of $|m\rangle$ orbitals only those orbitals $|e\rangle$ which are not occupied by electrons at $t = -\infty$ and use the values of $C_i^e(-\infty)$ as the initial values for the solution of (22). Because there are several empty orbitals, we obtain several sets of initial values and correspondingly several final solutions characterized by $C_i^e(+\infty)$ and $C_a^e(+\infty)$. The value of R^+ is then given by the sum over empty orbitals, i.e.

$$R^+ = \sum_e \langle C_a^{e+}(+\infty) \, C_a^e(+\infty) \rangle . \qquad (23)$$

Alternatively, one can choose the coefficients $C_i^o(-\infty)$ of occupied orbitals $|o\rangle$ at $t = -\infty$ as initial values and calculate R^+ by summing over occupied orbitals, i.e.

$$R^+ = 1 - \sum_o \langle C_a^{o+}(+\infty) \, C_a^o(+\infty) \rangle . \qquad (24)$$

In most cases R^+ is much smaller than one and (23) is more convenient. Some further details of the calculation are described in Ref.[28].

To make the calculation possible within an acceptable computing time we have chosen a simple model consisting of seven atoms shown in Fig. 1 [29]. All particles are assumed to have the same mass equal to 20 proton masses. The atom depicted in Fig. 1 by a shaded sphere will be sputtered away due to the impact of the fast impinging atom which is depicted by a black sphere. The kinetic energy of the bombarding atom is 400 eV. The other atoms shown by empty spheres form the model solid state substrate. The forces of the classical motion of atoms consist of two parts: weak, central-directed, attractive forces which hold the cluster together and repulsive forces. The central forces are given by the spatial derivatives of the potential function. This potential function is equal to $3.7 \, r^2$ (eV, Å) for $r \leq 4$ Å and has a constant values for $r > 4$ Å, where r is the distance from the center of the cluster. The repulsive forces are given by the spatial derivatives of the potential $1.45 \times 10^5 \exp(-3.5 \, r_{ij})$ (eV, Å), where r_{ij} is the distance between the atoms i and j. To simulate, at least partly, the fact that the substrate should be semiinfinite we assume that the next substrate atoms (not shown in Fig. 1) are fixed at their positions. They act only through their repulsive

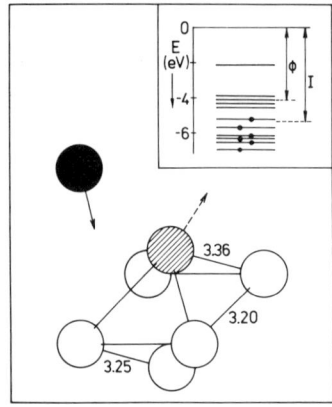

Fig. 1 The atomic model used in the computer simulation. The model consists of the atomic substrate (white spheres), the particle which is sputtered (shaded sphere), and the bombarding particle (black sphere). The interatomic distances are marked in Å. The electronic structure of the model is shown in the inset with electrons depicted by black dots. I is the ionization energy of the sputtered particle. ∅ marked the highest density of the empty electronic states.

forces and prevent forward sputtering. All these forces stabilize the interatomic distances at values marked in Fig. 1 in Å. The used interatomic repulsive potential is substantially harder than the potential in real systems. Consequently, interatomic distances never reach excessivly small values and approximate formulae for electronic transfer integrals can be used. It should be stressed that the basic conclusions, which stem from the calculation, do not depend critically upon the choice of the interatomic potential.

The substrate atoms have two electron energy levels separated by 2 eV, the deeper one being 6 eV below the zero (vacuum) level. The bombarding atom has no energy level (simulation of an inert gas ion) and the sputtered atom has only one level, separated by the energy ε_a from the vacuum level. The absolute value of ε_a, equivalent to the ionization energy I, has been varied to simulate different ionization energies of different sputtered particles. The electron transfer integrals between energy levels of two atoms are assumed to depend on the interatomic distance r_{ij} as $V^0_{ij} \exp[\gamma_{ij}(r^0_{ij}-r_{ij})]$, where r^0_{ij} is equal to 3.36 Å and V^0_{ij} to 0.29 eV. The energy levels of the cluster before the impact of the bombarding ion are shown in Fig. 1 in the inset. Also shown is the occupation by 6 electrons (we neglect the spin degeneracy).

By a proper choice of the impact point, far enough from the sputtered particle to avoid simple binary collision, the kinetic energy E_S of the sputtered particle can be varied from almost zero to energies greater than 100 eV. Using this method of changing E_S we have calculated the values of R^+ for different E_S keeping V^0_{ij} = 0.29 eV and γ = 1.5 Å^{-1} the same for all atomic pairs. The resulting dependencies of R^+ on E_S for various ionization energies I = $-\varepsilon_a$ are shown in Fig. 2. To demonstrate the dependence of R^+ on the ionization energy I, we have plotted in Fig. 3 the values of R^+ (at the fixed kinetic energy of E_S = 40 eV) as a function of I in a semilog plot. As seen, the data can be very well interpolated in a wide range of I by a straight line (the full line in Fig. 3). The line corresponds to the exponential dependence $\exp[-(I - \emptyset)kT_e]$ where T_e = 1800 K

288

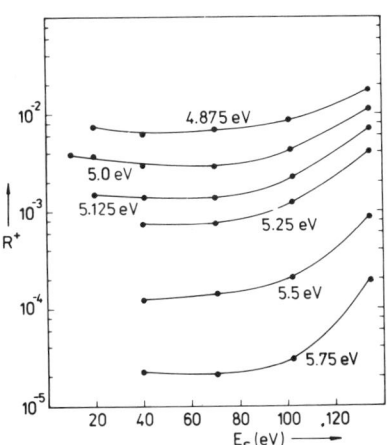

Fig. 2 The calculated ionization yield R^+ as a function of the kinetic energy E_S of the sputtered particle. The parameter is the ionization energy I. The calculation was carried out for $\gamma_a = \gamma_s = 1.5$ Å^{-1}

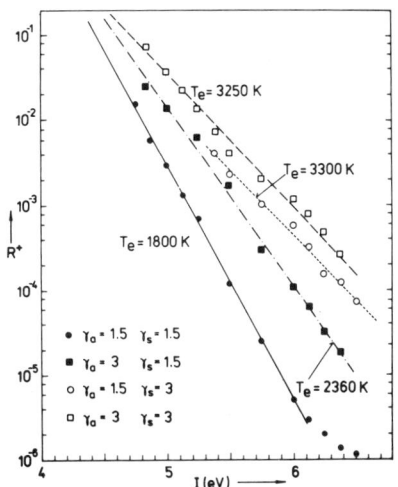

Fig. 3 The calculated ionization yield R^+ as a function of the ionization energy I for various combinations of γ (in Å^{-1}). The data are interpolated by straight lines corresponding to different temperatures T_e

and $\emptyset = 4.1$ eV. Inspecting the inset in Fig. 1 one can see that the value of $\emptyset$ falls just inside the group of the empty cluster levels.

In the next step we have changed various parameters of the cluster and studied the corresponding changes of R^+. We denote by γ_a the parameter γ between the substrate and the sputtered atom and by γ_s the parameter γ between the substrate atoms. Besides the case mentioned above ($\gamma_a = \gamma_s = 1.5$), we have analysed and plotted in Fig. 3 further cases: a) $\gamma_a = 1.5$; $\gamma_s = 3$ (interpolated by dotted line), b) $\gamma_a = 3$; $\gamma_s = 1.5$ (dashed-dotted line), c) $\gamma_a = 3$; $\gamma_s = 3$ (dashed line) (γ in Å^{-1}). The values of V_{ij}^0 are kept equal to 0.29 eV. The calculated data for these faster ionization processes are more scattered than those for the $\gamma_a = \gamma_s = 1.5$ case. Nevertheless, they still can be well interpolated in a reasonable range of I by straight lines and the effective temperatures can be estimated (see Fig. 3).

The model in Fig. 1 is still too crude to give results directly comparable with specific experiments. Also the fact that numerical techniques must be used prevents to obtain clear physical picture of the ionization process. Nevertheless, the calculated values of R^+ and their dependences on various parameters of the system can be favorably compared with experimental results described in the first chapter. Most notably, the ionization probability can be well described, in a large range of I, by the experimental relation $\exp[(\emptyset - I)/kT_e]$ with reasonable large value of T_e. This is in a way surprising because one can hardly expect that thermal equilibrium is established during such rapid, nonstationary processes. The parameter $\emptyset$ is not necessarily the work-function of the system but it rather coincides with the energy at which the density of unoccupied states is largest. As shown in Fig. 2, in agreement with experiments, the values of R^+ and of T_e increase smoothly and reasonably slowly with increasing E_S (T_e is 1800 K at E_S = 20 eV and increases to 3400 K at E_S = 120 eV). The effective temperature depends, through the values of γ, not only upon the electronic structure of the substrate but also upon the electronic structure of the sputtered atom itself.

Unfortunately, the interesting question whether the parameter T_e depends on the bulk properties of the substrate (i.e. on the band structure) or on the properties of individual substrate atoms cannot be addressed because of the limited physical size of the model. According to the results in Fig. 3, the parameter T_e depends mainly upon γ which characterizes the spatial dependence of the electron hopping integral and is not directly related to the band electronic structure. The band structure may be indeed entirely irrelevant because is smeared out in the collision cascade region [30, 31].

To correlate γ with electronic properties of individual colliding atoms is, at present, a difficult problem because V_{ij} is a phenomenological parameter which actually includes ill-defined complex microscopical processes of interatomic electron transfers and interatomic electron-electron interaction. For example, one can speculate that a generally lower temperature T_e of negative ions is caused by smaller γ associated with a large spread of negative ion wave functions. But other parameters of the substrate, such as the speed with which excitations spread in the substrate [see Eq. (9)] may also influence the value of T_e and must be carefully studied.

4. Conclusion

In this short review we have demonstrated the current status and trends in the theoretical description of ionization processes during sputtering. The status is clearly not satisfying because many conceptual questions are not yet resolved but some progress has been made in the general understanding of ionization. In particular, simple empirical formulas which seem to well describe ionization processes have been obtained theoretically on simplified models and thus chances of microscopical understanding of the processes are greater. For further progress of the theory, systematic experiments on well defined systems and the development of new unconventional experimental methods are urgently needed.

Acknowledgment

The author wishes to express his thanks to Dr. Zavadil and Dr. Ždánský for many stimulating discussions.

References

1. C.A. Andersen and J.R. Hinthorne, Anal.Chem. $\underline{45}$, 1421 (1973).
2. Z. Jurela, Int.J. Mass Spectr. and Ion Phys. $\underline{12}$, 13 (1973).
3. A.E. Morgan and H.W. Werner, Anal.Chem. $\underline{48}$, 699 (1976).
4. A.E. Morgan and H.W. Werner, Anal.Chem. $\underline{49}$, 927 (1977).
5. M.L. Yu, Phys.Rev.Lett. $\underline{40}$, 574 (1978).
6. Z. Šroubek, Solid State Comm. $\underline{32}$, 809 (1979).
7. Z. Šroubek, J. Zavadil and K. Žďánský, to be published.
8. V. Deline, C.A. Evans, Jr. and P. Williams, Appl.Phys.Letters $\underline{33}$, 578 (1978).
9. J.M. Schroeer, A. Dely and L.L. Deal, in Secondary Ion Mass Spectroscopy SIMS-II, Proc. 2nd Int. Conf., Stanford Univ., August 27-31, 1979, ed. by A. Benninghoven, C.A. Evans, Jr., R.A. Powell, R. Shimizu, H.A. Storms, Springer Series in Chemical Physics, Vol. 9 (Springer, Berlin, Heidelberg, New York 1979) p. 73
10. G. Blaise and G. Slodzian, Surface Sci. $\underline{40}$, 708 (1973).
11. Z. Šroubek, unpublished results.
12. M.A. Rudat and G.H. Morrison, Surface Sci. $\underline{82}$, 549 (1979).
13. H. Oechsner, Phys.Rev.Letters $\underline{24}$, 583 (1970).
14. N. Winograd, B.J. Garrison, T. Fleisch, W.N. Deglass and D.E. Harrison, Jr., J.Vac.Sci. and Technol., $\underline{16}$, 629 (1979).
15. K. Wittmaack, Surface Sci. $\underline{53}$, 626 (1975).
16. K. Wittmaack, Phys.Rev.Letters 43, 872 (1979).
17. D. Brochard and G. Slodzian, J.Phys. (Paris) $\underline{32}$, 185 (1971).
18. P. Joyes, J.Phys. $\underline{C5}$, 2192 (1972).
19. G. Blaise and A. Nourtier, Surface Sci. $\underline{90}$, 495 (1979).
20. P.W. Anderson, Phys.Rev. $\underline{124}$, 41 (1961).
21. D.M. Newns, Phys.Rev. $\underline{178}$, 1123 (1969).
22. Z. Šroubek, Surface Sci. $\underline{44}$, 47 (1974).
23. J.M. Schroeer, T.N. Rhodin and R.C. Bradley, Surface Sci. $\underline{34}$, 571 (1973).
24. L.I. Schiff, Quantum Mechanics (Mc Graw-Hill, New York, 1968), p. 282.
25. J.K. Nørskov and B.I. Lundqvist, Phys.Rev. B, $\underline{19}$, 5661 (1979).
26. A. Blandin, A. Nourtier and D. Hone, J.Phys. (Paris) $\underline{37}$, 369 (1976).
27. W. Bloss and D. Hone, Surface Sci. $\underline{72}$, 277 (1978).
28. Z. Šroubek, J. Zavadil, F. Kubec and K. Žďánský, Surface Sci. $\underline{77}$, 603 (1978).
29. Z. Šroubek, K. Žďánský and J. Zavadil, Phys.Rev.Letters $\underline{45}$, 580 (1980).
30. K. Wittmaack, Surface Sci. $\underline{85}$, 69 (1979).
31. P. Williams, Surface Sci. $\underline{90}$, 548 (1979).

Boundary Conditions for Models of Ion and Excited-State Formation in the Sputtering Process

R. Kelly

IBM Thomas J. Watson Research Center
Yorktown Heights, NY 10598, USA

Abstract

Although fully satisfactory models to explain the presence of ions and excited states in the sputtered flux do not exist, it is possible to indicate with fair detail what the models will have to describe. That is, what might be termed "boundary conditions" can be set up. For example, ions have significantly lower kinetic energies, and at the same time higher yields, than excited states. Ions are an order of magnitude smaller than excited states and appear to form in events resembling cascade sputtering, whereas excited states are large and appear to form in events resembling binary collisions. In a few respects ions are like excited states, including in the relative yields showing an approximately exponential dependence on the internal energy and in evidence for degeneracy playing a role.

In the case of ions the information available is not particularly restrictive and therefore does not point the way to developing a mechanism. The converse is true for excited states, especially the matter of the high kinetic energies and large sizes. We therefore pursue a model involving random, inelastic energy transfer. The model accomodates easily to high kinetic energies, to relative yields showing an approximately exponential dependence on the internal energy, to large sizes, and to most other details.

1. Introduction

Fully satisfactory models to explain the presence of ions and excited states in the sputtered flux are not at present available. Nevertheless, because of the existence of a variety of experimental results, it is possible to put certain restrictions on what the models will have to explain. Perhaps the most severe of these restrictions relate to the kinetic energies, absolute yields, and siz The kinetic energies, E, of sputtered ions, whether M^+, M^{2+}, or M^{3+}, are in general low, being maximizing functions resembling the total sputtered energy distribution [1, 2]. The energy distributions relevant to a bombardment situation, it will be recalled, have forms as follows [3]. The incident beam, Fig. 1a, is basically a delta function,

$$\text{beam flux} \propto \delta(E - E_1), \tag{1a}$$

where E_1 is the beam energy. The energy distribution of the internal recoils, Fig. 1b, has the form

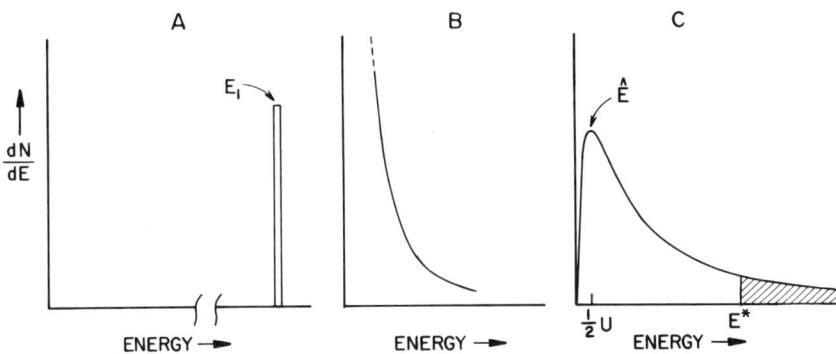

Fig.1 Schematic representation of the degradation of an incident beam having energy E_1 into a sputtered flux. (a) Energy distribution of incident ions according to (1a). (b) Energy distribution of recoiling atoms inside a target according to (1b). (c) Energy distribution of all sputtered atoms according to (1c). The maximum, $\hat{E}$, ideally occurs at $(1/2)U$, where U is the surface binding energy. E^* is the kinetic-energy threshold which is believed to be relevant to excited states (Section 2.1).

$$\text{internal recoil flux} \propto E_1/E^2. \quad (1b)$$

The total sputtered energy distribution, Fig. 1c, is given in normalized form by

$$\text{total sputtered flux} = dN(E)/dE = 2EU(E+U)^{-3}, \quad (1c)$$

where $N(E)$ is the number of sputtered particles of kinetic energy E and U is the surface binding energy. At the same time, the absolute yields of ions from oxidized surfaces are high, namely 0.05-1 for $Z_2 < 42$ [4-6], where Z_2 is the atomic number of the target. Concerning sizes, ions are small.

The kinetic energies of excited states, whether M^*, M^{+*}, or M^{2+*}, i.e. MI, MII, or MIII, appear to be very high [7-10] and, at the same time, the yields from oxidized surfaces are low [11]. Such results are rather unexpected, as it is natural to think of ions as being a limiting case of excited states and therefore having higher, rather than lower, energies and lower, rather than higher, yields. Concerning sizes, excited states are large.

Altogether, the following experimental details, which are of such a nature that they might be termed "boundary conditions", will be seen to exist:
(a) kinetic energies: low for ions but high for excited states, although with certain exceptions in the case of ions.
(b) relative yields: approximately exponential dependence on the internal energy for both ions and excited states, i.e. $\ln(\text{yield})$ $\propto$ (internal energy).
(c) absolute yields from oxidized surfaces: high for ions but low for excited states.
(d) degeneracy: tentatively relevant for both ions and excited states, although with qualifications.

(e) rate of decay: relevant for neither ions nor excited states, although with an important qualification in the case of excited states.
(f) sizes: ions are an order of magnitude smaller than excited states.
(g) resemblance to cascade sputtering: shown tentatively for ion formation.
(h) resemblance to binary collisions: shown tentatively for excited-state formation.

This is not all that is known about ions and excited states but the other aspects present various problems. In some cases the results are sufficiently reasonable that they do not constitute "boundary conditions", for example the correlation of yields with channeling [12]. In other cases, important information is presently being obtained but definite conclusions must await further experimentation, as in the correlation of ion yields from binary targets with electronegativity differences [13, 14]. In still a further case, concerning the drastic fall of some (but not all) yields when oxygen is removed from the target, extensive information is already available [2, 15] but its interpretation in terms of mechanisms has tended to obscure rather than aid setting up mechanisms. An example of the complexity encountered in the matter of surface oxygen includes the tendency for different species to show different responses:
(a) All AlI yields are strongly reduced when oxygen is removed from the target, whereas AlII and AlIII yields are largely unaffected [16]. With Si, SiI and SiII are reduced when oxygen is removed, but SiIII is increased [17].
(b) Most MgI yields are reduced when oxygen is removed, but MgII yields are variously decreased, not affected, or increased [18].
(c) The yields of Al^+, Al^{3+}, and Si^+ are reduced when oxygen is removed, but the yields of Al^{2+}, Si^{2+}, and Si^{3+} are increased [17, 19, 20].
(d) The velocities of species emitting AlI are greater in the absence of oxygen, and conversely for species emitting AlII and AlIII [16].

What follows will be divided into 3 sections. In Section 2 will be discussed the principle "boundary conditions" that models relating to the formation of ions and excited states will have to explain. It will emerge that the details relating to ions are so typical of the overall sputtering process that they do not point the way to developing a mechanism. For example, the details are such that neither molecular intermediaries,nor thermal events,nor formation within the solid can be ruled out. By contrast, the details relating to excited states are so atypical that they provide important hints. The mechanism must be one which involves high kinetic energies and this immediately rules out a role for molecular intermediaries [21] or thermal events [22]. It must also accomodate to large sizes and this rules out formation within the solid.

Section 3 will be devoted to a model, based on random, inelastic energy transfer, which accomodates naturally to high kinetic energies and large sizes. In some respects, it is the excited-state analog of the statistical model of ionization developed by Russek et al. [23, 24].

Section 4 will recapitulate the main points that have been made.

Although we will not be able to contribute too much to the problem of how ions are formed, it is worth recalling the recent progress of Sroubek et al. [25, 26]. They are able to predict both relative and absolute yields, and at the same time find, in agreement with experiment, only a weak dependence of ion yields on the kinetic energy of the sputtered particle.

2. Survey of Boundary Conditions

2.1 Kinetic Energies

Figure 2 shows kinetic-energy distributions for Al^+, Al^{2+}, and Al^{3+} [1] and serves to emphasize what has been the result of most energy-distribution studies for more than a decade: secondary ions have energies similar to sputtered atoms in general. The particular example is of interest in showing that Al^{2+} and Al^{3+} have energies nearly as low as Al^+. In other work, Yu [27] showed the distributions for Si^+ and Si^{2+} to be nearly superimposable, Williams [28] showed the 3 ions, Al^+, Al^{2+}, and Al^{3+}, to be even more similar than in Fig. 2, Blaise and Nourtier [1] confirmed low energies for Al_2^+ and Al_3^+, and Wittmaack [2] confirmed low energies for Nb^+, NbO^+, and NbO_2^+. A particularly broad survey of $\hat{E}$, the energy where $dN(E)/dE$ maximizes (cf. Fig. 1c), and $<E>$, the average ion energy, is given by Rudat and Morrison [29].

In fact, the behavior shown in Fig. 2 is probably better described as being usual but not universal. F^+ formed by bombard-

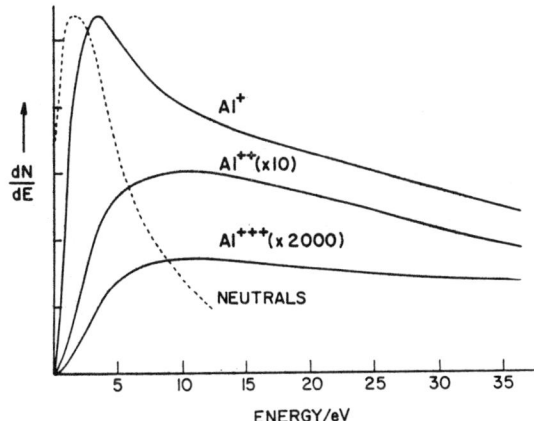

Fig.2 Solid curves: sputtered energy distributions of Al^+, Al^{2+}, and Al^{3+} emitted from polycrystalline Al bombarded with 6 keV Ar^+. Dashed curve: sputtered energy distribution of neutral Al obtained with incident 900 eV Ar^+ and normalized so as to have the same intensity at the maximum as the Al^+ curve. Ion curves due to Blaise and Nourtier [1] and neutral curve due to Oechsner [88].

ing HF-contaminated Si shows much lower energies [30], consistent with it being formed by <u>interatomic</u> Auger decay [31]. In other work there was evidence for higher energies with secondary ions coming from oxygen-free targets. In [32] the evidence took the form of a high-energy component which appeared either as a peak or a tail in the ion energy spectrum. In [33] the peak in the energy spectrum shifted to higher energies. We will, in what follows, have to overlook these examples of unusually low or high energies and assume that ions are adequately represented by Fig. 2. The possibility that unusual energies occur, especially unusually high values, still requires confirmation.

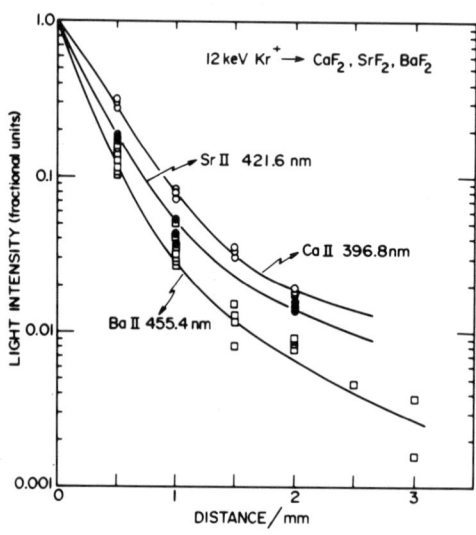

Fig.3 Differential measurements of "light-vs.-distance" for sputtered excited ions emitted when single-crystalline CaF_2, SrF_2, and BaF_2 are bombarded with 12 keV Kr^+. Due to Dzioba and Kelly [8].

Fig. 3 shows measurements of "light-vs.-distance", i.e. the light from sputtered, excited atoms or ions determined as a function of the perpendicular distance from the target [8]. The observed 1/e distances are 0.2-0.4 mm, whereas, given the particular lifetimes and a kinetic energy of 5 eV, one would expect 0.02-0.03 mm. The particular excited states thus have kinetic energies much higher than either ions or sputtered atoms in general. This result was first hinted in experiments by Gritsyna et al. [34, 35] and Braun et al. [36]. This work, which sometimes claimed impossibly high energies, might be criticized for lack of resolution and for use of a simplified mathematical analysis. More recent results, as in Fig. 3, seem to have overcome these problems, yet the inferred kinetic energies, while not <u>impossibly</u> high, are still distinctly higher than might be expected [7-10]. The basic premise of [34-36] regarding high energies thus appears to be correct.

The assignment of actual kinetic energies to excited states requires, apart from a correct mathematical analysis of information as in Fig. 3, an assumption about P(E), the probability of excitation for a given kinetic energy, E. For example, if we assume a distribution as in (1c) or Fig. 1c but starting at a kinetic-energy threshold, E^*, which is equivalent to $P(E) = P^*$ for $E > E^*$, results as in Table 1 are obtained. An explicit determination of $P(E)$ for HeI formed when He^+ is backscattered from Pb is in agreement with $P(E)$ being nearly independent of energy though does not establish a threshold [37]. Approximate energy-independence for $P(E)$ was also suggested in other work [38], even if on somewhat indirect grounds. We note that high values of E^* apply, not just to MI, but also to MII and MIII [39]; the molecule BH^*, on the other hand, has a relatively low E^* [40]. It is significant that high values of E^* continue to be obtained even when the superior resolving power of an optical multichannel analyzer is used to obtain the curves of light-vs.-distance [41].

Further indications that excited states may have unusually high kinetic energies also exist. Tolk et al. [42] found that the light yield as a function of the <u>incident</u> ion energy showed a threshold at 100-300 eV. MacDonald [10] found that measurements of Doppler broadening with a series of Ti compounds implied similar kinetic energies as did measurements of light-vs.-distance. Leung et al. [43] argued that the Doppler shift for various HI formed when H^+, H_2^+, or H_3^+ is backscattered from Mo corresponds to the excited states having energies near the incident ion energy.

Heiland et al. [44] and Bhattacharya et al. [38] observed that photon yields generally increased as the incident energy increased, with [38] in particular showing that the increasing yields lie in an energy regime where the sputtering coefficient <u>decreases</u>. A possible, even if not unique, interpretation of this result is that the excited states originate mainly from recoil-sputtered atoms. Since it is characteristic of recoil sputtering that kinetic energies increase as the incident energy increases [45, 46], there is a concomitant increase in the probability of the excited states surviving non-radiative processes.

<u>Table 1</u> Average values of the assumed threshold kinetic energy, $<E^*>$, for atoms (I) and ions (II) sputtered in excited states from LiF, Li, B, NaCl, MgF_2, CaF_2, SrF_2, BaF_2, and a mixed oxide glass

Type of transition	Number of examples	Average threshold energy, $<E^*>$ [eV]	Reference
LiI	9	30	[7]
BI	1	2200-3700	[40]
NaI	5	120	[7, 9, 41]
MgI	2[a]	900	[8]
MgII	1	1600	[8]
SiI	1	700	[9]
CaI	1	1600	[8]
CaII	2	1200	[8]
SrII	2	1800	[8]
BaI	3	1400	[8, 9, 41]
BaII	4	2000	[8, 9, 41]

[a]Exclusive of MgI at 285.2 nm [8].

We would suggest that, since ions and excited states have such different kinetic energies, with those of ions being consistently lower, ions and excited states have distinct origins. This conclusion follows, let it be emphasized, only if ion energies are lower than those of excited states. In the contrary case, with ions having energies higher than those of excited states, it would be reasonable to assume similar origins.

2.2 Relative Yields

Figure 4 [47] shows relative positive ion yields versus the ionization potential, I, for O_2^+ incident on various elemental targets plotted as

$$\log(\text{yield}) \text{ vs. } I.$$

The results are corrected for general matrix effects from a knowledge of the ion yields for 3 tracer species, H, P, and As, which served for normalization. They are not corrected for degeneracy. There is seen to be a well-defined linearity with a slope of 1.74 eV^{-1} or 6700 K.

A corresponding plot for particular excited states of some 34 different elements can be made using the absolute yields of Tsong and Yusuf [11] and is found to be quite erratic. This remains true even if corrections for degeneracy and branching are made by plotting

$$\ln(\text{yield}/g_i b_{fi}) \text{ vs. } \varepsilon_i. \tag{2}$$

Here g_i is the degeneracy, to be discussed in Section 2.4, b_{fi} is is the branching ratio, ε_i is the excitation energy, i stands for "initial", and f stands for "final".

But a plotting in terms of (2) of the yields of different excited states of the same element is well-known to show a reasonable linearity (Fig. 5a) [22]. Li is an important exception, to which we will return in Section 3.4 (Fig. 5b) [48].

The approximate linearity of plots of $\ln(\text{yield})$ vs. (internal energy) constitutes one of the most severe boundary conditions that models for the formation of ions or excited states must meet. We would emphasize that such linearity neither requires nor infers that thermal events play a role; it is simply a description of a rapid, monotonic fall in population as a function of internal energy. More specifically, we have argued elsewhere [22] that, since the slopes of plots of $\ln(\text{yield}/g_i b_{fi})$ vs. ε_i are largely independent of the ion and target mass, it is difficult to interpret them in terms of thermal spikes. Similarly, Sroubek et al. [26] have calculated ion yields using a quantum mechanical approach, plotted the results as $\ln(\text{yield})$ vs. I, and found slopes which, in temperature units, had values of 2000-3500 K.

2.3 Absolute Yields from Oxidized Surfaces

In connection with absolute yields, we will consider only oxidized surfaces. This is done in part because of the complexities of

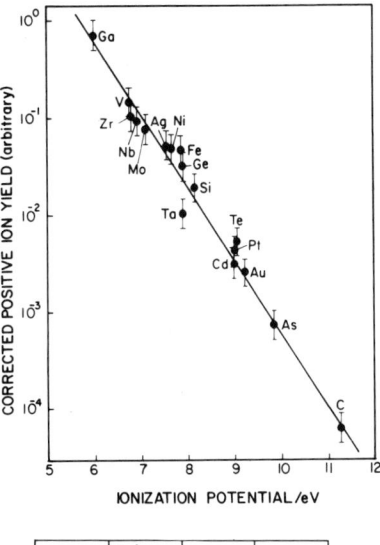

Fig. 4 Relative positive-ion yields versus the ionization potential for various elements bombarded with 25 keV O_2^+. Each target was implanted with a known amount of H, P, and As, the signals of which were used to normalize the matrix signals. The results <u>are not</u> corrected for degeneracy. Due to Williams et al. [47].

Fig. 5a Relative photon yields vs. the excitation energy plotted according to (2) for group III targets bombarded with 12 keV Kr^+. Due to Good-Zamin et al. [22].

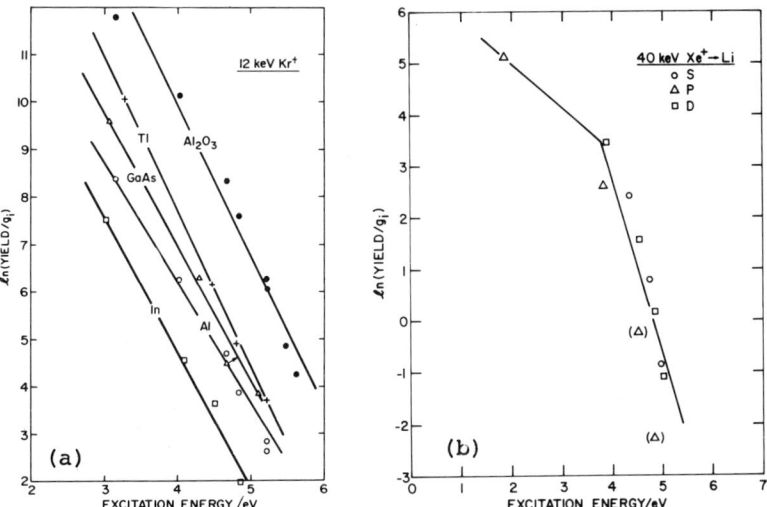

Fig. 5b Relative photon yields vs. the excitation energy plotted according to (2) for Li bombarded with 40 keV Xe^+. The 2 bracketed points involve such long lives that the signals **were probably** underestimated. Taken from Fig. 7a, which is in turn due to Jensen and Veje [48].

the responses which occur when oxygen is removed from the target (Section 1). In addition, there is the more general problem that results for truly oxygen-free systems are rare [14]. Even the use of ultra-high vacuum does not overcome the problem that oxygen is always contained in the target and sometimes contained in the ion beam.

The absolute yields of ions from oxidized surfaces are typically 0.05-1 for $Z_2 < 42$ [4-6], while those of excited states are typically 10^{-4}-10^{-2} [11]. The trend that emerges, that ions are more abundant than excited states, was subsequently confirmed for for Si and Ni, even if the ion yields were here rather low [49]. Ion yields can be difficult to measure, though many of the problems will be overcome in work initiated by Oechsner et al. [6], in which both neutrals and ions are detected with the same geometry.

We regard the trends in absolute yields to be equivalent to the trends in kinetic energies discussed in Section 2.1. For example, if excited states have high kinetic energies, then it is natural that the absolute yields be low.

Nevertheless, the apparent trends may be somewhat exaggerated. This is because, in a study of ions, all ions are in principle detectable [33] whereas, in a study of excited states, only those states which emit at appropriate wavelengths and have appropriate lifetimes are detectable. For example, the lowest lying states of group II and IV atoms, which should have the highest yields, in some cases emit at wave lengths not normally studied and in most cases are too long lived to emit within the detection volume.

2.4 Degeneracy

Morgan and Werner [50, 51] have shown that the correlation of ion yields with I for different ions sputtered from steel, halides, and oxides is such that the statistical weight, in these cases identified with the ratio of the ion to neutral partition functions, Q^+/Q^o, appears to enter (Fig. 6). If true, this means that, whatever the details of the model describing the origin of ions, we have

$$\text{population} \propto \text{partition function} = \Sigma g_i \exp(-\varepsilon_i/kT). \qquad (3)$$

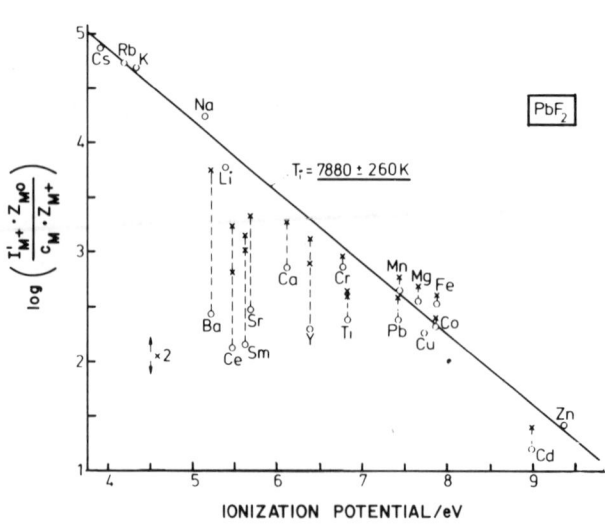

Fig.6 Relative positive-ion yields vs. the ionization potential for various dopants in PbF_2 bombarded with 6 keV Ar^+. "o" is the yield of the atomic ion, M^+; "x" is the sum of M^+ and MF^+; "x,x" is the sum of M^+, MF^+, and MF_2^+. The observed ion signals are divided by the atomic concentration, c_M, and multiplied by the electronic partition-function ratio, Z^o/Z^+, where $Z \equiv Q$. Due to Morgan and Werner [51].

We regard this result as problematical in two respects. Firstly, the partition-function ratios for the elements studied do not, except for groups I and III, differ significantly from unity (Table 2), so the linearity as in Fig. 6 would (except for points corresponding to groups I and III) be similar with or without the correction. Indeed, there has already been a good example of linearizing ion yields without the use of partition functions (Fig. 4), while Wittmaack [2] has shown explicitly that the results of [50] are equally well linearized with or without partition functions. Secondly, and more generally, the partition function is a thermal quantity and therefore has no place in a description of sputtering unless one is dealing explicitly with thermal sputtering.

Table 2 Partition-function and ground-state degeneracy ratios

Atom	Ratio of ion to neutral partition functions at 7000 K, Q^+/Q^o [86]	Ratio of ion to neutral ground-state degeneracies, g_o^+/g_o^o [78-82]
Li	0.44	0.50
K	0.37	0.50
Cu	0.41	0.50
Au	0.50	0.50
Mg	1.8	2.0
Ba	1.1	2.0
Zn	2.0	2.0
Cd	2.0	2.0
Al	0.17	0.17
In	0.22	0.17
C	0.62	0.67
Ge	0.56	0.67
Ti	1.5	1.3
Hf	0.96	0.48
P	1.8	2.3
Sb	0.80	2.3
V	0.87	0.89
Ta	1.2	1.3
S	0.48	0.44
Te	0.71	0.44
Cr	0.62	0.86
W	1.08	1.2
Fe	1.42	1.2
Ni	0.41	0.48
Pd	1.70	10.
Pt	0.57	0.67

It is still possible, however, that the various studies on secondary ions, including [50, 51], have been correct in demonstrating the validity of a relation having the general form of (3). We would suggest that, if this is so, it is an indication not that the partition function but rather that the ground-state degeneracy, g_o, is relevant:

population ∝ ground-state degeneracy = g_o. (4)

This avoids using a thermal quantity and is, at the same time, intuitively reasonable. There is no numerical problem, for, as is evident from Table 2, the partition-function ratio is in most cases similar to the ground-state degeneracy ratio, the ground state being here defined to include the sublevels differing in J of a state with a given L. (Pd is amongst the few exceptions.) It is clear that experiments which make explicit tests of (3) and (4) are needed.

Veje et al. [48, 52] have shown that a correlation of excited-state yields with the principle quantum number, n, for different excited states of Li and Mg^+ implies a qualified proportionality with g_i (Figs. 7a and 7b). The qualification is that the P states of Li and certain D and F states of Mg^+ have low yields. The Li result (Fig. 7a) can be understood in terms of the atoms in P states emitting in part beyond the detection volume by virtue of the long lives. With Mg^+ (Fig. 7b), the D states which are

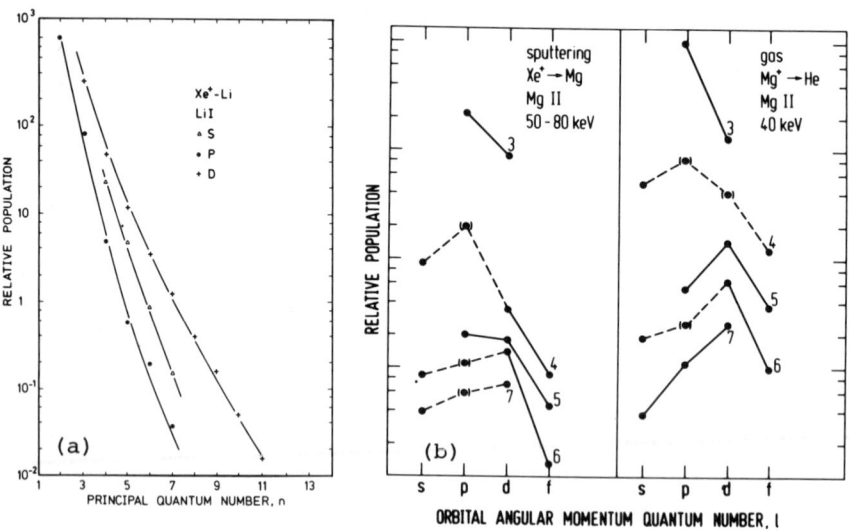

Fig.7a Relative LiI populations vs. the principal quantum number, n, for Li bombarded with 40 keV Xe^+. Due to Jensen and Veje [48].

Fig.7b Relative MgII populations vs. the orbital angular-momentum quantum number, ℓ, for solid Mg bombarded with 50-80 keV Xe^+ (left) and for gaseous He bombarded with 40 keV Mg^+ (right). Note how the F-state yields are low for both sputtering and binary collisions. Due to Andersen et al. [52].

low are those which have distinctly higher internal energies than the corresponding S and P states. The low signals are thus in accordance with (2). Only the F states of Mg^+ do not admit a simple interpretation. Perhaps they are disfavored owing to selection rules, for if the same selection rules as apply to the products of binary collisions applied, even if only weakly, to sputtering, such states would have low yields. Alternatively, these states are somehow disfavored by virtue of not being significantly broadened [53] (see also Section 3.4).

It would appear that degeneracy plays a role with excited states somewhat as in (4) but with some indication of levels with high L, such as the F states of Mg^+, being disfavored. As with secondary ions, it is clear that experiments to test the role of degeneracy are needed.

2.5 Rate of Decay

The rate of decay is never a relevant quantity with ground-state ions but may be very important with excited states. There is, first of all, the problem of the physical nature of the region in which the excited states emit radiation. If the emission occurs in a region resembling an equilibrium plasma, the population will be proportional to the transition "probability", A_{fi}:

$$\text{population} \propto A_{fi}. \tag{5a}$$

If the emission occurs in a region resembling a vacuum, the population will be independent of A_{fi} and, instead, proportional to the branching ratio, b_{fi}:

$$\text{population} \propto b_{fi} = A_{fi}/\Sigma A_{f'i}. \tag{5b}$$

In practice there are numerous examples in which both (5a) and (5b) are used. The problem was explored experimentally by Morrow et al. [54], to which we refer also for an enumeration of examples of the use of the two equations. Typical results, as in Fig. 8, show that a forbidden transition which has a low yield from a plasma can have a very high yield in a sputtering geometry. Thus, the CdI feature at 326.1 nm is about 200 times stronger than plasma data [89] suggest. This result is readily understood if (5b) is valid.

Nevertheless, it is not always correct to argue that excited-state yields are independent of A_{fi}. If the decay rate constant,

$$\gamma_i = \Sigma A_{f'i},$$

is sufficiently small, e.g. $< 0.004 \times 10^8$ s^{-1} for the geometry of [54], significant decay occurs beyond the reach of the detector. Indeed, for a viewing geometry with a lens so placed that the detection volume is minimized [39], the populations of most excited states will scale as γ_i, so that (5b) becomes

$$\text{population} \propto b_{fi}\gamma_i = A_{fi} \tag{5c}$$

and thus takes on the same form as (5a).

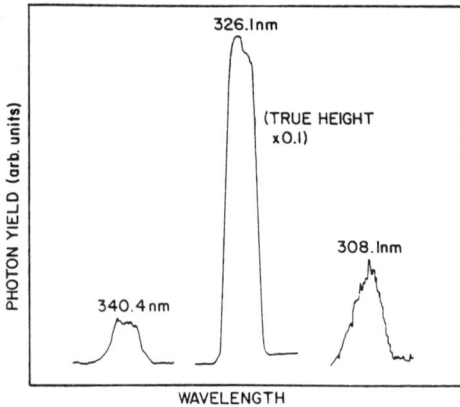

Fig.8 Reproductions of strip-chart recordings of photon yields for Cd bombarded with 12 keV Kr^+. The line at 326.1 nm is here by far the most intense (10 times more than shown), whereas with Cd-containing plasmas [89] it is very weak. The inference is that (5b) rather than (5a) is valid. Due to Morrow et al. [54].

2.6 Sizes

The size of a positive ion can be identified with the conventional crystallographic ion radius. The sequence Li^+ to Cs^+ thus encompasses 0.68-1.67 Å, Be^{2+} to Ba^{2+} encompasses 0.35-1.34 Å, and B^{3+} to Tl^{3+} encompasses 0.23-0.95 Å. Such radii will be seen to be very small compared with those of excited states.

The size of an excited state would, ideally, be identified with the mean radius of the electron density, $<r>$, as evaluated for the particular electronic configuration but we are unable to locate a source for this information. An alternative is to use the following formula for $<r>$ of a hydrogenic entity [55]:

$$<r> = (3n^2 a_o/2Z^*)\{1 - \ell(\ell + 1)/3n^2\}$$
$$= 0.7937n^2\{1 - \ell(\ell + 1)/3n^2\}, \quad (6)$$

where n is the principal quantum number, ℓ is the orbital angular-momentum quantum number, a_o is the Bohr radius, 0.5292 Å, and Z^* is the charge number seen by the excited electron, here taken as 1. Still a further alternative [56] is to deduce an effective size, $<r^*>$, by using (6) but with n replaced by the effective principal quantum number [57], n^*, defined by the relation

$$\varepsilon_i = I - 13.60/n^{*2}. \quad (7)$$

Values of $<r>$ and $<r^*>$ for Na are compared in Table 3. It is clear that excited states are an order of magnitude larger than ions.

Because of $<r>$ and $<r^*>$ for excited states being so large, a severe restriction arises for any mechanism purporting to explain excitation. The act of excitation must occur at or beyond the target surface rather than within the target, a point recognized also by Wright and Gruen [58].

Table 3 Sizes of excited states of Na

State	Hydrogenic size, $<r>$, from (6) [Å]	Effective quantum number, n^*, from (7)	Effective size, $<r^*>$, from (6) with n^* [Å]
4s	12.7	2.64	5.5
7s	38.9	5.62	25.1
4p	12.2	3.13	7.2
7p	38.4	6.15	29.5
4d	11.1	3.98	11.0
7d	37.3	6.97	37.0

2.7 Resemblance of Ion Formation to Cascade Sputtering

Two features of secondary-ion formation already considered suggest a similarity to cascade sputtering. These are the maximizing energy distributions (Section 2.1) and the high yields (Section 2.3).

A further aspect relates to the surface binding energy. The surface binding energy, U, governs the overall sputtering process as

sputtering coefficient $\propto 1/U$,

and governs the sputtered energy distribution, at least in ideal cases, as in (1c). Here we see that $dN(E)/dE$ maximizes for $\hat{E} = (1/2)U$. An explicit test of the role of U in the overall sputtering process lies in the close correlation of the sputtering

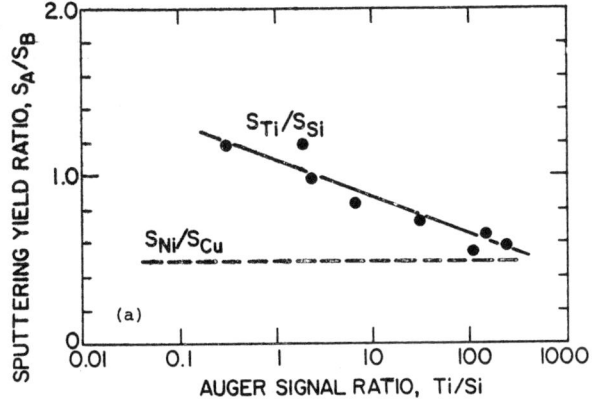

Fig.9a Sputtering-yield ratios, S_{Ni}/S_{Cu} and S_{Ti}/S_{Si}, versus the surface concentration ratio. The latter was determined by Auger analysis, including a sensitivity correction, after prolonged Xe^+ bombardment. The Ni-Cu data, for 500 eV Ar^+, are due to Shimizu et al. [90]. The Ti-Si data, for 1.5 keV Xe^+, are due to Narusawa et al. [62].

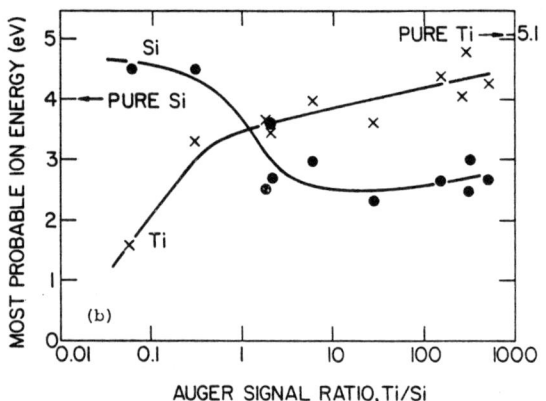

Fig.9b Most probable ion energies, $\hat{E}$, versus the surface concentration ratio. Note how, for those compositions where Ti is lost preferentially (Fig.9a), $\hat{E}_{Ti}$ is less than $\hat{E}_{Si}$. Due to Narusawa et al. [62].

coefficient with the inverse of the heat of sublimation [59, 60]. Corresponding tests with ions, on the other hand, are rare. This may be assumed to be in part because of the difficulty in obtaining absolute yields which are both accurate [33] and free of matrix effects [47]. In part it will be because the maximum in dN(E)/dE is often not well defined, if observed at all [e.g. 61]. One instance where the role of U appears to have been demonstrated with ions is in work by Narusawa et al. [62]. They showed that for compositions of Ti-Si such that Ti is lost preferentially (Fig. 9a), $\hat{E}_{Ti}$ is less than $\hat{E}_{Si}$ (Fig. 9b). Another instance is work by Richards and Kelly [63]. They obtained values of $\hat{E}$ for cations from single-crystal halides which showed both the expected ordering and the expected magnitudes (Table 4).

Table 4 Observed ($2\hat{E}$) and expected ($[5/3]\Delta H_a$) surface binding energies (U) for single-crystal halides[a]

Target	Energy, $\hat{E}$, at which dN(E)/dE maximizes, from [63] [eV]	U = $2\hat{E}$ [eV]	$(5/3)\Delta H_a$ [eV]
LiF	2.95	5.9	7.4
NaCl	2.9	5.8	5.6
KCl	3.0	6.0	5.6
KBr	2.6	5.2	5.1
KI	2.0	4.0	4.5

[a]The heat of atomization, ΔH_a, is for a "half-space" surface atom, hence one with 3 bonds for an NaCl-type substance. $(5/3)\Delta H_a$ applies to an in-surface atom with 5 bonds. Note that there is good precedent for describing surface energies with halides in terms of bonds even if formal bonds do not exist [87].

Similar information is totally unavailable for excited states. We would suggest, however, on the basis of the kinetic energies being so high (Section 2.1) and of the possibility that recoil sputtering is involved (Section 2.8) that the answer would be that U plays no role at all.

2.8 Resemblance of Excited-state Formation to Binary Collisions

One of the most characteristic features of excitation in binary collisions is the existence of well-defined energy thresholds. Such thresholds are interpreted in terms of "inner" curve crossing [64, 65] and may possibly be equivalent to the high kinetic energies discussed in Section 2.1. On the other hand, the similarities may be misleading, as high kinetic energies are also to be expected due to the general inefficiency of inelastic energy transfer (Section 3.3).

A more explicit example relating to binary collisions is that of Snowdon et al. [66, 67]. They showed that the excited states from solid Zn sputtered with 50 keV Ar^+ gave a comparable fit to a plot according to (2), i.e. $\ln(yield/g_i b_{fi})$ vs. ε_i, as did those from gaseous Ar or Zn bombarded with Zn^+.

Andersen et al. [52] compared $Xe^+ \to Mg$(solid) with $Mg^+ \to He$(gas) and found similar relative populations (Fig. 7b). They concluded that such results suggest either that excitation is governed by the wave functions of the free atoms or else that it results from binary collisions. These are really equivalent conclusions.

The possibility that excited states originate from recoil-sputtered atoms, thence in events resembling binary collisions, was considered in [46]. Expressions were developed for the recoil-sputtering coefficient of both matter, S_M, and energy, S_E. As seen in Table 5, the values of S_M are of order 0.03-0.3 and therefore substantially larger than the area under the high-energy (i.e. $E > E^*$) tails of cascade-sputtering energy distributions (cross hatching in Fig. 1c). Moreover, the mean energy

Table 5 Recoil-sputtering quantities for 12 keV Kr^+ incident on various targets [46]

Target	Coefficient for recoil-sputtering of matter, S_M [atoms/ion]	Coefficient for recoil-sputtering of energy, S_E [fraction/ion]	Mean energy per recoil-sputtered atom, $<E> = S_E E_1/S_M$ [eV/atom]
LiF	0.34	0.0069	240
Li	0.57	0.012	270
NaCl	0.19	0.0095	610
MgF_2	0.11	0.0064	700
CaF_2	0.076	0.0066	1050
SrF_2	0.041	0.0046	1360
BaF_2	0.027	0.0031	1370

per recoil-sputtered atom,

$$\langle E \rangle = S_E E_1 / S_M,$$

is consistently either larger than, or similar to, $\langle E^* \rangle$ as in Table 1. The conclusion is that, from a numerical point of view, recoil sputtering is better able than cascade sputtering to account for excited states.

Taken together, the information of this and the preceding Section shows that, just as ions seem to originate in events resembling cascade sputtering, so excited states are formed in events which are more nearly like binary collisions. Explicit experiments are, however, relatively rare.

3. Excited-state Formation

3.1 The Boundary Conditions

As stated in Section 1, there are still no fully satisfactory models to explain the presence of ions and excited states in the sputtered flux. It was then shown in Section 2 that any models that are proposed must meet certain restrictions relating to kinetic energies, relative yields, absolute yields from oxidized surfaces, degeneracy, rate of decay, sizes, resemblance to cascade sputtering, and resemblance to binary collisions.

The restrictions on models of ion formation are not particularly severe, as ions appear to arise in typical, sputtering-like events. But the restrictions on models of excited-state formation are very severe indeed, especially the evidence for high kinetic energies and large sizes.

Under these circumstances, it is natural to attempt to set up a mechanism for excited-state formation which meets the various restrictions. We note first of all that a role for molecular intermediaries [21] or thermal events [22] can be immediately ruled out owing to the high kinetic energies. Models which involve excitation within the solid can also be ruled out owing to the large sizes. There is one approach, however, which accomodates naturally to high kinetic energies and large sizes, and which we will attempt to develop. It will be based on random, inelastic energy transfer. The transfer will be assumed to take place at the outer surface in essentially binary collisions and with the transferred energy distributed statistically amongst the accessible excited states. The details of the inelastic energy transfer will be taken as in Firsov's theory [68-70].

The model will be seen to succeed fairly well in explaining the high kinetic energies, the relative yields, the role of degeneracy, and the resemblance to binary collisions. The problem raised by large sizes, while not included explicitly in the model, is accomodated by virtue of the assumption that the collisions leading to excitation are at the outer surface. We have not yet applied the model to absolute yields except to confirm the feasibility.

The main failing of the model is that, by virtue of being statistical, it contains no chemical information such as is needed to explain why some excited-state yields fall when oxygen is removed from the target (Section 1). The model would also be unable to explain any correlations with electronegativity differences analogous to those found with ions [13, 14] which may emerge in future work.

3.2 Inelastic Energy Transfer, ΔE_e, According to the Formalism of Firsov

When 2 atoms collide at low enough velocities, the behavior is said to be adiabatic, "not going through". This means that the electrons are at all times relaxed to the correct diatomic molecular energies. Such processes are comparatively simple to treat for inner shells or for small atoms. For outer shells the number of relevant states becomes unwieldy, while with the inclusion of excited levels the number of relevant states becomes impossibly high. Fig. 10 gives a hint as to what is involved when even a very small number of excited states is included.

Firsov [68] has suggested the use of the Thomas-Fermi atc to handle the complexities of collisions where inner shells, outer shells, together with excited levels are involved. This approximation is valid whenever the value of $\Delta\varepsilon$, the level separation, is much smaller than ΔE_e, the inelastic energy transfer:

$$\Delta\varepsilon \ll \Delta E_e.$$

His final result is an expression containing the atomic numbers, the velocity of the moving atom, and the impact parameter. We are interested in large-angle events and will use the approximation [cf. 23] that the impact parameter be replaced by the distance of closest approach. Doing this, and letting the colliding atoms have the same atomic number, Z, we obtain

$$\Delta E_e \simeq \frac{0.190(E/M)^{1/2}Z^{5/3}}{(1 + 0.391Z^{1/3}\check{r})^5}, \tag{8}$$

where E is the kinetic energy of the moving atom in eV, M is the mass of the moving atom in u, and $\check{r}$ is the distance of closest approach in Å.

Kishinevski [70] has attempted to treat large-angle events explicitly, the result being an expression containing the atomic numbers, the velocity, and $\check{r}$. We consider the special case that the colliding atoms have the same atomic number and approach head-on, so that, if V(r) is the interatomic potential, we have [cf. 71]

$$V(\check{r}) = (1/2)E. \tag{9}$$

ΔE_e is then given by

$$\Delta E_e \simeq \frac{0.110(E/M)^{1/2}Z^{5/3}}{(1 + 0.713Z^{1/3}\check{r})^3}. \tag{10}$$

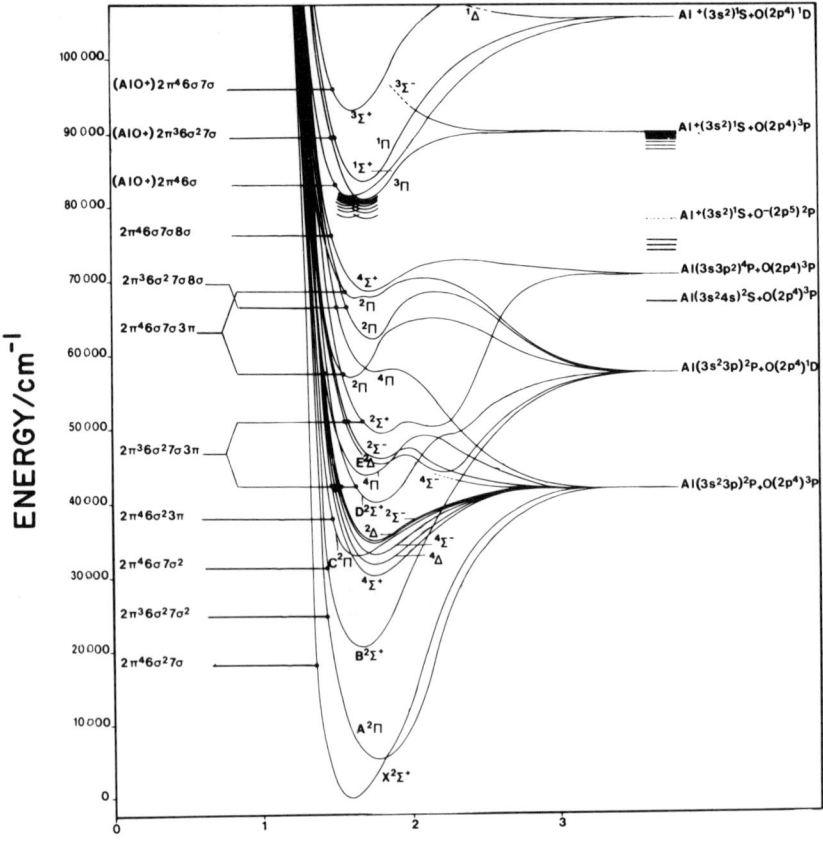

Fig.10 Calculated potential-energy curves for a limited number of electronic states of the Al-O system. Only one excited state for each of Al and O is shown yet the number of molecular states, thence the number of curve crossings, is very high. Due to Schamps [91].

To evaluate $\check{r}$, (9) should be expressed with $V(r)$ as for Thomas-Fermi atoms,

$$\frac{14.4 Z^2 \Phi(\check{r}/a)}{(\check{r}/a)a} = (1/2)E,$$

where $\underline{a}$ is the Thomas-Fermi screening length in Å (tabulated in [72]) and Φ is the Thomas-Fermi screening function (tabulated in [73]).

It is important to realize that the use of Firsov's theory, thence (8) and (10), does not constitute a restriction on what follows. One could just as well have taken ΔE_e according to the model of Oen and Robinson [74] or even regarded it as an empirical phenomenon. The most relevant experiments, in this regard, are probably those of Eckstein et al. [75] and Heiland and Taglauer [76], which involve an evaluation of the inelastic energy transfer at particularly low incident energies, 0.2-15 keV.

3.3 General Properties of ΔE_e

Examples of ΔE_e calculated from (8) and (10) are given in Table 6 and Fig. 11. Two important features will be noted. Firstly, the values of E required to make ΔE_e exceed typical excitation energies, 2-5 eV, are relatively large, several 100 eV. We suggest that this is the reason for the experimental kinetic energies, $<E^*>$ of Table 1, being so large. Moreover, it follows that it is best to describe E^* as being large not because of a true threshold effect [64, 65] but because inelastic energy transfer is an inefficient process, especially when it is <u>random</u>.

Secondly, the plots of ΔE_e vs. E are approximately linear at low enough energies, so that one can write

$$\Delta E_e = KE \qquad (11)$$

rather than, for example, $\Delta E_e = K'E^{1/2}$. Such a dependence is at first sight problematical in view of the fact that $(dE/dx)_e$, the electronic stopping, scales as velocity, i.e. $E^{1/2}$, in most experimental situations. Eckstein et al. [75] and Heiland and Taglauer [76] have deduced ΔE_e from low-energy ion-surface scattering events. They identified ΔE_e with the slight shift of the

Table 6 Examples of the evaluation of the inelastic energy transfer, ΔE_e

Collision pair	Kinetic energy, E [eV]	Distance of closest approach, ř [Å]	ΔE_e from (8) [eV]	ΔE_e from (10) [eV]
Li→Li	100	0.53	1.2	0.71
	300	0.30	3.6	2.0
	600	0.20	6.4	3.6
	1000	0.15	9.6	5.4
K→K	100	1.6	0.31	0.37
	300	1.1	1.6	1.4
	600	0.85	4.2	3.2
	1000	0.70	8.3	5.9
Cs→Cs	100	2.4	0.063	0.17
	300	1.7	0.39	0.71
	600	1.4	1.1	1.7
	1000	1.2	2.6	3.2

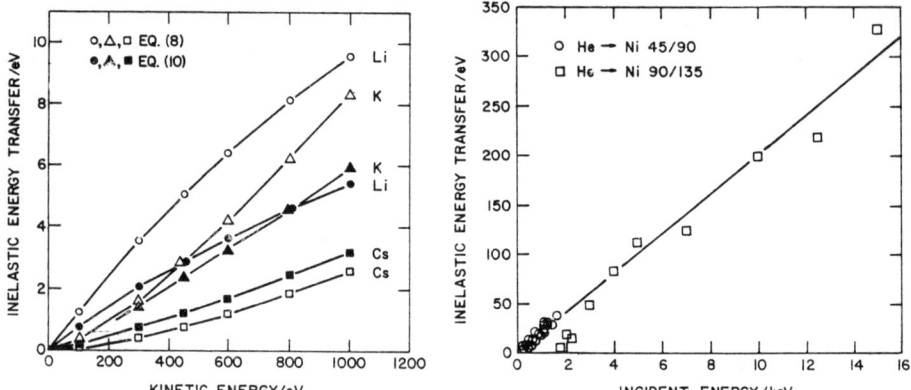

Fig.11 ΔE_e calculated according to the formalism of Firsov, using (8) and (10), for the 3 collision pairs Li→Li, K→K, and Cs→Cs. The kinetic energies required to make ΔE_e exceed typical excitation energies, 2-5 eV, are relatively large.

Fig.12 ΔE_e determined experimentally for $^4\text{He}^+$ scattered from Ni at either ψ = 90° (normal) incidence with respect to the surface and a laboratory scattering angle θ = 135° or at ψ = 45°, θ = 90°. Note the factor-of-two numerical similarity to the Li→Li curve of Fig.11. Due to Eckstein et al. [75] and Heiland and Taglauer [76].

scattered energy from the value expected on the basis of conserving energy and momentum and found a linear dependence as in (11) up to 15 keV (Fig. 12). The expressions of Oen and Robinson [74] and Firsov [68] for ΔE_e were evaluated in [76] for He→Ni and found to be approximately linear in E for low enough energies. Measurements of $(dE/dx)_e$ by Blume et al. [77] showed proportionality to $E^{1/2}$ above about 10 keV but a stronger dependence on E for lower energies. The E-dependence of both ΔE_e and $(dE/dx)_e$ thus appears to depend on the energy regime and will be taken as in (11) in the present work.

3.4 Relative Populations

Perhaps the most important test of the significance of ΔE_e is to use it to deduce relative populations and then examine plots of $\ln(\text{population}/g_i)$ vs. ε_i, where ε_i is, as before, the excitation energy. Such plots must be approximately linear and, at the same time, have slopes with expected values. The slopes will, for historical reasons, be expressed in units of "temperature" with no intent to mean "temperature" in the conventional sense.

Numerical values for the ε_i will be required, a problem which is not straightforward in spite of the existence of comprehensive tables [78-82]. The first complication is that the ε_i will be shifted when an atom is near a surface [53]. To the lowest approximation this shift is equal to the sum of the electron-(electron image) and electron-(ion image) energies, namely $\delta\varepsilon \simeq 3.6/x$

eV for a metal, with x the distance from the surface in Å. The shift, $\delta\varepsilon$, is thus approximately uniform for all levels, and neither the order nor separation of the levels should be changed in a significant way. We will subsequently overlook $\delta\varepsilon$ on the grounds that it should not be important to <u>relative</u> populations.

The second complication is that the ε_i will be broadened. Some broadening, $\Delta\varepsilon_s$, is due to the proximity of the surface. For example, for a distance of about 2 Å $\Delta\varepsilon_s$ for an S state is $\sim$1.0 eV, for a P state is $\sim$0.3 eV, and for a D state is $\sim$0.1 eV [53]. Even larger broadenings for S states were deduced in [83]. Additional broadening, $\Delta\varepsilon_u$, is due to the uncertainty principal [23]:

$$\Delta\varepsilon_u \simeq \hbar v/<r>,$$

where v is the velocity of the moving atom and $<r>$ is, as before, the excited-state size. We are unable to comment on whether broadening is significant in the present context, for example, whether it changes the order or separation of the ε_i in an important way or whether it is in any way relevant to the problem of states with large L being disfavored (Section 2.4). We will therefore have to overlook broadening, but recognize that this is probably the main oversight in the model.

To calculate relative populations, five assumptions will be made. Firstly, a sputtered atom is assumed to be an outer layer atom which is struck from behind. This is in reasonable agreement with the simulations of Harrison et al. [84] which showed 80-100% of all atoms sputtered from single-crystal Cu to come from the outer layer. The second assumption, necessary to accomodate to the large sizes of excited states (Section 2.6), is that the same collision which creates a sputtered atom is also the one which leads to excitation. The third assumption is that the probability of the striking atom having kinetic energy E is given by (1b), namely

$$\text{probability} \propto E^{-2}dE. \tag{12}$$

The problem here is that an E^{-2} dependence is strictly correct only for cascade sputtering and there is some evidence that recoil sputtering may play a role (Section 2.8). The fourth assumption is that a given, singly excited S, P, D, or F state forms with a probability scaling with the degeneracy, g_i, whereas states of the type G, H, ..., as well as doubly excited states, do not form at all. This proposal is in part supported by experiment (Section 2.4).

Only the fifth assumption is problematical. In order to accomodate to the experimental result that states with increasing ε_i have a rapidly decreasing population (Section 2.2), we will assume that, for a given value of ΔE, all excited states with $\varepsilon_i \leq \Delta E_e = KE$ can form. This would in principle be rigorous if the energy discrepancy, $\Delta E_e - \varepsilon_i$, were readily converted into kinetic or some other form of energy but little is known about such a possibility. One could also devise situations in which this assumption becomes rigorous based on the broadening being so extensive that many states overlap. Russek and Thomas [23], for example, made such a proposal.

The fourth and fifth assumptions taken together introduce a factor

$$\text{weighting factor} = \frac{g_i}{g_0 + g_1 + g_2 + \ldots + g_i} \equiv g_i/G_i, \qquad (13)$$

where g_0 is, as before, the ground-state degeneracy and G_i might be called the <u>cumulative</u> degeneracy. To simplify the calculations, levels <u>differing</u> only in J will be combined, so that g_i is given by

$$g_i = \Sigma(2J + 1), \qquad (14)$$

where J is the total angular-momentum quantum number.

These five assumptions are sufficient to permit an expression for the relative population of the ith state to be written. It is clear that, if KE < ε_i, the ith state cannot form at all. If ε_i < KE < ε_{i+1}, then the ith state should form proportionally to

$$(g_i/G_i) \int_{\varepsilon_i/K}^{\varepsilon_{i+1}/K} E^{-2} dE.$$

If ε_{i+1} < KE < ε_{i+2}, then the ith state should form proportionally to

$$(g_i/G_{i+1}) \int_{\varepsilon_{i+1}/K}^{\varepsilon_{i+2}/K} E^{-2} dE,$$

and so on. The final result is easily seen to be

$$\text{population}/g_i \propto \sum_{k=i}^{\infty} (1/G_k)(1/\varepsilon_k - 1/\varepsilon_{k+1}). \qquad (15)$$

This relation is seen to disfavor higher levels for 2 reasons: because G_i increases rapidly and because the level separation, $\Delta\varepsilon_k = \varepsilon_{k+1} - \varepsilon_k$, decreases rapidly. Both have as a consequence that there are more and more sinks for the available energy.

Had the fifth assumption not been made, the final result would have been

$$\text{population}/g_i \propto (1/\varepsilon_i - 1/\varepsilon_{i+1}).$$

This relation varies erratically as ε_i increases.

To evaluate (15), the first 50 excited states of each element to be considered were taken from tables [78-82]. In compiling the list, the few states of the type G and H, as well as doubly excited states, were excluded and the rather more numerous missing states of the type S, P, D, and F were interpolated. States with similar energies were grouped together.

Table 7 shows the sort of primary data needed for the particular case of Al. Figs. 13 and 14 show plots of $\ln(\text{population}/g_i)$ vs. ε_i for a selection of elements and Table 8 compares the slopes with what is observed experimentally. The model is seen to succeed essentially as required: it predicts approximate linearity in plots of $\ln(\text{population}/g_i)$ vs. ε_i and it yields slopes which are within a factor of 2 of typical experimental values.

Table 7 Example of the primary data needed to deduce the populations of sputtered excited states of Al according to (15)

Excited-state number	Excitation energy, ε_i [eV]	Degeneracy, g_i, from (14)	Cumulative degeneracy, G_i, from (13)
0	0	6	6
1	3.14	2	8
2	4.02	10	18
3	4.08	6	24
4	4.67	2	26
5	4.83	10	36
6	4.99	6	42
7	5.12	14	56
8, 9	5.24	12	68
10, 11	5.43	20	88
12-14	5.57	18	106

The behavior of Li is interesting, as a discontinuity is evident in both theory (Fig. 13) and experiment (Fig. 5b). Likewise, with Al we find a convex-up shape in both theory (Fig. 14) and experiment (Fig. 5a).

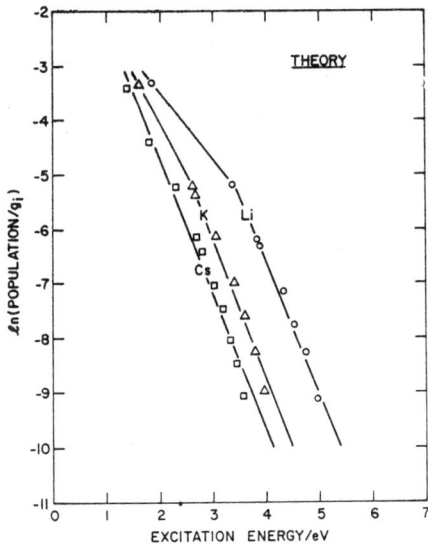

Fig.13 Relative excited-state populations (arbitrary units) vs. the excitation energy as calculated with (15) for the 3 collision pairs Li→Li, K→K, and Cs→Cs. The slopes, 2.44-2.50 eV^{-1} or 4600-4800 K, are within a factor of 2 of typical experimental values. The discontinuity with Li is closely duplicated in the experimental results of Jensen and Veje (Fig. 5b).

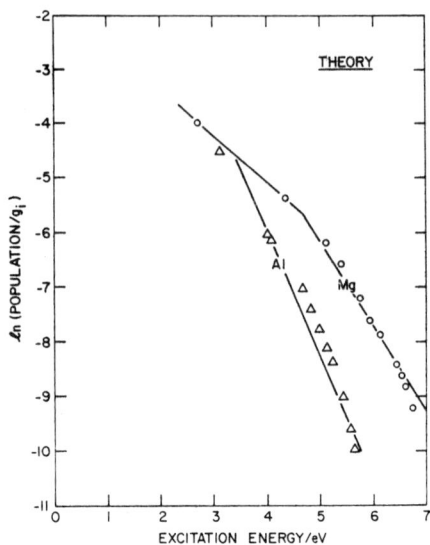

Fig.14 Relative excited-state populations (arbitrary units) vs. the excitation energy as calculated with (15) for the 2 collision pairs Mg→Mg and Al→Al. The slopes, 1.55 and 2.45 eV^{-1} or 7500 and 4700 K, are again within a factor of 2 of typical experimental values. The convex-up shape with Al is apparent also in experiment (Fig. 5a). The discontinuity with Mg is not experimentally accessible owing to the state at 2.71 eV being metastable.

Table 8 Comparison of the slopes of calculated plots of $\ln(\text{population}/g_i)$ vs. ε_i with those observed experimentally

Target	Calculated slopes from Figs. 13 and 14		Experimentally observed slopes [K]
	(a) [eV^{-1}]	(b) [K]	
Li	2.44	4800	3500 (Fig. 5b)
K	2.55	4600	...
Cs	2.50	4600	...
Mg	1.55	7500	4700 for Mg; 4000 for Be [22]
Al	2.45	4700	4600 for Al; 4200 for GaAs; 4100 for In [22]

4. Concluding Remarks

4.1 Ions as a Limiting Case of Excited States

It is natural to think of sputtered ions as being a limiting case of sputtered excited states. This view is supported by such work as that which shows ion and excited-state yields to vary in a similar way under channeling conditions [12] and to fall to similar extents when the target is freed of oxygen [17, 85]. Also indicative of a similarity is the result that the yields of both ions and excited states vary approximately as $g_i \exp(-[\text{internal energy}]/kT)$.

A further requirement if ions are a limiting case of excited states is that ions should have higher kinetic energies, lower yields, and similar sizes. Information on the kinetic energies of ions has been available since at least 1968 [61] and on the kinetic energies of excited states has been available, at least in principal, since 1971 [34]. The data analysis in the latter was somewhat oversimplified, however, so the evidence, that excited states had high kinetic energies, tended to be overlooked. Improved data analysis became available in 1980 [7-10, 41] and confirmed that excited states have unusually high kinetic energies. An extension of the theory of recoil sputtering showed that the necessary population of energetic sputtered atoms exists [46]. Information on the yields of ions [4-6] and excited states [11] from oxidized surfaces is also now available, and confirms that excited states have unusually low yields. The evidence on sizes has been available for half a century but was frequently overlooked: ions are an order of magnitude smaller than excited states [55].

A final aspect of ions and excited states is that the former appear to originate in events resembling cascade sputtering and the latter in events which are more nearly like binary collisions.

The picture that emerges is that it is wrong to regard ions as a limiting case of excited states.

4.2 Mechanistic Implications

The details relating to ion formation, most of which were summarized in Section 4.1, have in common that they are typical of the overall sputtering process. This means that they do not give strong indications about the underlying mechanism. For example, one cannot exclude a role for molecular intermediaries, or thermal events, or formation *within* the solid.

By contrast, the details that relate to excited states are so atypical of the overall sputtering process that they provide important mechanistic hints. For example, the mechanism must be one which involves high kinetic energies and this immediately rules out a role for molecular intermediaries or thermal events. It must also accomodate to large sizes and this rules out formation *within* the solid.

It is easily shown that inelastic energy transfer is a very inefficient process, especially when it is random. Kinetic energies of several 100 eV give inelastic transfers, ΔE_e, of only 2-5 eV, which is of the order of an excitation energy. This inefficiency is an obvious and simple explanation for excited states having high kinetic energies and suggests that it would be fruitful to pursue random, inelastic energy transfer in greater detail.

The following 5 assumptions are introduced:
(i) a sputtered atom is an outerlayer atom which is struck from behind.
(ii) the same collision which creates a sputtered atom is also the one which leads to excitation.

(iii) the probability of the striking atom having a kinetic energy E scales as E^{-2}.
(iv) a given, singly excited S, P, D, or F state forms with a probability which scales as the degeneracy, $g_i = \Sigma(2J + 1)$.
(v) for a given value of ΔE_e, all excited states with $\varepsilon_i \leq \Delta E_e$ can form.

A further important detail is that, for low enough kinetic energies, ΔE_e appears to scale as E rather than, for example, $E^{1/2}$. The final result is that one can write the following expression for the yield of excited states from a sputtered surface:

$$\text{population}/g_i \propto \sum_{k=i}^{\infty} (1/G_k)(1/\varepsilon_k - 1/\varepsilon_{k+1}),$$

where G_i is the cumulative degeneracy (13). This relation gives approximate linearity in plots of $\ln(\text{population}/g_i)$ vs. ε_i and yields slopes which are within a factor of 2 of typical experimental values. Even the discontinuity found with Li and the convex-up shape found with Al have experimental precedents.

4.3 Needed Experiments

We have indicated at various points where additional experimental information is needed. One obvious problem relates to the question of whether ion yields scale with the partition function or with the ground-state degeneracy (Section 2.4). Closely related is whether excited-state yields in general scale with the degeneracy, or do so only for states with low L (Section 2.4). The approach taken by Veje et al. [48, 52] and here reproduced in Figs. 7a and 7b is particularly direct.

An explicit demonstration that the surface binding energy plays a role in the overall sputtering process lies in the well-known correlation of the sputtering coefficient with the inverse of the heat of atomization [59, 60]. Corresponding tests with ions are rare, one of the best being that of Narusawa et al. [62], which is here reproduced in Figs. 9a and 9b. There is no information at all with excited states.

Various indications exist that the events leading to excited states resemble binary collisions (Section 2.8). Those based on the existence of **high kinetic energies, as well as comparisons** of the yields from solid and gas targets, are important but somewhat circumstantial. We regard the approach taken by Andersen et al. [52] and here reproduced in Fig. 7b as a far better one.

Even such basic questions as whether or not excited states have unusually high kinetic energies (Section 2.1) and low yields (Section 2.3) are not established beyond doubt. The possibility of feeding from long-lived cascades will distort curves of light vs.-distance as in Fig. 3 in such a way as to simulate high kinetic energies, though this problem was probably overcome by Yusuf and Tsong [41] in their recent high-resolution experiments. The lowness of excited-state yields may be in part exaggerated owing to the existence of low-lying states which are not observable. This problem does not occur, however, with groups I and III and, since the relevant yields are typical, the implication is that yields are indeed low.

References

1. G. Blaise and A. Nourtier, Surface Sci. 90 (1979) 495.
2. K. Wittmaack, Nucl. Instr. Meth. 168 (1980) 343.
3. R. Kelly, in: Proc. Intern. Conf. on Ion-beam Modification of Mater. (Budapest, 1978) p. 1465.
4. A. Benninghoven, in: Chem. and Phys. of Sol. Surfaces (CRC Press, Cleveland, 1976) p. 207.
5. K. Wittmaack, in: Inelastic Ion-surface Collisions (Academic Press, New York, 1977) p. 153.
6. H. Oechsner, W. Rühe, and E. Stumpe, Surface Sci. 85 (1979) 289.
7. S. Dzioba, O. Auciello, and R. Kelly, Rad. Effects 45 (1980) 235.
8. S. Dzioba and R. Kelly, Surface Sci. 100 (1980) 119.
9. I.S.T. Tsong and N.A. Yusuf, Nucl. Instr. Meth. 170 (1980) 357.
10. R.J. MacDonald, in: Proc. 3rd Intern. Workshop on Inelastic Ion-surface Collisions (this book).
11. I.S.T. Tsong and N.A. Yusuf, Appl. Phys. Lett. 33 (1978) 999.
12. P.J. Martin and R.J. MacDonald, Rad. Effects 32 (1977) 177.
13. J.J. Cuomo, R.J. Gambino, J.M.E. Harper, J.D. Kuptsis, and J.C. Webber, J. Vac. Sci. Technol. 15 (1978) 281.
14. M.L. Yu and W. Reuter, J. Vac. Sci. Technol. (in press).
15. G.E. Thomas, Surface Sci. 90 (1979) 381.
16. M. Braun, Phys. Scripta 19 (1979) 33.
17. P.J. Martin, A.R. Bayly, R.J. MacDonald, N.H. Tolk, G.J. Clark, and J.C. Kelly, Surface Sci. 60 (1976) 349.
18. C.B. Kerkdijk and R. Kelly, Rad. Effects 38 (1978) 73.
19. D. Brochard and G. Slodzian, J. Phys. 32 (1971) 185.
20. P. Williams and C.A. Evans, Surface Sci. 78 (1978) 324.
21. G.E. Thomas, Rad. Effects 31 (1977) 185.
22. C.J. Good-Zamin, M.T. Shehata, D.B. Squires, and R. Kelly, Rad. Effects 35 (1978) 139.
23. A. Russek and M.T. Thomas, Phys. Rev. 109 (1958) 2015.
24. J.B. Bulman and A. Russek, Phys. Rev. 122 (1961) 506.
25. Z. Sroubek, J. Zavadil, F. Kubec, and K. Zdansky, Surface Sci. 77 (1978) 603.
26. Z. Sroubek, K. Zdansky, and J. Zavadil, Phys. Rev. Lett. 45 (1980) 580.
27. M.L. Yu (IBM, Yorktown Heights) unpublished.
28. P. Williams (University of Illinois, Urbana) unpublished.
29. M.A. Rudat and G.H. Morrison, Intern. J. Mass Spect. Ion Phys. 30 (1979) 197.
30. P. Williams, Phys. Rev. Lett. (submitted).
31. M.L. Knotek and P.J. Feibelman, Phys. Rev. Lett. 40 (1978) 964.
32. A.R. Bayly and R.J. MacDonald, Rad. Effects 34 (1977) 169.
33. A.R. Krauss and D.M. Gruen, Appl. Phys. 14 (1977) 89.
34. V.V. Gritsyna, T.S. Kiyan, R. Goutte, A.G. Koval', and Ya. M. Fogel', Bull. Acad. Sci. USSR 35 (1971) 530.
35. T.S. Kiyan, V.V. Gritsyna, and Ya. M. Fogel', Nucl. Instr. Meth. 132 (1976) 435.

36. M. Braun, B. Emmoth, and I. Martinson, Phys. Scripta 10 (1974) 133.
37. E. Taglauer, W. Heiland, R.J. MacDonald, and N.H. Tolk, J. Phys. B 12 (1979) L533
38. R.S. Bhattacharya, D. Hasselkamp, and K.-H. Schartner, J. Phys. D 12 (1979) L55.
39. R. Kelly and N. Penebre (IBM, Yorktown Heights) unpublished.
40. R. Kelly, S. Dzioba, N.H. Tolk, and J.C. Tully, Surface Sci. (in press).
41. N.A. Yusuf and I.S.T. Tsong, Surface Sci. (in press).
42. N.H. Tolk, D.L. Simms, E.B. Foley, and C.W. White, Rad. Effects 18 (1973) 221.
43. S.Y. Leung, N.H. Tolk, W. Heiland, J.C. Tully, J.S. Kraus, and P. Hill, Phys. Rev. A 18 (1978) 447.
44. W. Heiland, J. Kraus, S. Leung, and N.H. Tolk, Surface Sci. 67 (1977) 437.
45. S. Dzioba and R. Kelly, J. Nucl. Mat. 76 (1978) 175.
46. S. Dzioba and R. Kelly, Nucl. Instr. Meth. (in press).
47. P. Williams, W. Katz, and C.A. Evans, Nucl. Instr. Meth. 168 (1980) 373.
48. K. Jensen and E. Veje, Z. Phys. 269 (1974) 293.
49. P. Williams, I.S.T. Tsong, and S. Tsuji, Nucl. Instr. Meth. 170 (1980) 591.
50. A.E. Morgan and H.W. Werner, Anal. Chem. 48 (1976) 699.
51. A.E. Morgan and H.W. Werner, J. Chem. Phys. 68 (1978) 3900.
52. N. Andersen, B. Andresen, and E. Veje, in: Proc. Symp. on Sputtering, eds. P. Varga, G. Betz, and F.P. Viehböck (Inst. für Allgemeine Physik, Technische Univ., Vienna, 1980) p. 726.
53. J.W. Gadzuk, Phys. Rev. B 1 (1970) 2110.
54. M.R. Morrow, O. Auciello, S. Dzioba, and R. Kelly, Surface Sci. 97 (1980) 243.
55. L. Pauling and E.B. Wilson, Introduction to Quantum Mechanics (McGraw-Hill, New York, 1935) p. 144.
56. T. Andersen and A. Lindgård, J. Phys. B 10 (1977) 2359.
57. C. Candler, Atomic Spectra and the Vector Model (Adam Hilger, Institute of Physics, Bristol) p. 16.
58. R.B. Wright and D.M. Gruen, J. Chem. Phys. 72 (1980) 147.
59. D. Rosenberg and G.K. Wehner, J. Appl. Phys. 33 (1962) 1842.
60. P. Sigmund, Phys. Rev. 184 (1969) 383.
61. J.-F. Hennequin, J. Phys. 29 (1968) 655.
62. T. Narusawa, T. Satake, and S. Komiya, J. Vac. Sci. Technol. 13 (1976) 514.
63. J. Richards and J.C. Kelly, Rad. Effects 19 (1973) 185.
64. S. Dworetsky, R. Novick, W.W. Smith, and N. Tolk, Phys. Rev. Lett. 18 (1967) 939.
65. N.H. Tolk, J.C. Tully, C.W. White, J. Kraus, A.A. Monge, D.L. Simms, M.F. Robbins, S.H. Neff, and W. Lichten, Phys. Rev. A 13 (1976) 969.
66. K.J. Snowdon, G. Carter, D.G. Armour, B. Andresen, and E. Veje, Surface Sci. 90 (1979) 429.
67. K.J. Snowdon, G. Carter, D.G. Armour, B. Andresen, and E. Veje, Rad. Effects Lett. 43 (1979) 201.

68 O.B. Firsov, Soviet Phys.-JETP 36 (1959) 1076.
69 E.S. Parilis and L.M. Kishinevskii, Soviet Phys.—Solid State 3 (1960) 885.
70 L.M. Kishinevskii, Bull. Acad. Sci. USSR, Phys. Ser. 20 (1962) 1433.
71 C. Lehmann, Interaction of Radiation with Solids and Elementary Defect Production (North-Holland, Amsterdam, 1977) p. 10.
72 K.B. Winterbon, Chalk River Report AECL-3194 (1968).
73 P. Gombas, in Atoms II, Encyclopedia of Physics, Vol. 36 (Springer, Berlin, Heidelberg 1956) p. 109
74 O.S. Oen and M.T. Robinson, Nucl. Instr. Meth. 132 (1976) 647.
75 W. Eckstein, V.A. Molchanov, and H. Verbeek, Nucl. Instr. Meth. 149 (1978) 599.
76 W. Heiland and E. Taglauer, in: The Physics of Ionized Gases, ed. R.K. Janev (Institute of Physics, Beograd, Yugoslavia, 1978) p. 239.
77 R. Blume, W. Eckstein, and H. Verbeek, Nucl. Instr. Meth. 168 (1980) 57.
78 C.E. Moore, Atomic Energy Levels, vol. 1 (U.S. Govt. Printing Office, Washington, 1971) p. 8.
79 ibidem, vol. 3, p. 124.
80 W.C. Martin and R. Zalubas, J. Phys. Chem. Ref. Data 9 (1980) 1.
81 W.C. Martin and R. Zalubas, J. Phys. Chem. Ref. Data 8 (1979) 817.
82 C. Corliss and J. Sugar, J. Phys. Chem. Ref. Data 8 (1979) 1109.
83 M. Rémy, C.R. Acad. Sci. Paris C 287 (1978) 235.
84 D.E. Harrison, P.W. Kelly, B.J. Garrison, and N. Winograd, Surface Sci. 76 (1978) 311.
85 R. Shimizu, T. Okutani, T. Ishitani, and H. Tamura, Surface Sci. 69 (1977) 349.
86 L. de Galan, R. Smith, and J.D. Winefordner, Spectrochim. Acta 23B (1968) 521.
87 R.A. Swalin, Thermodynamics of Solids, 2nd ed. (Wiley, New York, 1972) p. 230.
88 H. Oechsner, Appl. Phys. 8 (1975) 185.
89 A.N. Zaidel', V.K. Prokof'ev, S.M. Raiskii, V.A. Slavnyi, and E.Ya. Shreider, Tables of Spectral Lines (IFI/Plenum, New York, 1970).
90 H. Shimizu, M. Ono, and K. Nakayama, Surface Sci. 36 (1973) 817.
91 J. Schamps, Chem. Phys. 2 (1973) 352.

On the Velocity Measurements of Sputtered Excited Atoms[*]

M. Szymoński, A. Paradzisz, and L. Gabła
Jagellonian University, Institute of Physics, Reymonta 4
30-059 Kraków, Poland

Measurements of the intensity of light emitted from sputtered excited Mg atoms as a function of the distance from the surface have been performed under an oxygen environment. It is shown that: 1/ inclusion of the cascade transitions plays a crucial role in analysis of the experimental data, 2/ form of the velocity dependent excitation probability is much less important, 3/ application of this experimental method for determination of the velocity spectrum of excited particles is very limited.

1. Introduction

Measurements of the intensity decay of light emitted by sputtered atoms as a function of the distance from the surface have recently been used for determination of excited particle velocities. Several attempts were made for extracting mean [1-5] or threshold [4,6,7] energies of the sputtered excited atoms. None of them, however, avoided the following disturbing factors which are crucial for proper analysis of the experimental data: 1/ the intensity decay function for a given transition is usually strongly disturbed by the cascading transitions from upper long-lived levels which are particularly important at large distances from the surface, 2/ the E^{-2} expression usually used as description of the differential flux of sputtered particles is not correct either for low-eV energies (unimportant for excited state production) or for energies in a keV region (very important for excitation) [8], 3/ the dependence of the excitation probability on the velocity of sputtered atoms has been greatly oversimplify.
In this communication we will present one particular measurement of the intensity decay function of the MgI $3\ ^1P_1^o - 3\ ^1S_o$ line (285.2 nm) emitted by sputtered magnesium atoms. For this example an attempt will be made to account for the above disturbing factors in analysis of the experimental data.

2. Experimental

A polycrystalline Mg target was bombarded with a 4 keV Ar^+ beam directed along the surface normal. The ion current den-

---

[*]This work was carried out as a part of Research Project M.R.I/5

sity was about 40 $\mu A/cm^2$. A partial pressure of oxygen during the experiment was 1 x 10^{-6} torr, thus the surface was covered by an oxide layer. Thanks to low vapour pressure of the pumping oil, special liquid-nitrogen cooled traps and elevation of the target temperature up to 200 °C, the surface was free of hydro-carbide contaminants.

The light emitted above the target was detected through the two 0.1 mm wide slits parallel to the surface plane.

By moving the sample along the normal, the intensity decay function for a given transition was measured with an accuracy of about 0.1 mm.

3. Results and Discussion

The experimental intensity distribution $I(x)$ of MgI 285.2 nm line ($3\ ^1P_1^o - 3\ ^1S_0$ transition) is plotted as a function of $\log(I(x))$ versus distance x in Figure 1. GRITSYNA and coworkers [1] interpreted such plots as due to presence of two dis-

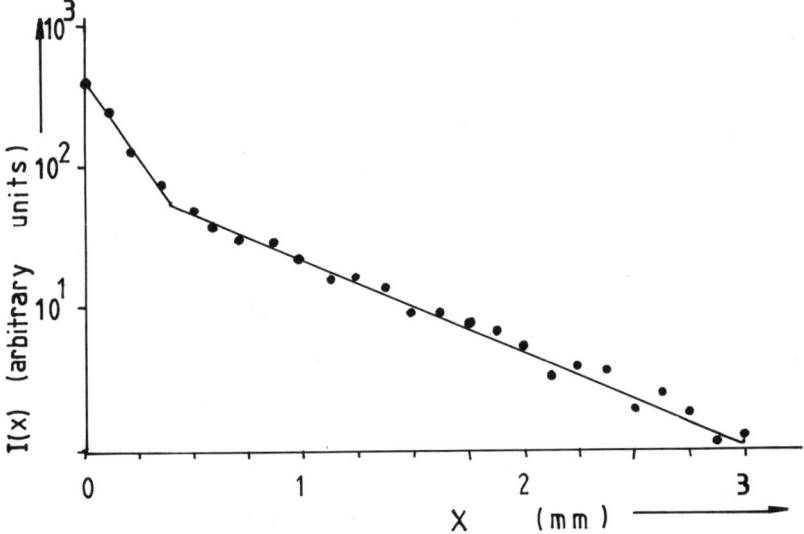

Fig.1 The experimental intensity distribution of MgI 285.2 nm line

tinct groups of the excited atom velocities (slow and fast) represented by the mean values $\langle v_i \rangle$. The intensity decay as a function of distance x was assumed to be:

$$I(x) = \sum_{i=1,2} I_{oi}\ \exp\left(-x/\langle v_i\rangle \tau_o\right), \qquad (1)$$

where I_{oi} is the proportionality factor, $\langle v_i \rangle$ is the mean velocity of the slow $(i=1)$ and the fast $(i=2)$ excited particles, and $1/\tau_o$ is the decay constant for the given transition. A possible filling up of the upper level of the transition via cascade process was ignored in this treatment. From the slopes of the two straight lines drawn through the experimental points in Figure 1 we calculated $v_1 = 7.45 \times 10^6$ cm/s and $v_2 = 2.4 \times 10^7$ cm/s which correspond to kinetic energies $E_1 = 700$ eV and $E_2 = 7$ keV respectively. Since the primary beam energy was only 4 keV it is quite obvious that this type of analysis is entirely wrong, primarily due to ignoring the cascade transitions.

In reality one should use a much more complicated approximation for $I(x)$ dependence:

$$I(x) = I_o \int_{E_{min}}^{E_{max}} \int_0^{\pi/2} F(E) \cdot P(E,\theta) \cdot D(E,\theta,x) \sin\theta \cos\theta \, dE \, d\theta, \quad (2)$$

where $F(E)$ is the kinetic energy distribution of the sputtered atoms (the angular distribution is assumed to be isotropic, i.e. $\sim \cos\theta$, θ is the sputtering angle), $P(E,\theta)$ is the probability of excitation of the sputtered atom and $D(E,\theta,x)$ is the decay function describing the dependence of the excited level population on the distance x.
KELLY and coworkers [6] assumed that $F(E) \sim 1/E^2$, $P(E) = 0$ for $E < E^*$ and $P(E) = P^*$ for $E \geq E^*$, and $D(E,\theta,x) \sim$
$\sim \exp\left(-\frac{a + x\sqrt{m/\tau_o}}{\sqrt{2E} \cos\theta}\right)$, where "a" is the survival probability constant for non-radiative decay ("a" for oxidized surface is negligible [6]), and "m" is the mass of the sputtered atom. Integration was carried out from threshold energy E^* to infinity and differential form of equation (2) was used, since the target was viewed through a sufficiently narrow slit. No cascade correction was made in this approach.
Using the above form of F, P and D functions, however, we were not able to reproduce our experimental data except for the first few points. Apparently the cascade transitions play a dominant role at large distance x. In order to include this process we solved differential equation describing the time dependence of the population of our excited level $3\ ^1P_1^o$:

$$dN(t)/dt = -1/\tau_o N(t) + \sum_i 1/\tau_i \, N_i(t) \quad ,$$

where summation is carried out over all cascade transitions from levels "i" which lifetimes τ_i are considerably larger than τ_o. The solution can be approximated by:

$$N(t) \cong N_o \exp(-t/\tau_o)\left[1 + \sum_i a_i \exp(b_i t)\right] , \quad (3)$$

where $t = x\sqrt{m/2E} / \cos\theta$, $a_i = N_i(0)\tau_o / N(0)(\tau_i - \tau_o)$ and

$b_i = (\tau_i - \tau_o)/\tau_i \tau_o$. Since the lifetimes of the most important cascade transitions are about 10 times larger than $\tau_o \approx 2.3$ ns we can further simplify the equation (3):

$$N(t) \sim \exp(-t/\tau_o) \left[1 + A \exp(bt) \right], \quad (4)$$

where A will be taken as fitting parameter. Neglecting "a" we can write:

$$D(E, \vartheta, x) \approx N\left[t(E, \vartheta, x) \right], \quad (5)$$

and using equation (2) with KELLY's form of F and P we can fit the experimental intensity curve. The agreement now is rather good (see Figure 2) but the value of $E^* = 2100$ eV $\pm$ 100 eV seems much too high compare to primary beam energy 4 keV.

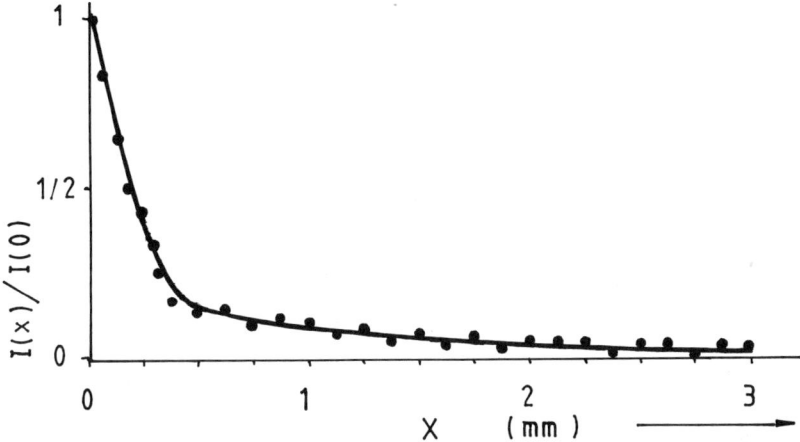

Fig.2 The intensity decay function of MgI 285.2 nm line. The curve calculated according to equation (2) is plotted (solid line) together with the experimental points. The differences between various forms of P(E) are not greater than thickness of the solid line

Furthermore we can almost arbitrary change value E^* and still obtain the same agreement with the experiment by changing the shape of $P(E, \vartheta)$ function. Generally the slower increase of P function, the smaller value of E^* can be achieved. For example $P(E) = (1 - E^*/E)^{\frac{1}{2}}$ was used and from the fit $E^* = 1500$ eV was determined. In both cases, however, the number of sputtered particles with energy higher than E^* is so small in comparison to the total flux (3-4 %) that even assuming P = 1 for $E \gg E^*$

the absolute photon yield would be smaller than that observed experimentally for this line [9]. We think, therefore, that these types of excitation probability function are physically unrealistic and no greater physical meaning should be ascribed to the values of E^*.
SCHROEER et al. [10] proposed the following excitation probability function:

$$P(E,\Theta) \sim \exp(-\alpha/\sqrt{E}\cos\Theta) \quad , \qquad (6)$$

where α is the factor not dependent on energy. Since in this case the excitation does not require threshold energy, a more accurate expression for energy distribution should be used [11]:

$$F(E) \sim E/(E + E_b)^3 \quad , \qquad (7)$$

where the surface binding energy for oxidized magnesium E_b = 8 eV [12]. Substitution of (5), (6) and (7) into expression (2) allows again for accurate reproduction of the experimental curve. The fitted value of α = 85 $(eV)^{1/2}$. The advantage of the last form of P relies upon a much more realistic energy range for which the sputtered atoms have non-zero excitation probability. Thus a much higher photon yield can be expected.
A comparison of the various probability functions P used in this work is presented in Figure 3.

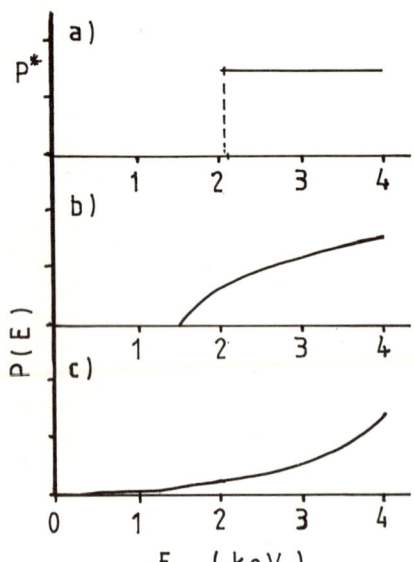

Fig.3 Comparison of the various probability functions $P(E)$, a/ $P(E)$ = 0 for $E < E^*$ and $P(E) = P^*$ for $E \geq E^*$, b/ $P(E) = (1 - E^*/E)^7$, c/ $P(E) \sim \exp(-\alpha/\sqrt{E})$

References

1. V. V. Gritsyna, T. S. Kijan, R. Goutte, A. G. Koval and Ya. M. Fogel, Bull. Acad. Sci. USSR, 35, 530 (1970).
2. M. Braun, B. Emmoth and J. Martison, Physica Scripta, 10, 133 (1974).
3. M. Braun, Physica Scripta, 19, 33 (1979).
4. I.S.T. Tsong and N.A. Yusuf, Nucl.Instr.Meth., 170, (1980).
5. C.M. Loxton, R.J. MacDonald and P.J. Martin, Surf.Sci., 93, 84 (1980).
6. S. Dzioba, O. Auciello and R. Kelly, Rad. Effects, 45, 235 (1980).
7. S. Dzioba and R. Kelly, Surf.Sci., 100, at press (1980).
8. P. Sigmund, Phys.Rev., 184, 383 (1969).
9. I.S.T. Tsong, this Proceedings.
10. J.M. Schroeer, T.N. Rhodin and R.C. Bradley, Surf.Sci., 34, 571 (1973).
11. M. Szymoński, Acta Physica Polonica, 56 A, 289 (1979).
12. R. Kelly and N.Q. Lam, Rad.Effects, 19, 39 (1973).

Index of Contributors

Baragiola, R.A. 38
Boiziau, C. 48
Burgdörfer, J. 207

Conrad, H. 73

Datz, S. 142

Eckstein, W. 157
Eicher, S. 138
Ertl, G. 73

Feibelmann, P.J. 104
Feldman, L.C. 112

Gabła, L. 322
Graser, W. 211

Haberland, H. 73

Kelly, R. 292
Kleber, M. 184
Kraus, J.S. 112,196
Küppers, J. 73

Laubert, R. 120
Loxton C.M. 224

MacDonald, R.J. 224
Madey, T.E. 80
Martin, P.J. 224
Morris, R.J. 112,196

Pian, T.R. 112
Paradzisz, A. 322

Rau, C. 138,196

Schröder, H. 207
Sellin, I.A. 120
Sesselmann, W. 73
Sigmund, P. 2,251
Sroubek, Z. 277
Szymónski, M. 322

Tolk, N.H. 112,196
Tougaard, S. 2
Traum, M.M. 112
Tsong, I.S.T. 258
Tully, J.C. 112,196

Varelas, C. 211

Winter, H. 216

Zwiegel, J. 184

M. A. Van Hove, S. Y. Tong

Surface Crystallography by LEED

Theory, Computation and Structural Results

1979. 19 figures, 2 tables. IX, 286 pages
(Springer Series in Chemical Physics, Volume 2)
ISBN 3-540-09194-7

Contents: Introduction. – The Physics of LEED. – Basic Aspects of the Programs. – Symmetry and Its Use. – Calculation of Diffraction Matrices for Single Bravais-Lattice Layers. – The Combined Space Method for Composite Layers: by Matrix Inversion. – The Combined Space Method for Composite Layers: by Reverse Scattering Perturbation. – Stacking Layers by Layer Doubling. – Stacking Layers by Renormalized Forward Scattering (RFS) Perturbation. – Assembling a Program: The Main Program and the Input. – Subroutine Listings. – Structural Results of LEED Crystallography. – Appendices. – References. – Subject Index.

Secondary Ion Mass Spectrometry SIMS-II

Editors: A. Benninghoven, C. A. Evans, Jr., R. A. Powell, R. Shimizu, H. A. Storms

1979. 234 figures, 21 tables. XIII, 298 pages
(Springer Series in Chemical Physics, Volume 9)
ISBN 3-540-09843-7

Contents: Fundamentals. – Quantitation. – Semiconductors. – Static SIMS. – Metallurgy. – Instrumentation. – Geology. – Panel Discussion. – Biology. – Combined Techniques. – Postdeadline Papers.

Sputtering by Particle Bombardment I

Physics and Applications

Editor: R. Behrisch

1981. 117 figures, 24 tables.
Approx. 280 pages
(Topics in Applied Physics, Volume 47)
ISBN 3-540-10521-2

Contents: R. Behrisch: Introduction and Overview. – P. Sigmund: Sputtering by Ion Bombardment: Theoretical Concepts. – M. T. Robinson: Theoretical Aspects of Monocrystal Sputtering. – H. H. Anderson, H. Bay: Sputtering Yield Measurements. – H. E. Roosendaal: Sputtering Yields of Single Crystalline Targets.

Vibrational Spectroscopy of Adsorbates

Editor: R. F. Willis
With contributions by numerous experts

1980. 97 figures, 8 tables. XII, 184 pages
(Springer Series in Chemical Physics, Volume 15)
ISBN 3-540-10429-1

Contents: Introduction. – Theory of Dipole Electron Scattering from Adsorbates. – Angle and Energy Dependent Electron Impact Vibrational Excitation of Adsorbates. – Adsorbate Induced Optical Phonons. – Inelastic Electron Tunnelling Spectroscopy. – Inelastic Molecular Beam Scattering from Surfaces. – Neutron Scattering Studies. – Reflection Absorption Infrared Spectroscopy: Application to Carbon Monoxide on Copper. – Raman Spectroscopy of Adsorbates at Metal Surfaces. – Vibrations of Monoatomic and Diatomic Ligands in Metal Clusters and Complexes – Analogies with Vibrations of Adsorbed Species on Metals. – Coupling Induced Vibrational Frequency Shifts and Island Size Determination: Co and Pt {001} and Pt {111}.

Springer-Verlag
Berlin
Heidelberg
New York

Electron Spectroscopy for Surface Analysis
Editor: H. Ibach

1977. 123 figures, 5 tables. XI, 255 pages
(Topics in Current Physics, Volume 4)
ISBN 3-540-08078-3

Contents: H. Ibach: Introduction. – D. Roy, J. D. Carette: Design of Electron Spectrometers for Surface Analysis. – J. Kirschner: Electron-Excited Core Level Spectroscopies. – M. Henzler: Electron Diffraction and Surface Defect Structure. – B. Feuerbacher, B. Fitton: Photoemission Spectroscopy. – H. Froitzheim: Electron Energy Loss Spectroscopy.

Interactions on Metal Surfaces
Editor: R. Gomer

1975. 112 figures. XI, 310 pages
(Topics in Applied Physics, Volume 4)
ISBN 3-540-07094-X

Contents: J. R. Smith: Theory of Electronic Properties of Surfaces. – S. K. Lyo, R. Gomer: Theory of Chemisorption. – L. D. Schmidt: Chemisorption: Aspects of the Experimental Situation.– D. Menzel: Desorption Phenomena. – E. W. Plummer: Photoemission and Field Emission Spectroscopy. –E. Bauer: Low Energy Electron Diffraction (LEED) and Auger Methods. – M. Boudart: Concepts in Heterogeneous Catalysis.

H. Raether
Excitation of Plasmons and Interband Transitions by Electrons
1980. 121 figures, 17 tables. VIII, 196 pages
(Springer Tracts in Modern Physics, Volume 88)
ISBN 3-540-09677-9

Contents: Introduction. – Volume Plasmons. – The Dielectric Function and the Loss Function of Bound Electrons. – Excitation of Volume Plasmons. – The Energy Loss Spectrum of Electrons and the Loss Function. – Experimental Results. – The Loss Width. – The Wave Vector Dependency of the Energy of the Volume Plasmon. – Core Excitations. – Application to Microanalysis. – Energy Losses by Excitation of Cerenkov Radiation and Guided Light Modes. – Surface Excitations. – Different Electron Energy Loss Spectrometers. – Notes Added in Proof.– References. – Subject Index.

Springer-Verlag
Berlin
Heidelberg
New York